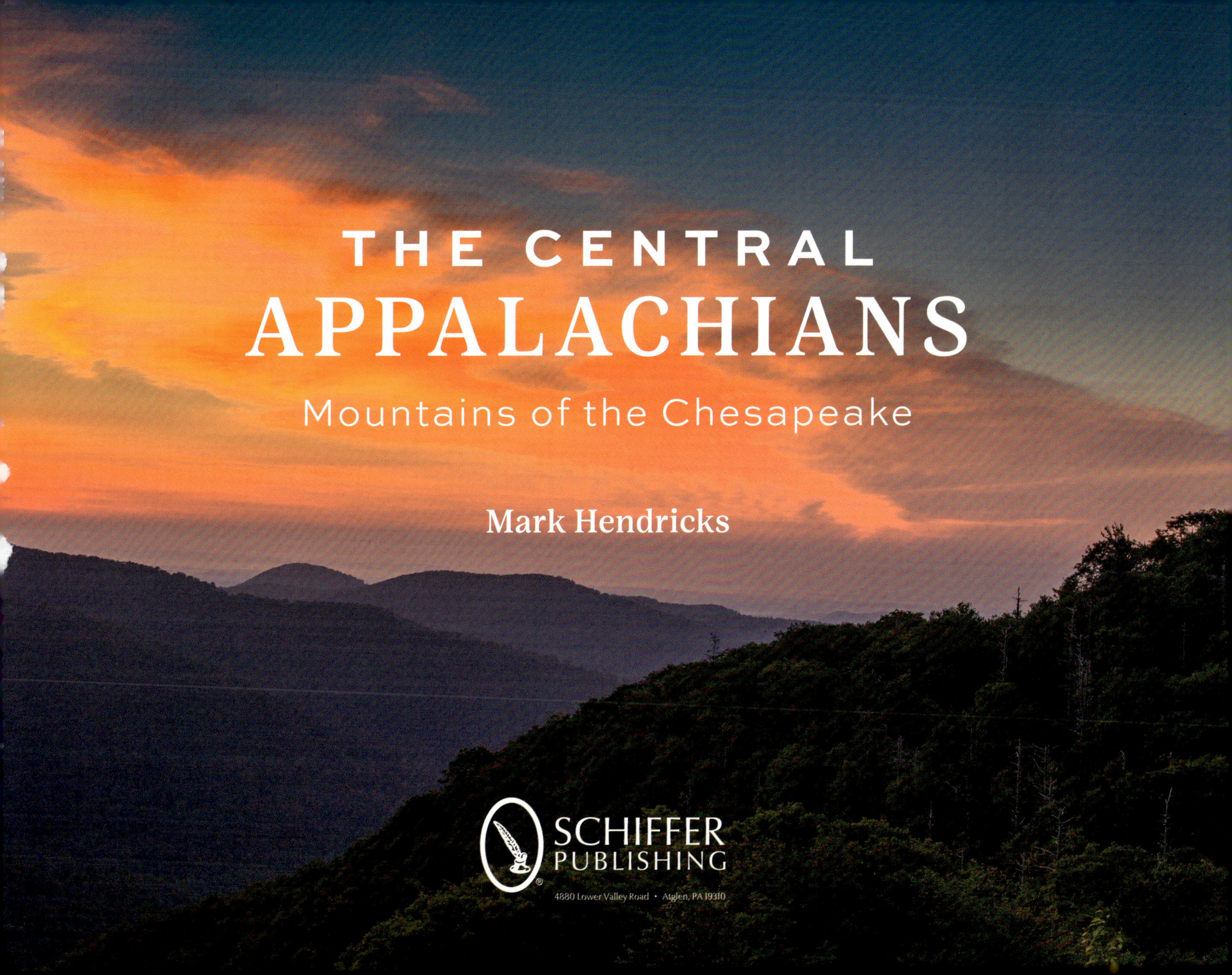

THE CENTRAL APPALACHIANS

Mountains of the Chesapeake

Mark Hendricks

SCHIFFER PUBLISHING
4880 Lower Valley Road • Atglen, PA 19310

Other Schiffer Books by the Author:

Natural Wonders of Assateague Island, 978-0-7643-5340-6

Other Schiffer Books on Related Subjects:

Choptank Odyssey: Celebrating a Great Chesapeake River,
Tom Horton and David W. Harp, 978-0-7643-5000-9

Chesapeake Wildlife: Stories of Survival and Loss,
Pat Vojtech, 978-0-87033-536-5

Delaware Water Gap National Recreation Area,
Michael P. Gadomski, 978-0-7643-5753-4

Library of Congress Control Number: 2023941104

Designed by Molly Shields
Type set in Sweet Sans Pro/Cambria

ISBN: 978-0-7643-6737-3
Printed in China

Published by Schiffer Publishing, Ltd.
4880 Lower Valley Road
Atglen, PA 19310
Phone: (610) 593-1777; Fax: (610) 593-2002
Email: info@schifferbooks.com
Web: www.schifferbooks.com

Previous page: Colorful clouds illuminate the Hemlock Springs Overlook in Virginia's Shenandoah National Park. The dead hemlocks on the mountain are a consequence of the hemlock woolly adelgid, an invasive insect that has decimated eastern hemlock trees in some areas of the Central Appalachians.

FSC
www.fsc.org
MIX
Paper from responsible sources
FSC® C167893

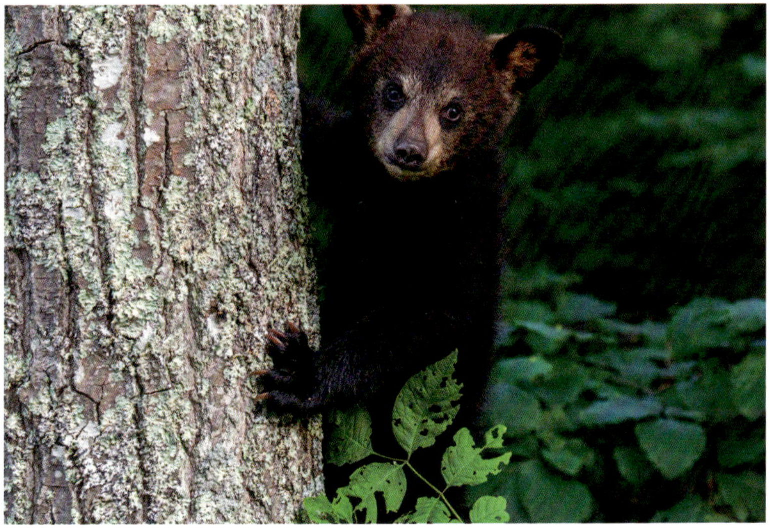

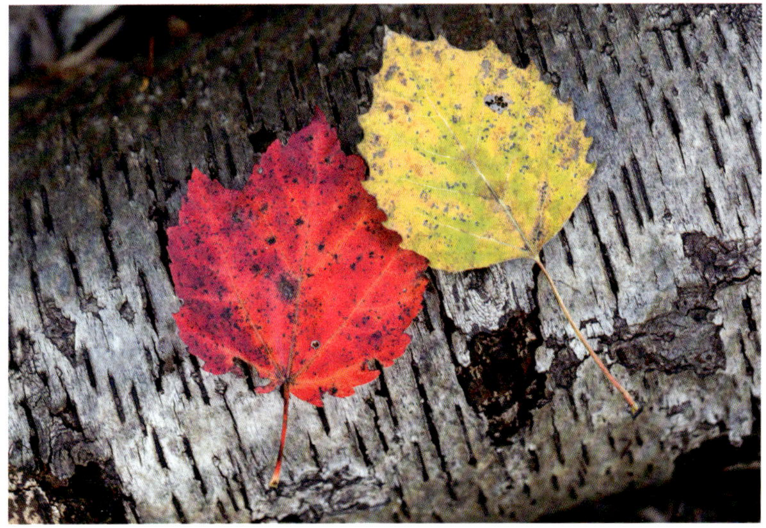

A black bear cub. George Washington National Forest, Virginia.

Maple and bigtooth aspen leaf in the Blue Ridge Mountains. Cunningham State Park, Maryland.

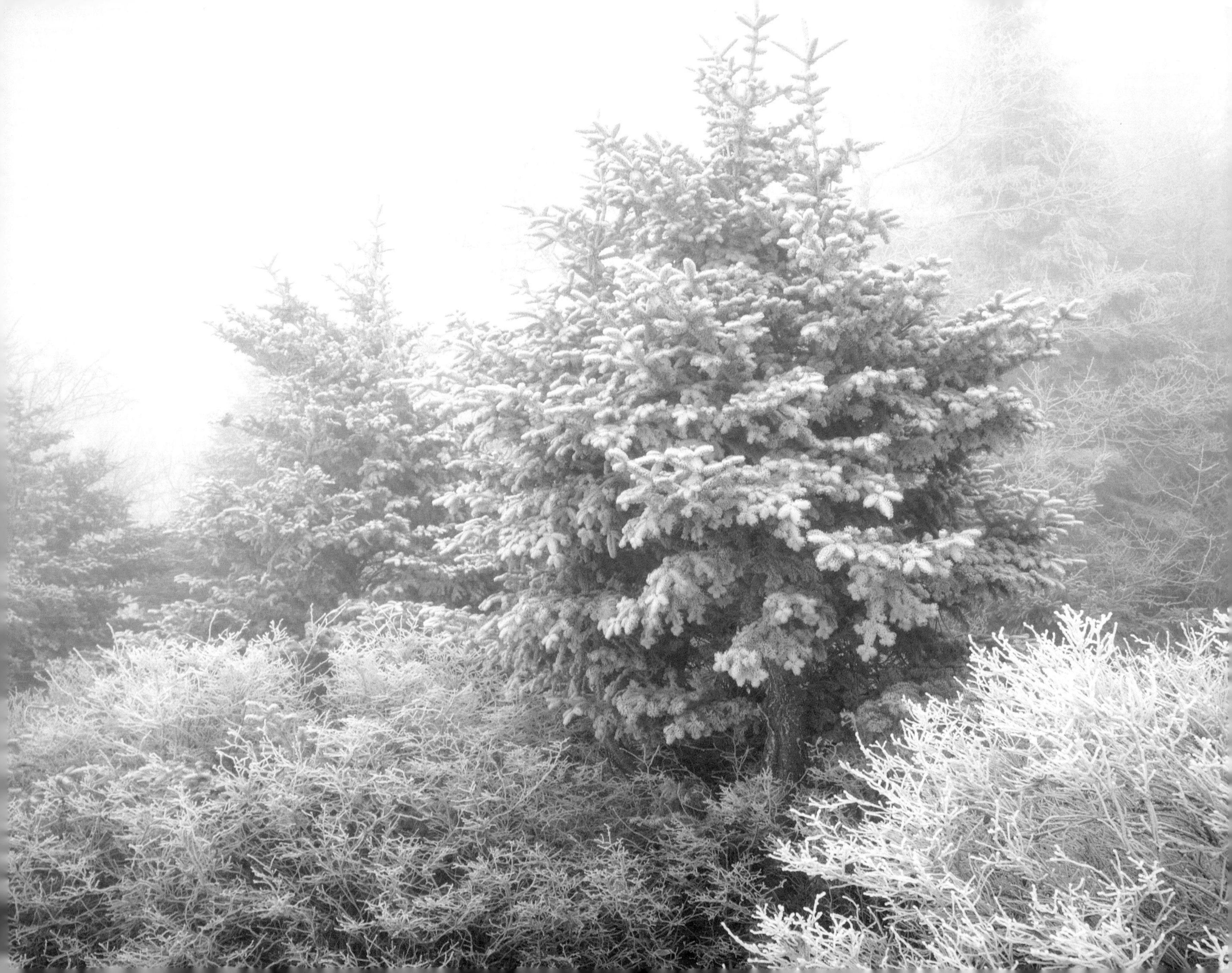

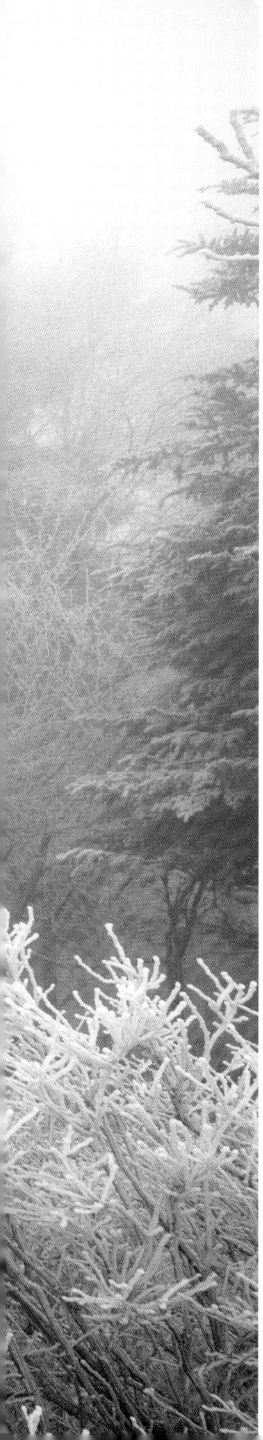

DEDICATION

To Liliana & Adrian
You give my work a new sense of purpose.
Papa loves you.

And to Dan Feller
Without your skills, expertise, patience, and friendship, this book would simply not be.

Red spruce in a winter storm. More common in New England and eastern Canada, these conifers are still found in the cooler, higher elevations of the Central Appalachians. Dolly Sods Wilderness, West Virginia.

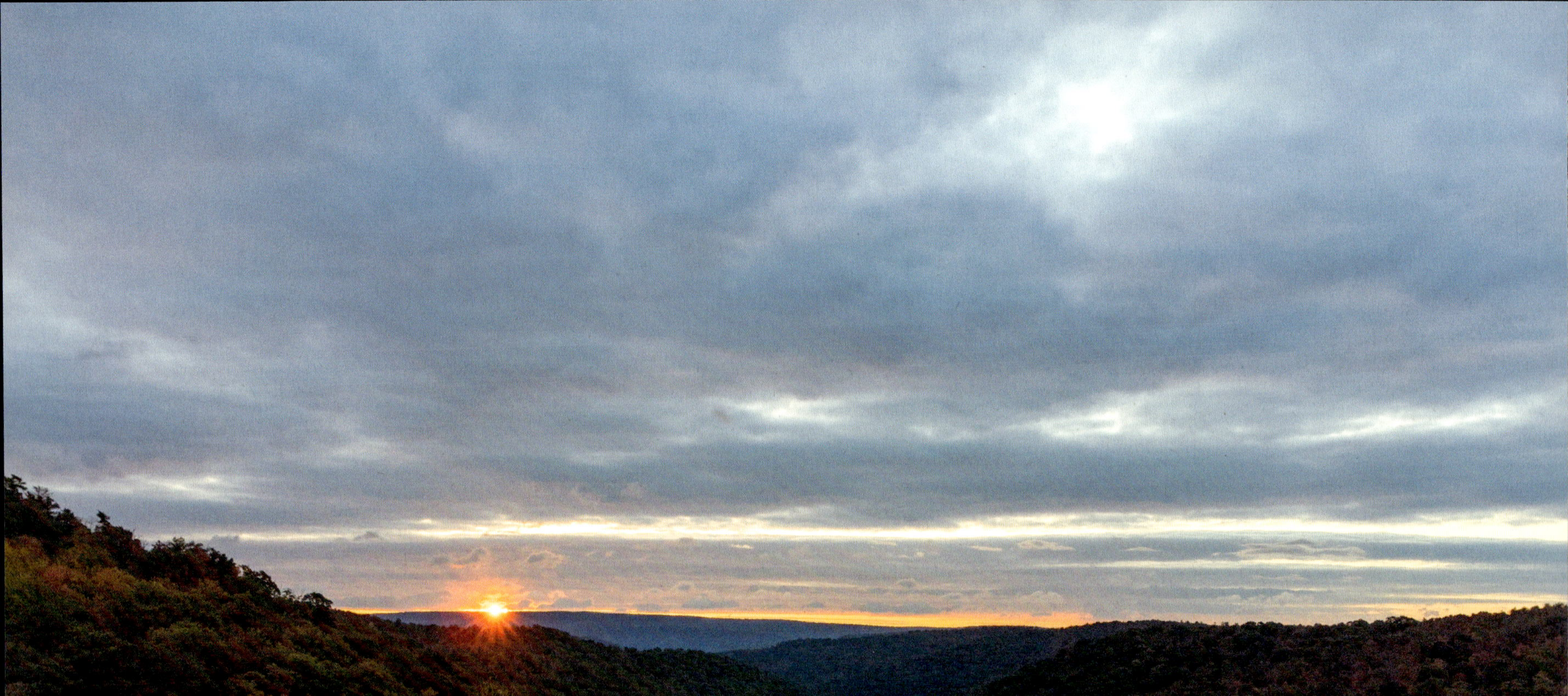

CONTENTS

Aerial view of the Monroe Valley in Garrett County, Maryland

FOREWORD

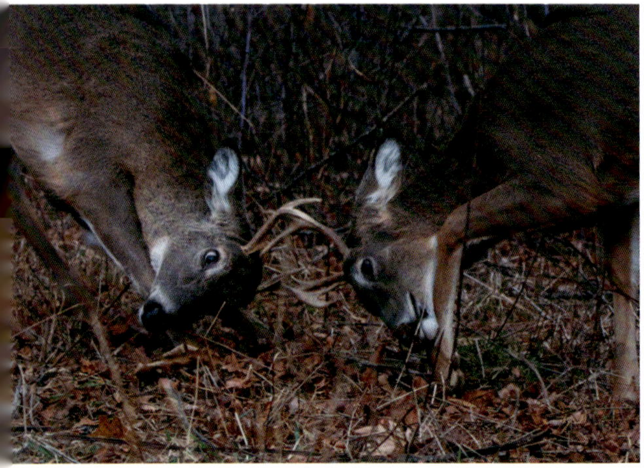

Young whitetail deer spar in late autumn

I have spent a lifetime wandering the forests and wild places in my home state of Virginia, and I've seen many things in my travels. Some were beautiful, some were ugly, and some were even comical. But these were passing visions, now fading in my memory as I get older. They were snapshots in time that only I was blessed to witness. In a state with both mountains and ocean coastline, the forms of life here are diverse and ever changing throughout the seasons. Surrounded by this abundant life, there are plenty of interesting things to see. But are you in the right place at the right time to see them? Are you willing to brave the mountain storms or crawl through frigid muddy streams to glimpse these sights, both great and small?

With the eye of an artist and the soul of a woodsman, Mark Hendricks has done the hard work of capturing these moments for us—preserving them and sharing with us the majesty of nature, in epic landscapes and small, quiet scenes. From elusive beasts and rarely seen plants, to colorful characters and the struggle to save species, Mark leads us through wild places and shows us where to look.

So, look deeply at these photos, lose yourself in these stories, and try to imagine the hardships and patience it took to capture these images and stories. They aren't just a fading moment in someone's memory. They are now our vision to share, and with thoughtful conservation of the natural resources they highlight, future generations can still see these photographs as things that are out in the wild—and not just as a record of things that are gone.

Tim MacWelch

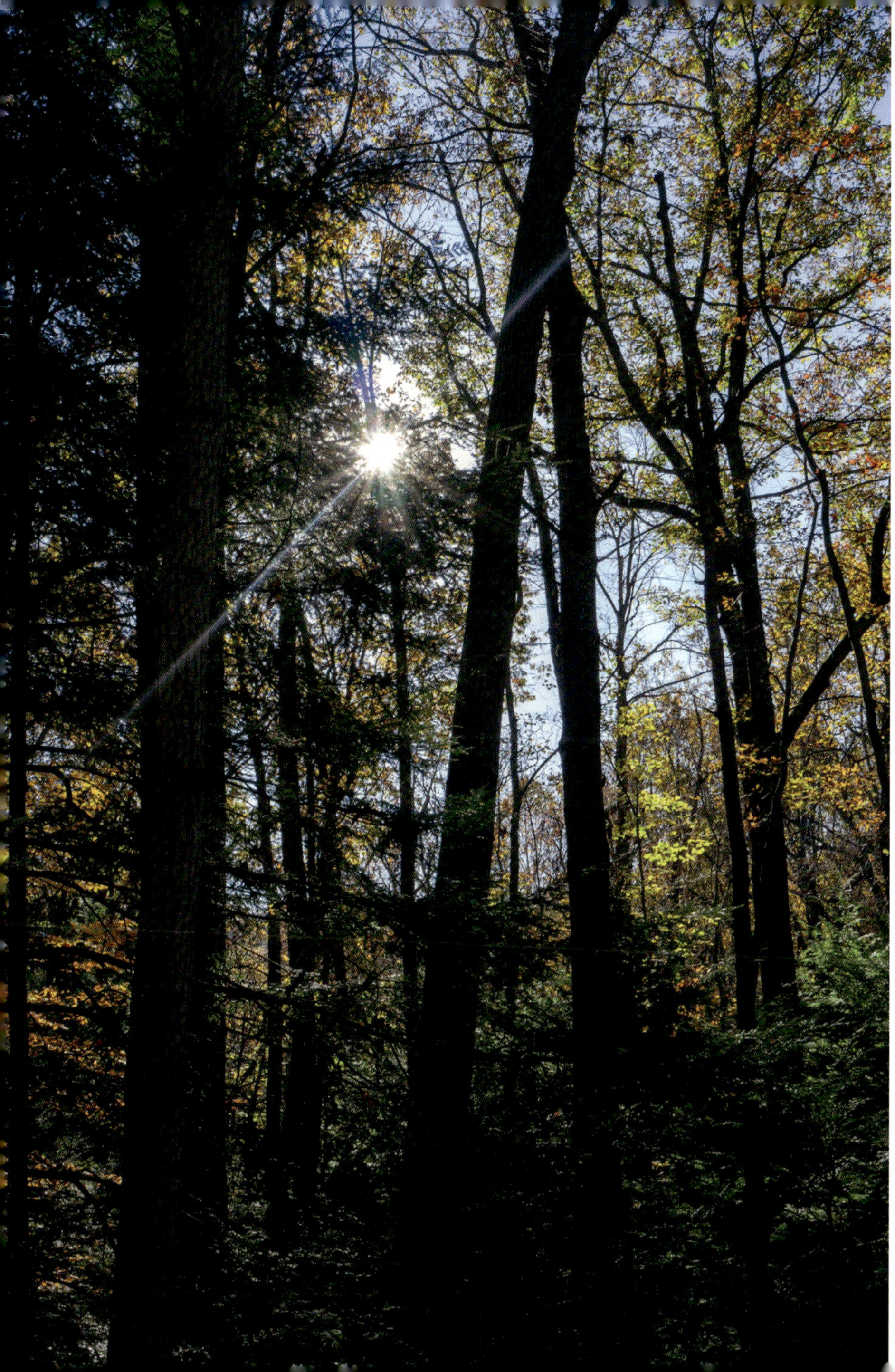

Tim MacWelch is the *New York Times* bestselling author of *Prepare for Anything*, *Hunting & Gathering*, and *How to Survive Anything*. He is *Outdoor Life* magazine's survival blogger. Tim is the founder and head instructor of Advanced Survival Training School as well as its sister school, Earth Connection, where for over two decades he has taught ordinary people how to detach from the grid and live closer to nature, and how to use those skills in any number of emergencies.
www.AdvancedSurvivalTraining.com

Eastern hemlocks and beech trees in autumn.
Frederick Municipal Watershed, Maryland.

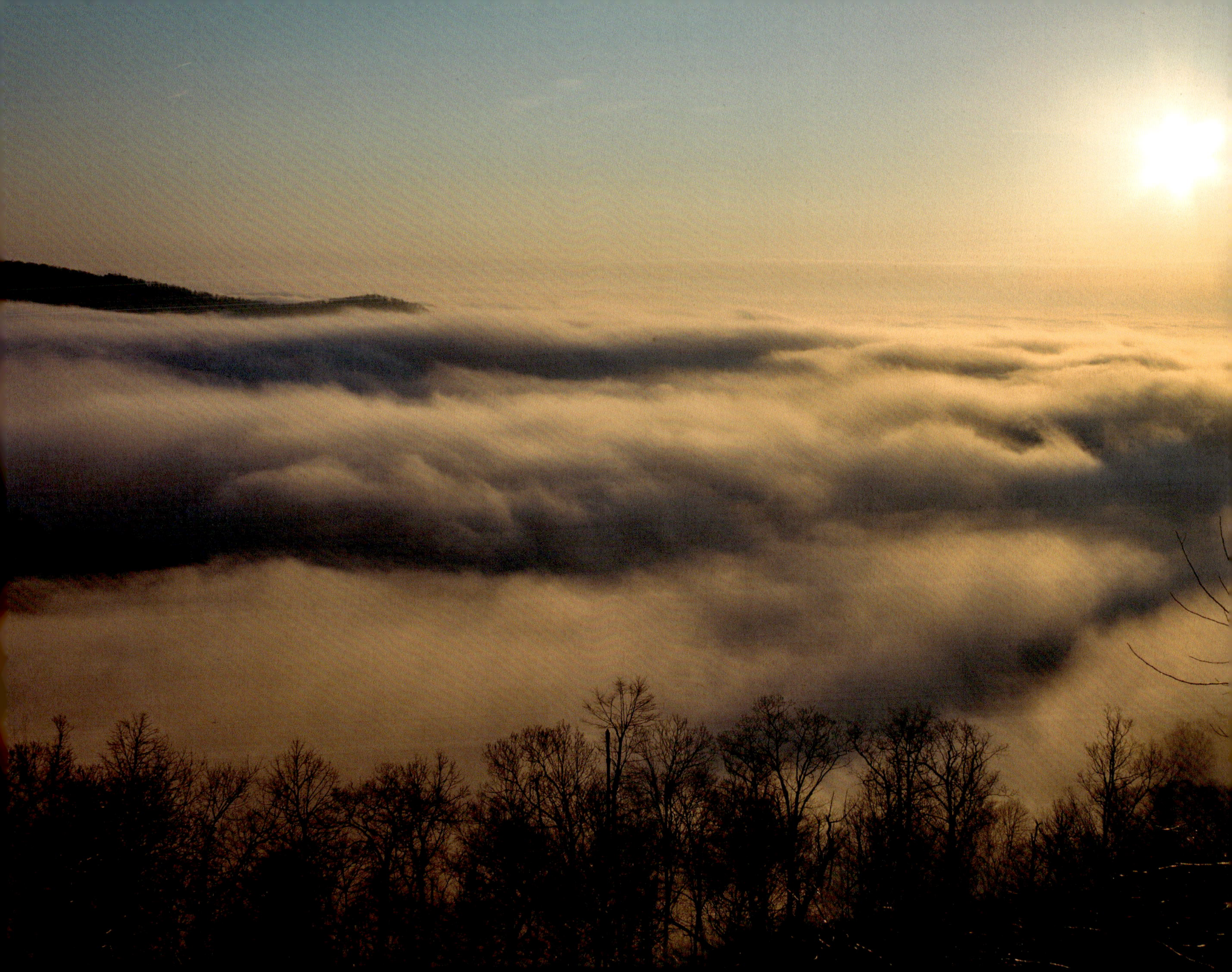

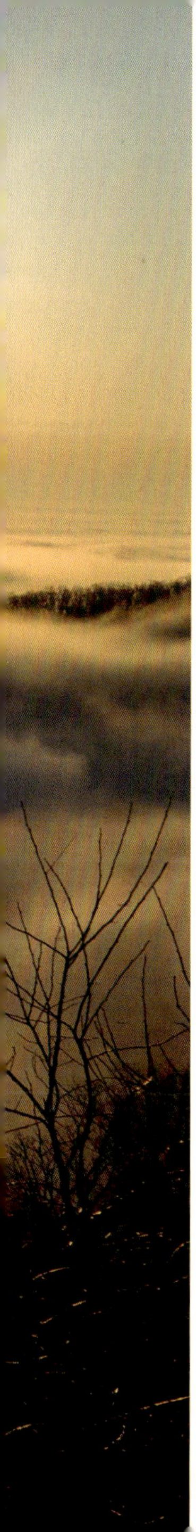

PREFACE

Rolling fog is a mesmerizing phenomenon. The first time I experienced it was from atop Skyline Drive in Virginia's Shenandoah National Park. I was watching an ancient ritual play out: a collision between the night's cool mountain air and the ample moisture from the valley below. The scene was ostensibly more maritime than mountain, as rushing clouds danced in a plunging repetition of hypnotic arcs and crashing waves. The high, soft peaks, engulfed in clouds, became islands in the oceanic trance. The ethereal scene was further complemented by the warm glow of the rising sun. While science could explain the process, it was magic to my eyes.

There is something mysterious about the Appalachians. A mystical past of colliding continents and violent volcanic activity birthed this most ancient mountain chain—a past of endless forests, wild beasts, and mythical lore.

I grew up in Baltimore City, Maryland, where, in my formative schooling years I often learned about the Chesapeake Bay and its vast watershed. And within those boundaries are the Appalachians. I was impressed by the geology of the watershed; the idea that a single drop of rain from the mountains would eventually flow into the Chesapeake Bay felt worthy of appreciation, much as the world's great animal migrations feel worthy of appreciation.

I work as a freelance photojournalist. This affords me opportunities to host presentations on my work. One project of mine, titled *Visualizing the Chesapeake*, is a public-speaking presentation of the watershed and my work within it. It takes the audience through the diverse habitats within the watershed. And every time I presented on the topic, there were attendees unaware that the Appalachians were connected to the Chesapeake Bay; hence the impetus for this book. Books highlighting the Chesapeake Bay abound, but rarely about the mountainous areas within the watershed.

My focus on this book would take me through the Appalachian regions of four states: Maryland, Virginia, West Virginia, and Pennsylvania, broadly defined as the Central Appalachians. New York, where the watershed begins at Cooperstown, was omitted from the book. The reasoning behind this decision was centered on the geographic continuity that makes up the Central Appalachians. Additionally, the agriculture-rich Piedmont Province, which extends eastward from the foot of the mountains to the Atlantic coastal plain, was excluded, since the Piedmont is not normally included within the main Appalachian provinces. Historically, even after Piedmont was extensively farmed, the Appalachians were still very much a wilderness and a barrier to early westward settlement.

The project would find me in well-known destinations such as Shenandoah National Park and the Blue Ridge Parkway, but also in lesser-known locales. Many of these lesser-known areas are home to rare species, and some exist solely to protect rare species. The Central Appalachians are not a simple monolith; one day I would be bouldering up a steep, rocky talus slope to a ridge, and the next day, hiking into a forest rich with red spruce and carpeted in moss. However, this project would not focus on listing and photographing mountains alone. Of course, summits would be included, but more important to me was to highlight the rich biodiversity of these Appalachians.

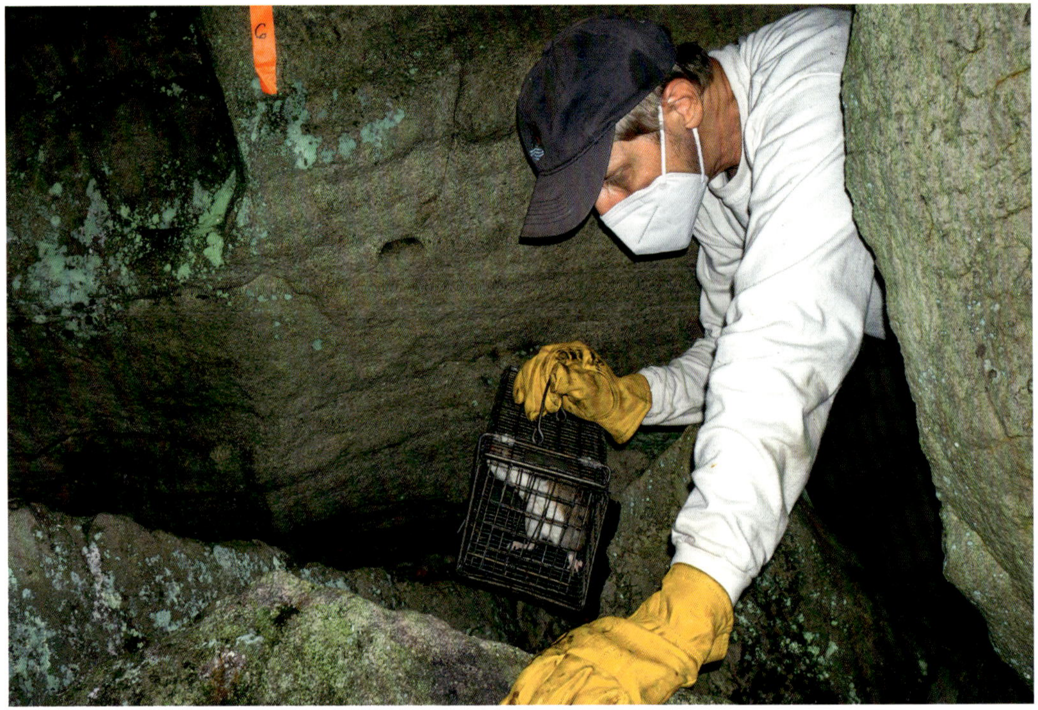

There were many surprises. From spelunking into the dark realm of caves to submerging into rivers, bogs, and vernal pools, I encountered an incredible diversity of life. I learned of species I was not previously aware of but that have since become most endeared to me, such as the Allegheny woodrat.

The best surprise of all, though, are the people: the biologists, conservationists, naturalists, and residents who opened up their hearts, homes, and land to assist me in these efforts.

A surprise I never could have anticipated was the COVID-19 pandemic. How could I continue my travels throughout the region when I had a toddler and newborn at home? Not only did the pandemic cause a delay in my work but also in research on imperiled species, since the risk of zoonosis was much too great. Eventually the project would move forward, albeit with caution.

When I began this project, it was just me and my wonderful wife. Now we have two amazing children who inspire my work in ways I could not have fathomed. It was no longer about me and a career, but about them. I hope that the places highlighted in this book will be further protected and supported, expanded, and reclaimed; that my children can experience these same wonders as I have, while being good stewards of the land.

For you, dear reader, I hope this book inspires you to visit and explore the Central Appalachians, to support the communities and those who work tirelessly to conserve it, to embrace their soft wind on an October morning, to feel the warmth of the sun after a summer's rain, to sit and listen to the warblers sing, and to understand the interconnectedness of these special places. The health of the mountains, the health of the water, and the health of the Bay. It is all but one and the same.

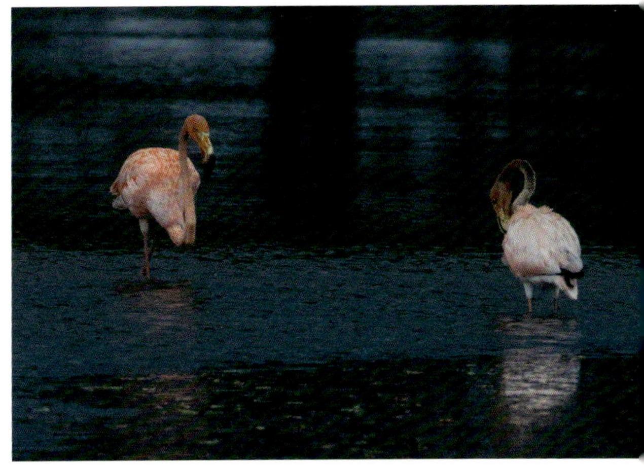

On the last day of this project, the author had a rare sighting of American flamingos, blown off course by Hurricane Idalia and foraging in a pond in south-central Pennsylvania. But with more-extreme storms could rare bird sightings like this become more common?

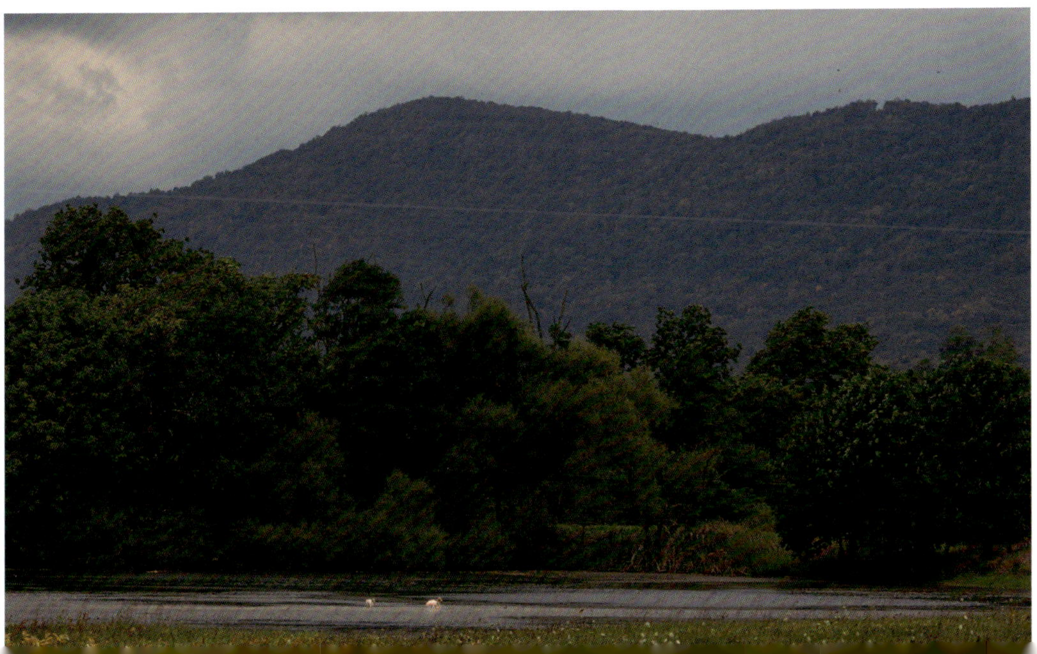

Biologist Dan Feller carries a live trapped Allegheny woodrat as part of a biological study on this threatened species. Garrett County, Maryland. *Adjacent page, top right*

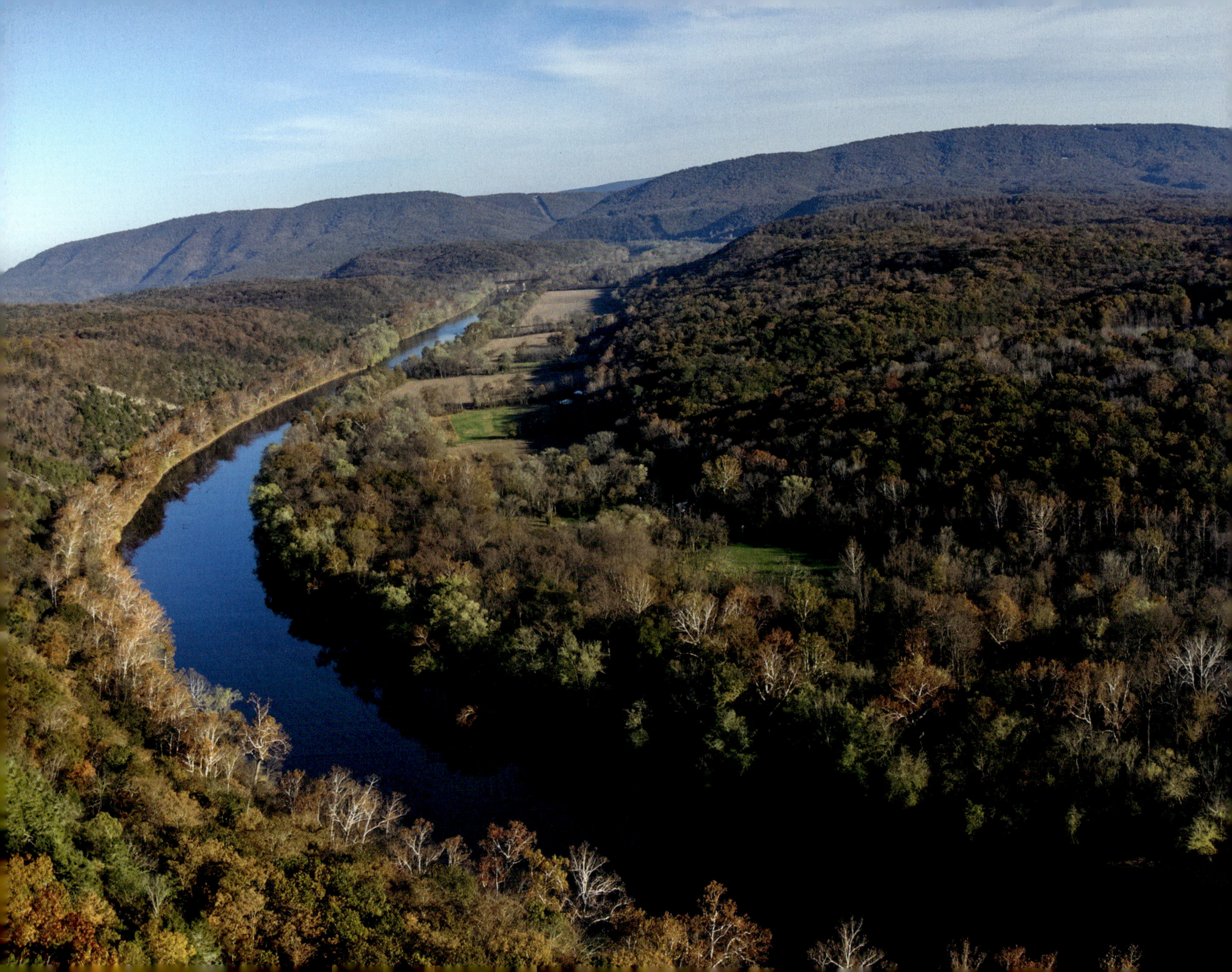

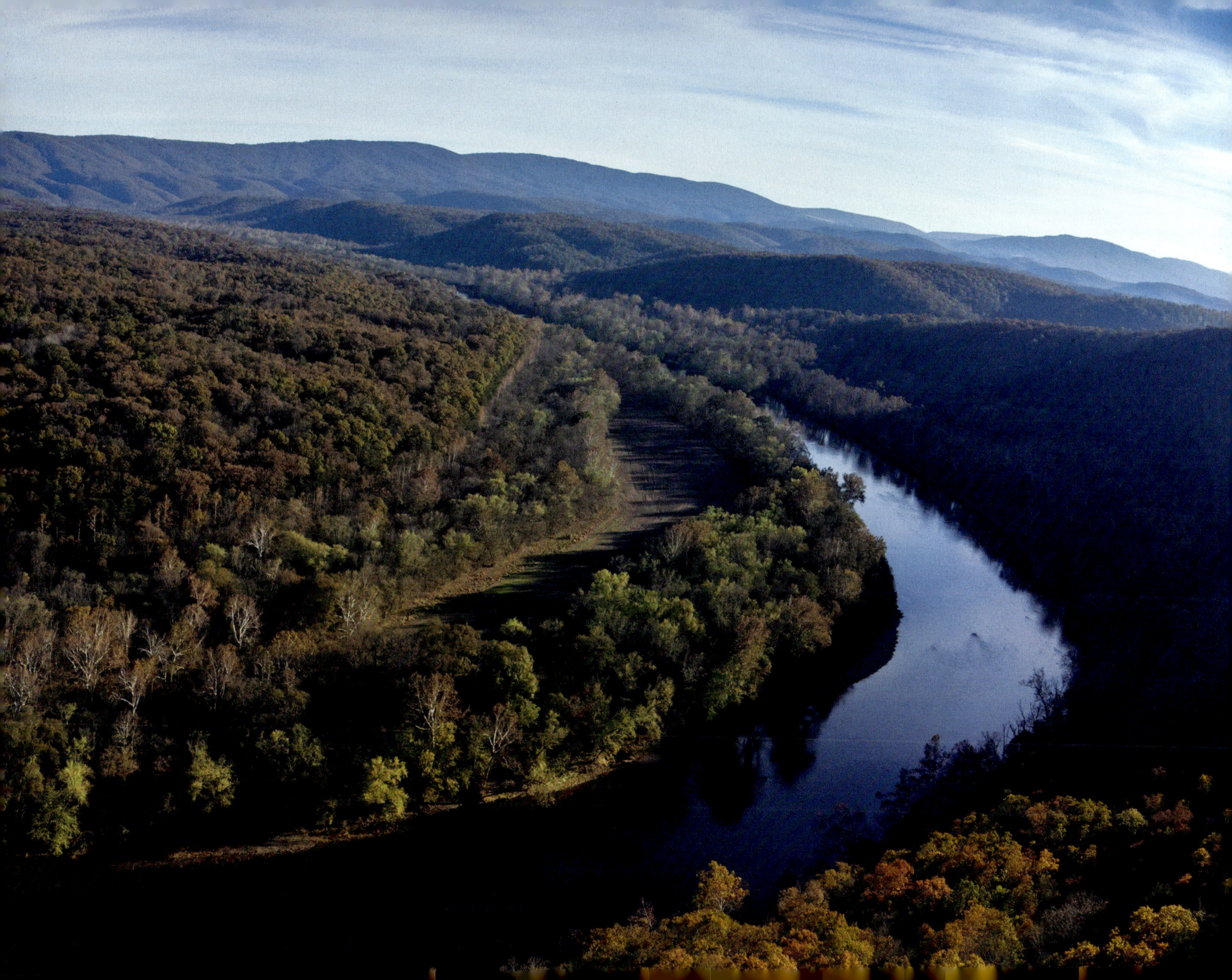

ABOUT THE CENTRAL APPALACHIANS & THE CHESAPEAKE BAY WATERSHED

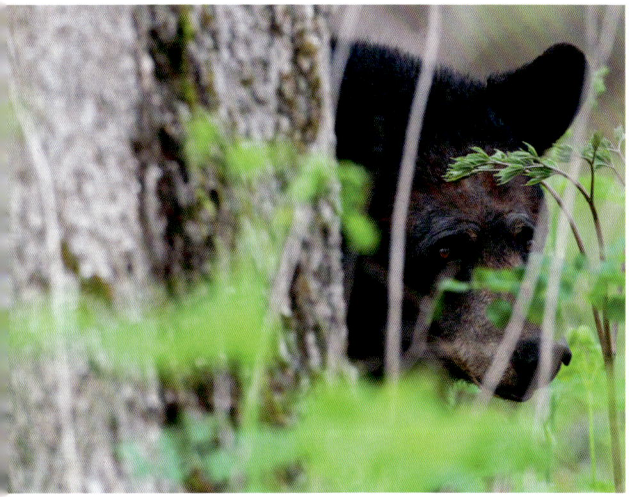

A curious black bear sow. The black bear population has increased significantly in the Central Appalachians.

Ancient. Majestic. Biodiverse. These are just some of the superlatives that describe the Appalachian mountain chain, a pantheon of the natural world. A genesis that began over 480 million years ago, when the tectonic plates of ancestral North America and Africa began a slow approach to one another during a time period known as the Ordovician, part of the Paleozoic era. Over 200 million years later, the continents collided, resulting in rocks being pushed westward into North America and piling up to form the mountains that we know today. In previous epochs the peaks towered as high as the Rocky Mountains, the Swiss Alps, and possibly the Himalayas, but millions of years of erosion has since transformed their outward appearance into the soft, almost gentle, forested look we know today.

Stretching over 1,500 miles from Newfoundland to Alabama, the Appalachians are one of the most biodiverse regions in a temperate environment. A rich topography of hills and ridges, valleys, and plateaus contributes to a stunning array of plant and animal life as well as important migration corridors.

Rich in natural resources, the Appalachians have fueled a nation.

The Chesapeake Bay is big.
Holding over 18 trillion gallons of water, it the largest estuary in North America and third largest on the planet. The equally impressive Chesapeake Bay watershed spans over 64,000 miles, with a land-to-water ratio of 14:1. With so much land draining into so little water, what happens terrestrially has a big impact on the water. Six states and all of Washington, DC, are found within its borders,

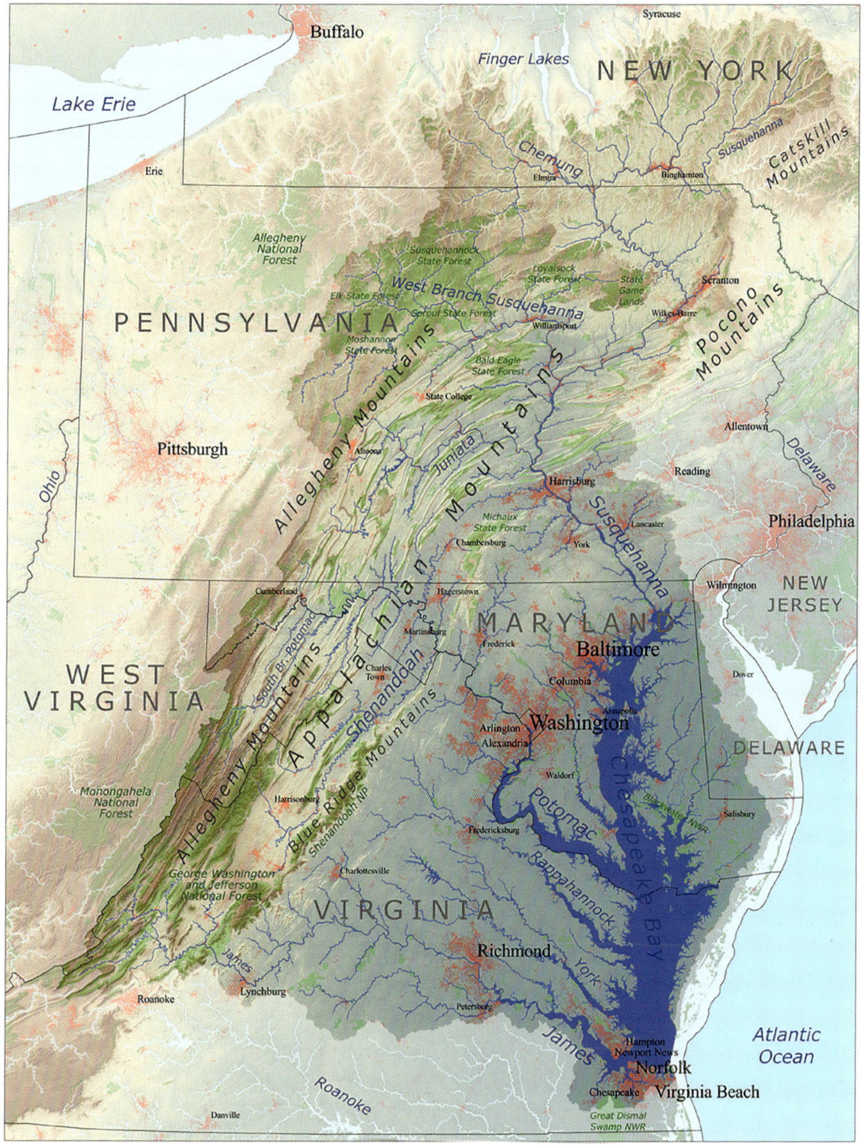

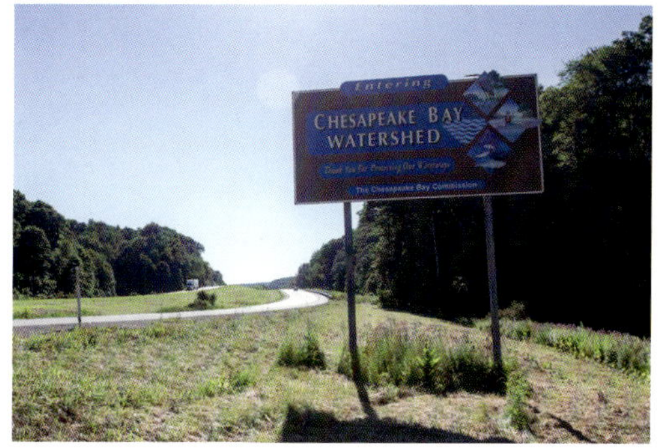

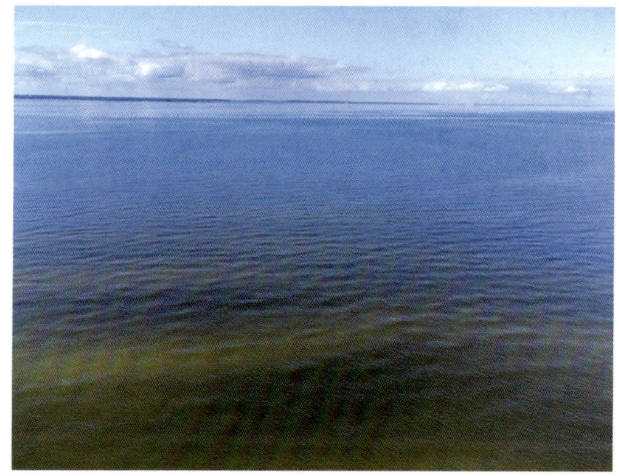

The Chesapeake Bay watershed. *Courtesy of Karl Musser*

Welcoming motorists to the watershed on Interstate 80, headed east in northern Pennsylvania, a reminder of how vast the watershed is

Aerial view of the Chesapeake Bay

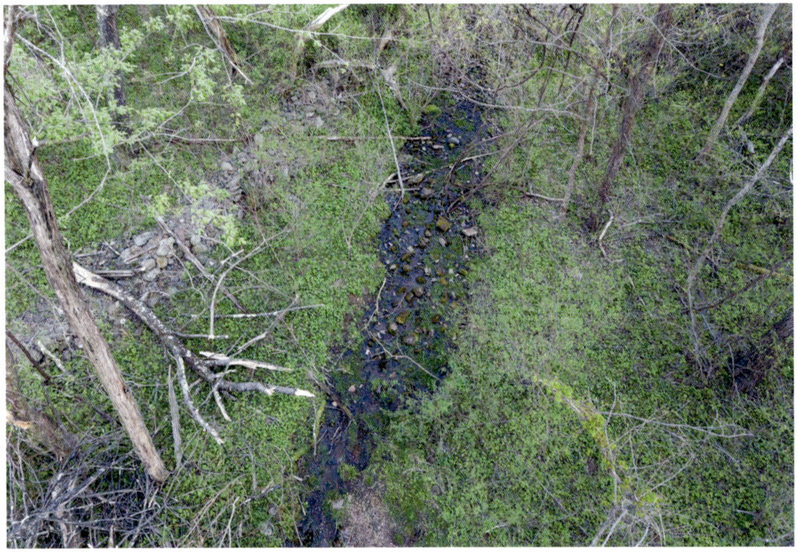

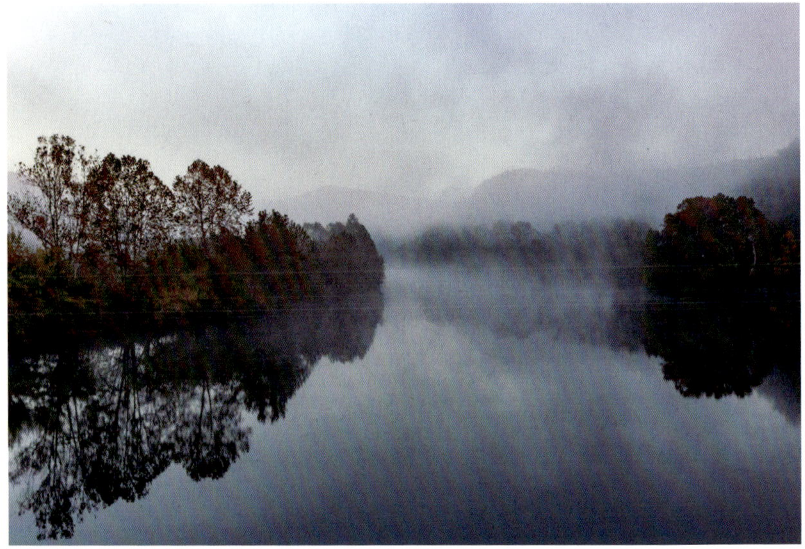

and over eighteen million people call it home. The total length of all the tributaries in the watershed is over 100,000 miles, and these are responsible for 51 billion gallons of water that flow into the bay each day.

The Appalachians are one of the geological provinces that make up the watershed and provide its dramatic northern and western boundaries. Four of the five largest tributaries that feed the bay have headwaters in the Appalachians: the Susquehanna, Rappahannock, Potomac, and James Rivers. The Susquehanna River alone accounts for half the fresh water the flows into the bay. When the James and the Potomac Rivers are included, that figure rises to 80 percent.

A mosaic of forest and habitat types exist in the Central Appalachians, which contribute to the rich biodiversity of the watershed.

The headwaters of the Rappahannock River during spring in the Blue Ridge Mountains at Chester Gap in Virginia

Autumn fog encompasses the James River seen from the Blue Ridge Parkway in Virginia

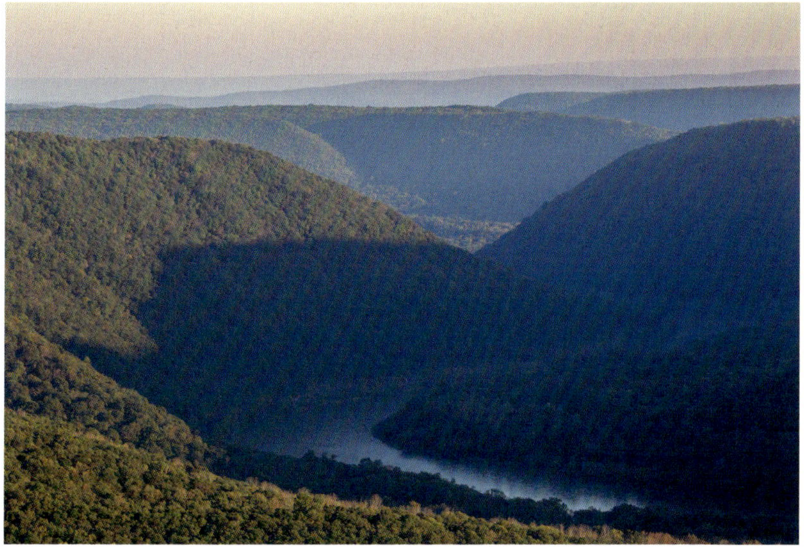

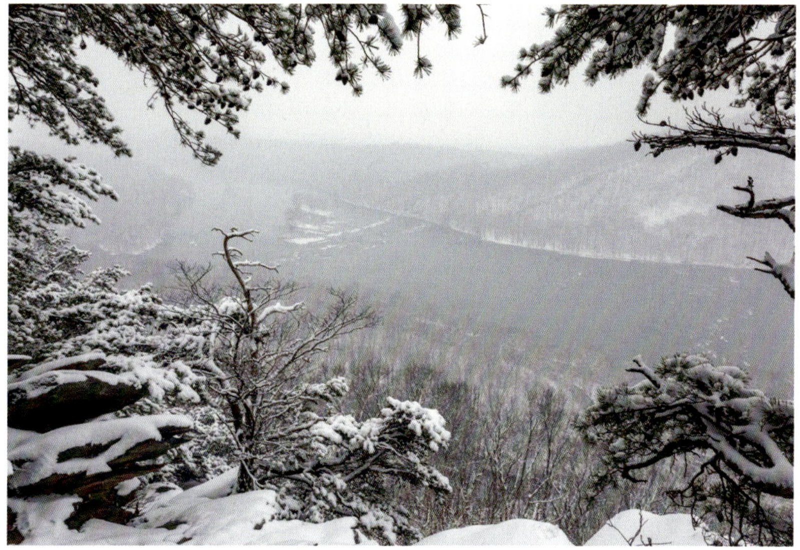

A summer's evening near the West Branch of the Susquehanna River, seen from Hyner View State Park in Pennsylvania

Winter comes to the Potomac River, seen from Weverton Cliffs off the Appalachian Trail in Washington County, Maryland.

The predominant forest types of the mountains are oak and hickory, and in some areas a combination of oak, hickory, and also pine. American chestnuts were once a dominant species but have heavily declined in the wake of the chestnut blight. Today, blight-resistant chestnut trees are being planted. Other forest types include northern hardwood forests, which may be found in northern Pennsylvania as well as at the higher elevations of above 3,000 feet along the Allegheny Front in western Maryland, West Virginia, and western Virginia. This cool, mesic environment may also be found along north-facing slopes and cool ravines. Dominant tree species include sugar maple, birch, black cherry, and beech, interspersed with conifers such as eastern hemlock, white pine, and even red spruce and balsam fir in some areas. Red spruce forest exists in West Virginia above 4,000 feet, and strands of spruce exist on Virginia's Jack Mountain at over 4,200 feet, as well as in mountain bogs in Maryland and Pennsylvania.

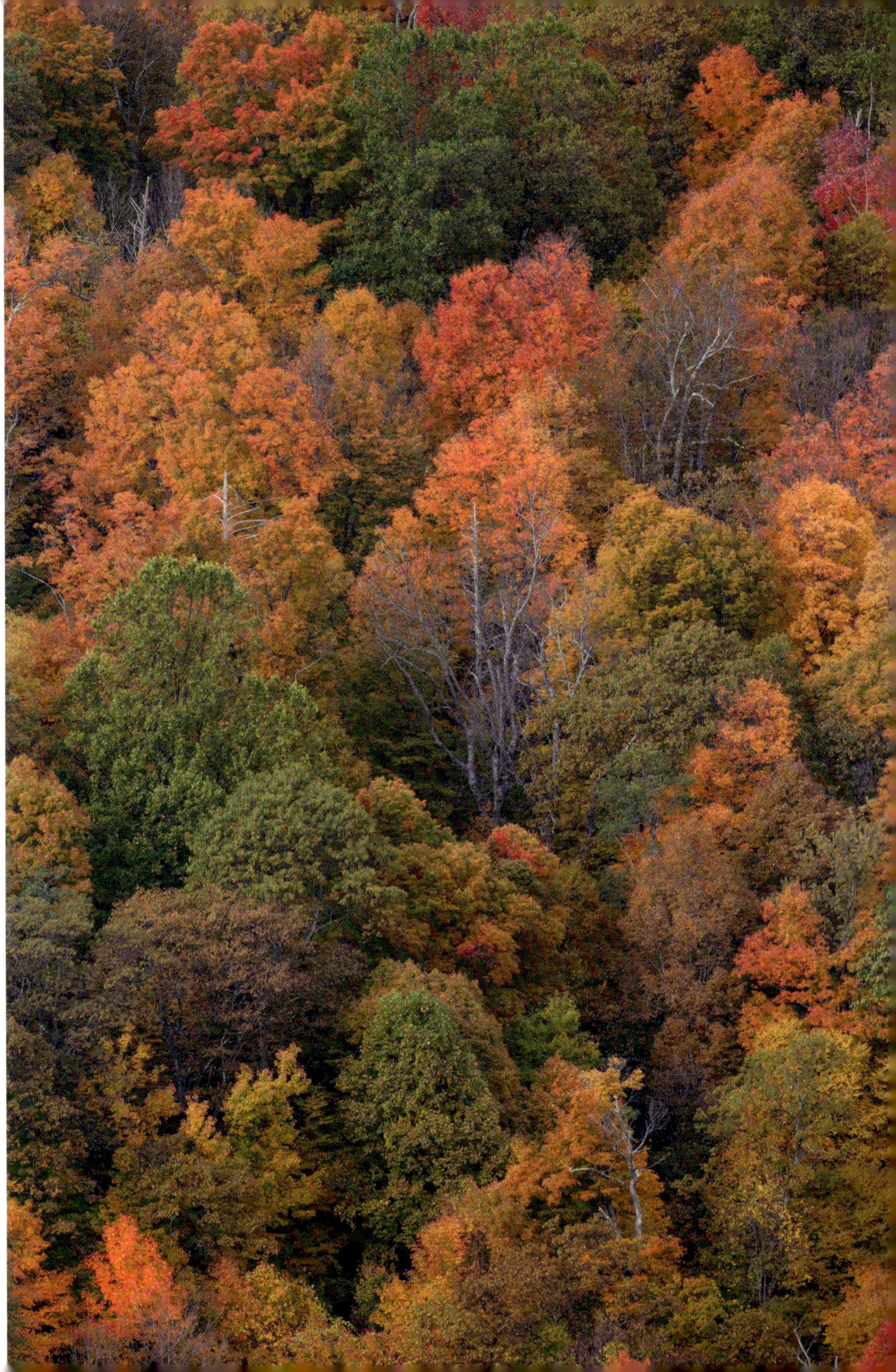

Oak-hickory forest in autumn, Virginia

Spring forest in Garrett County, Maryland

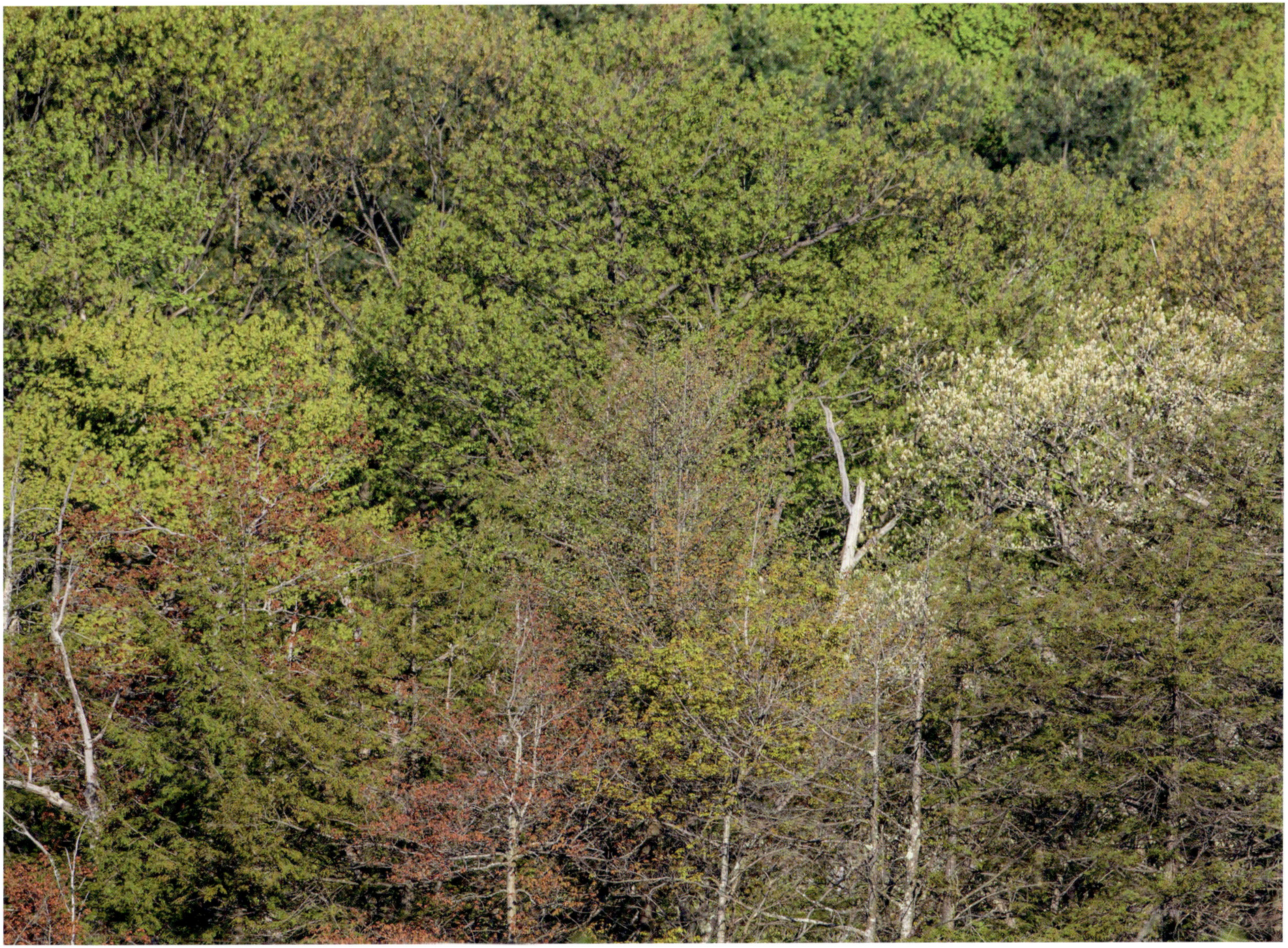

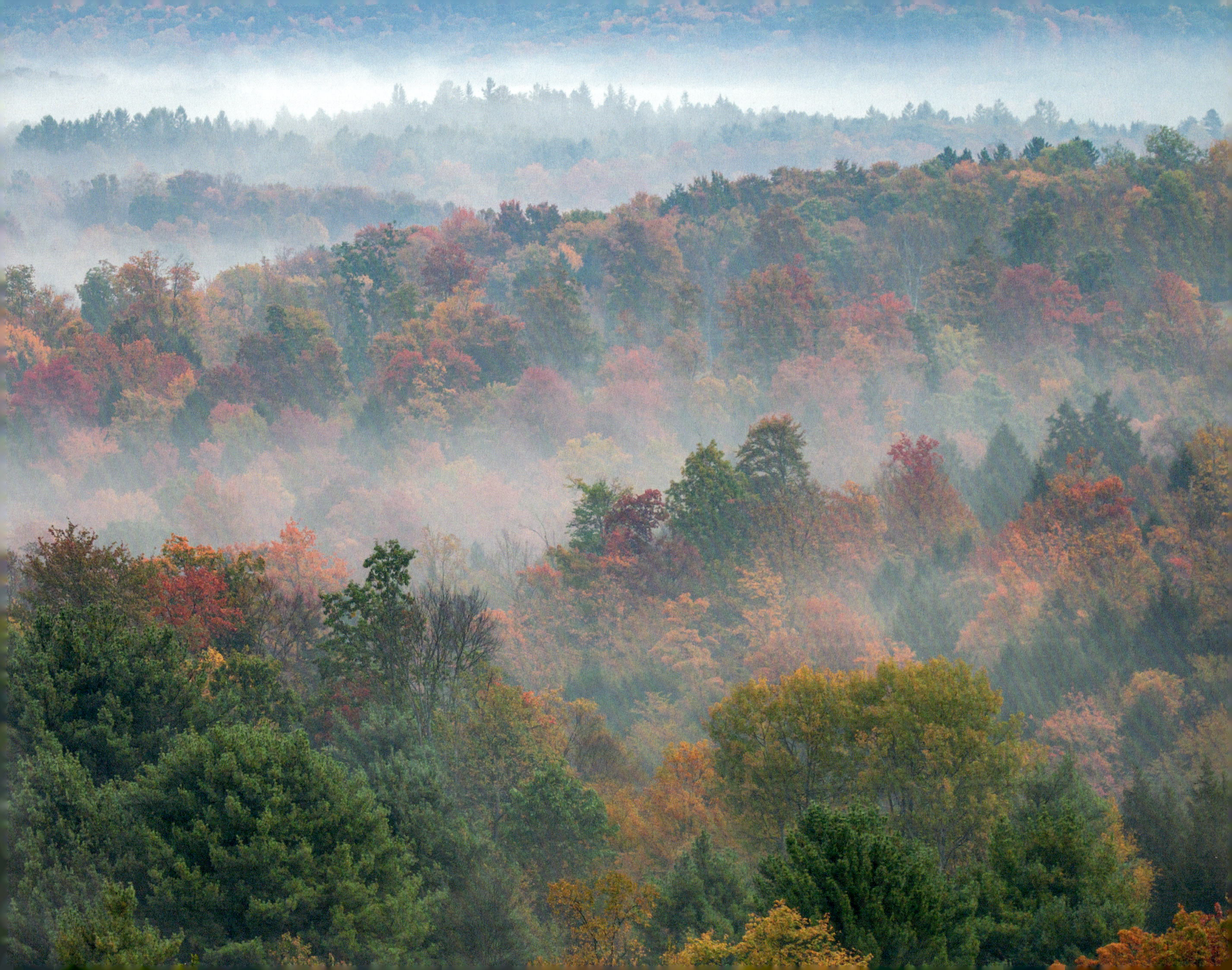

Today, much of the Appalachians may be viewed as an altered landscape with a legacy of poor forest management. While some old-growth forest still exists, much of it was cleared away by the early twentieth century. The altered landscape has since given way to secondary forest, which has forever changed the region. Today, tracts of land are being managed to allow the forest to regenerate so that it can eventually become old growth. These forests are important not only for wildlife habitat but also for carbon sequestration and water quality. Much of the forest of the central Appalachians exists on private land.

Besides forests, a myriad of habitats are also found in the Central Appalachians. Boreal bogs, fens, wetlands, shale barrens, and rock outcrops all contribute to the region's biodiversity.

Glaciers never reached as far south as the Central Appalachians but were present in the northern ridges of the glaciated Allegheny Plateau, a province within the Appalachian Plateau of Pennsylvania. The effects of glaciation, however, are still felt today. It was when the glaciers melted that the Susquehanna River was flooded and formed the Chesapeake Bay. In the mountains, boreal species typically found much farther north still make their home in the high peaks, frost pockets, and wetlands.

All four states that form these Central Appalachians of the Chesapeake Bay watershed include national or state parks and forests, nature preserves, and wildlife management areas that provide recreation and study for visitors. A rich hiking culture is present, since the world's most famous path, the 2,000-mile-plus Appalachian Trail, transverses through the Blue Ridge Mountains of the region.

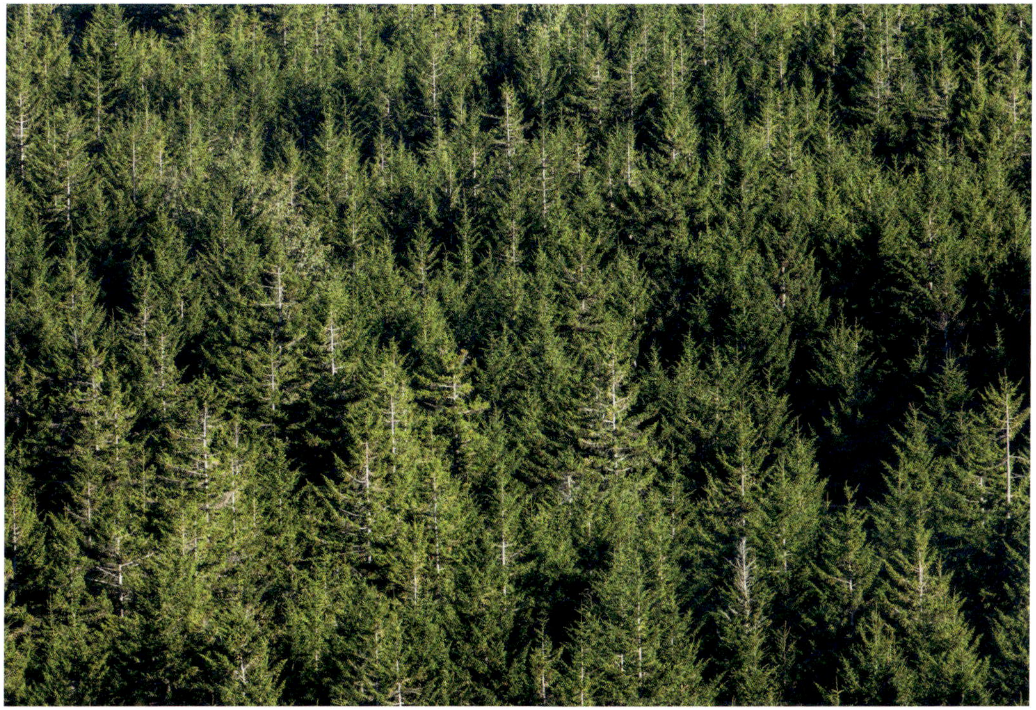

Northern hardwood forest in Elk State Forest, Pennsylvania

Red spruce forest in Dolly Sods Wilderness, West Virginia

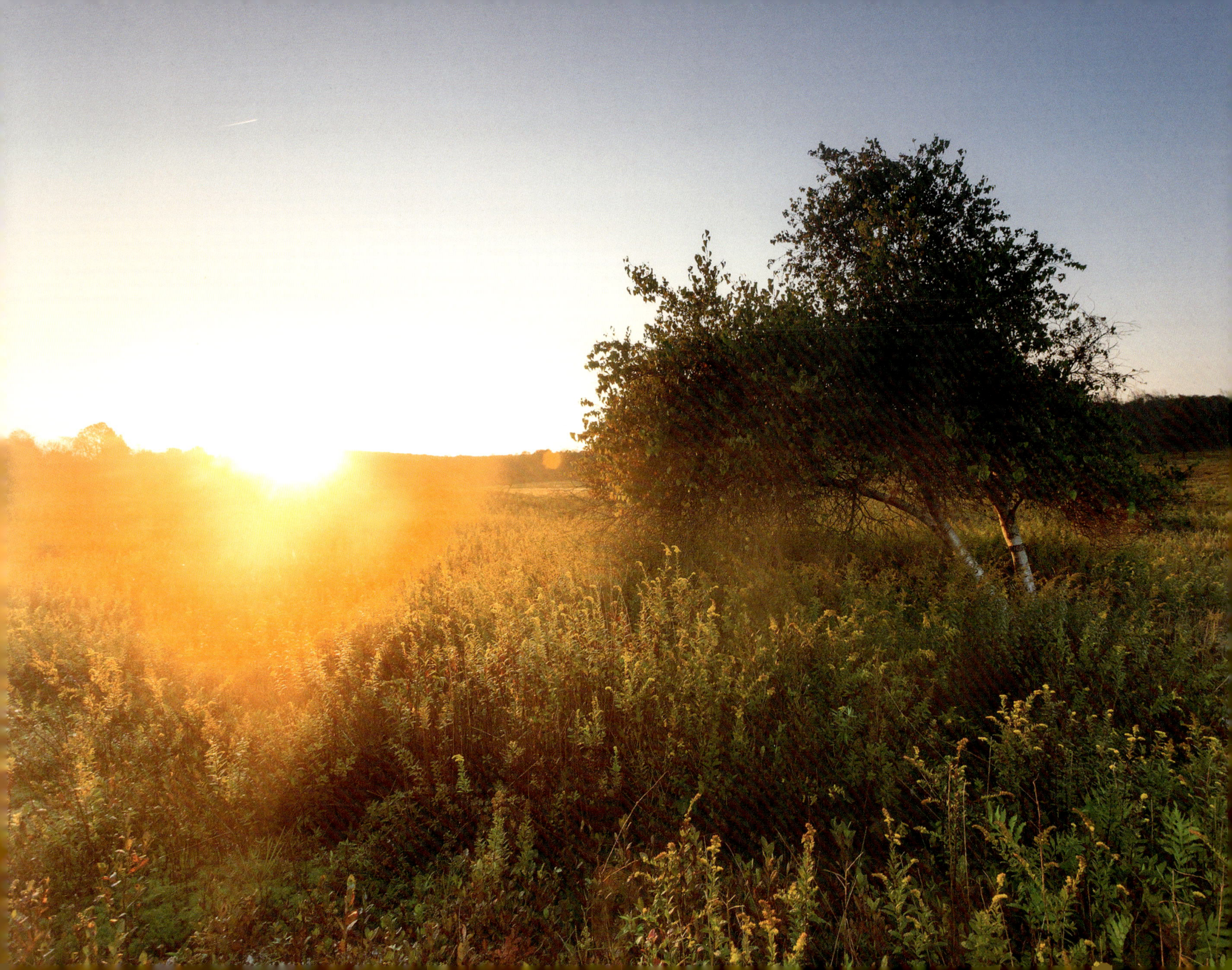

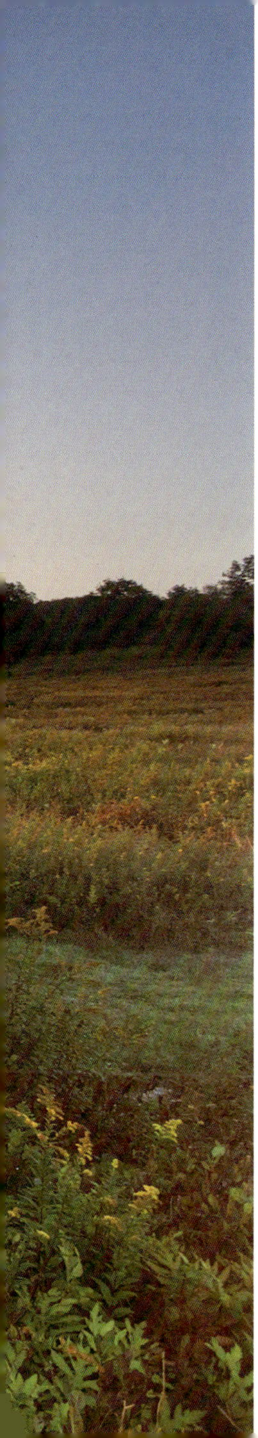

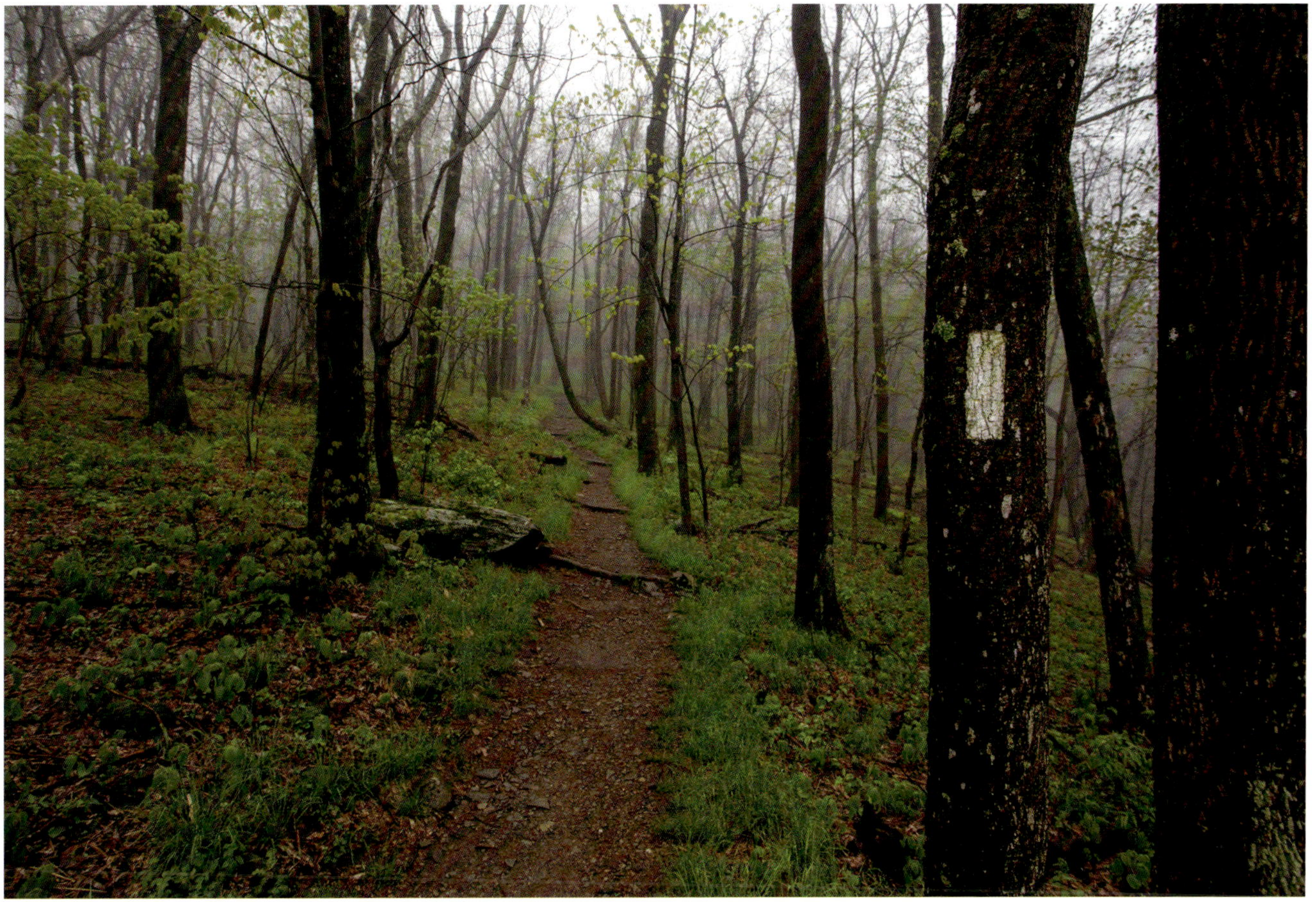

A gray birch tree in Shenandoah National Park. A species of the North, a disjunct population still survives in the park and is a relic of the much-cooler past of the mountains.

The ultimate symbol of the region's rich hiking culture: the white blaze of the Appalachian Trail

Appalachian Provinces in the Chesapeake Bay Watershed

The Blue Ridge

Rising at the western base of the Piedmont Province, the Blue Ridge Mountains form the first mountain province in the watershed and are bordered by the Great Valley to the west. Its famed, namesake blue haze comes from isoprene, a hydrocarbon produced by trees as a response to excessive heat. Oak trees produce a significant amount of isoprene and are a dominant species throughout the forests of the Blue Ridge.

The province is made up of deformed metamorphic rocks such as gneisses, slates, and quartzite. It is also mineral rich. Within the Chesapeake Bay watershed, the province extends from southern Virginia northward to the South Mountain range in Maryland and Pennsylvania. Here it is narrow, being about 10 miles wide at its widest. Outside the watershed, the mountain belt becomes wider and reaches all the way to Georgia. The Southern Appalachians of the Blue Ridge host most of the highest peaks found in the Appalachians and also contain Great Smoky Mountains National Park.

Shenandoah National Park and the Blue Ridge Parkway are the most-famous public lands within this province. Shenandoah National Park features a section containing a very rare biome called the Northern Blue Ridge Mafic Fen. Found in the Big Meadows area, this wetland sits on top of soil weathered from mafic igneous rocks, and hosts many rare plants.

Catoctin Mountain, which forms the easternmost ridge in the Blue Ridge Province, is home to the US presidential retreat Camp David.

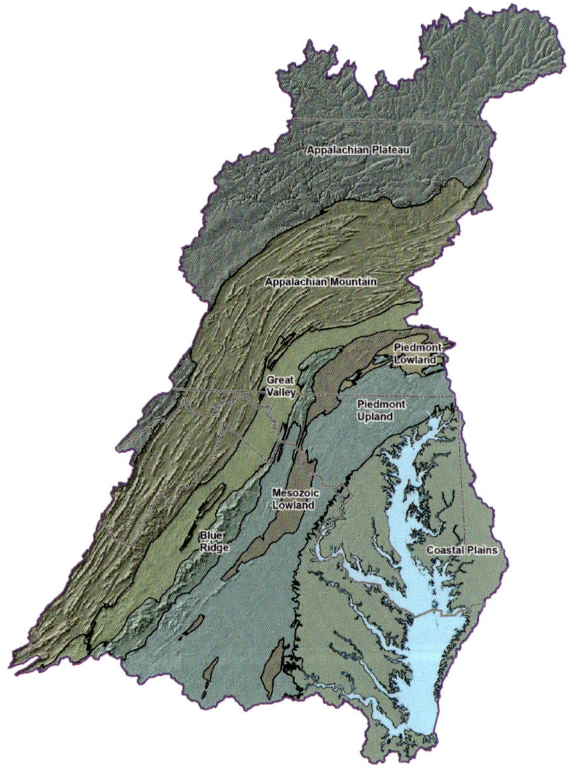

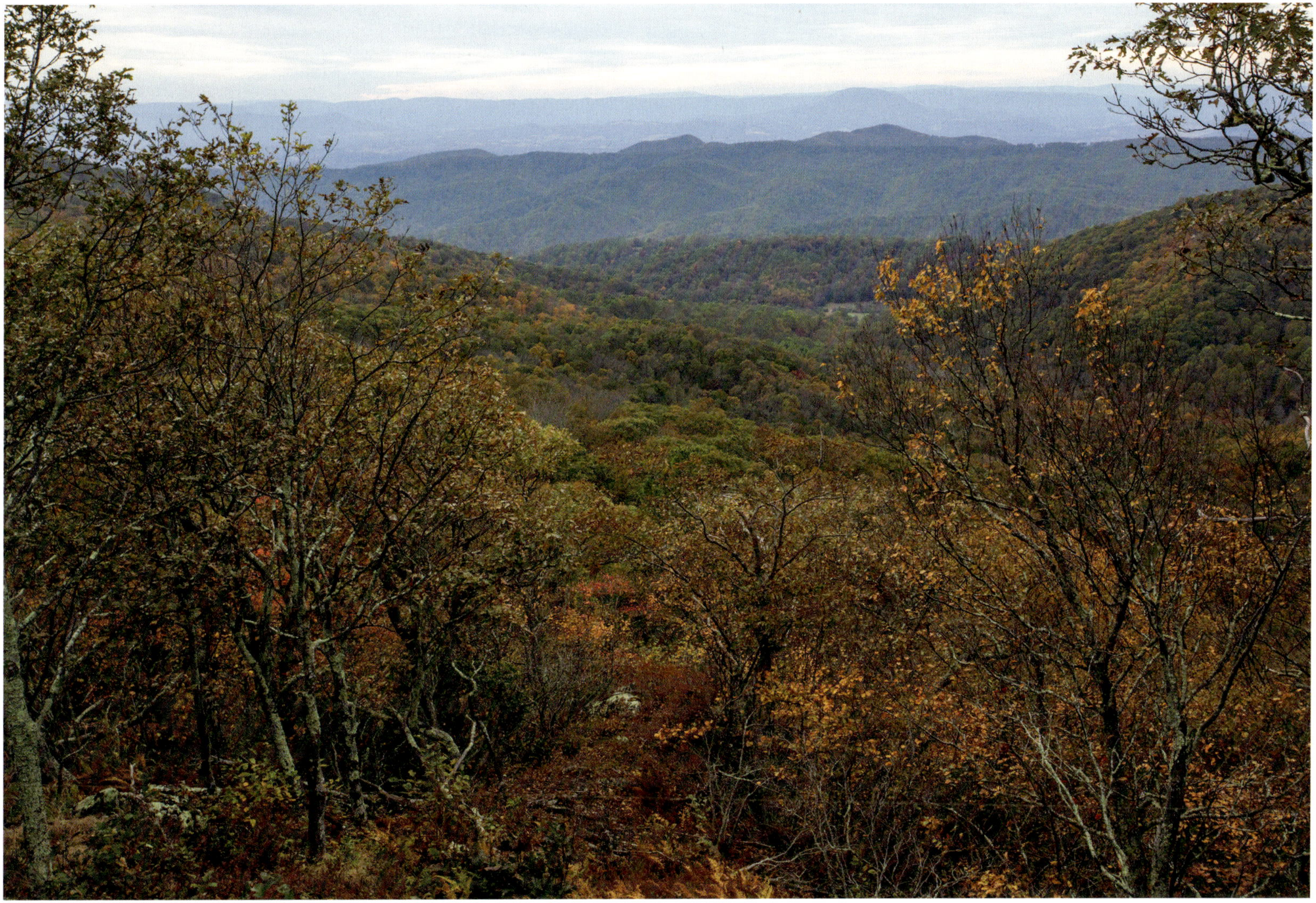

The geological provinces of the Chesapeake Bay watershed.
Map courtesy of Chesapeake Bay Program

Looking toward the Blue Ridge from the Appalachian Trail. George Washington National Forest, Virginia.

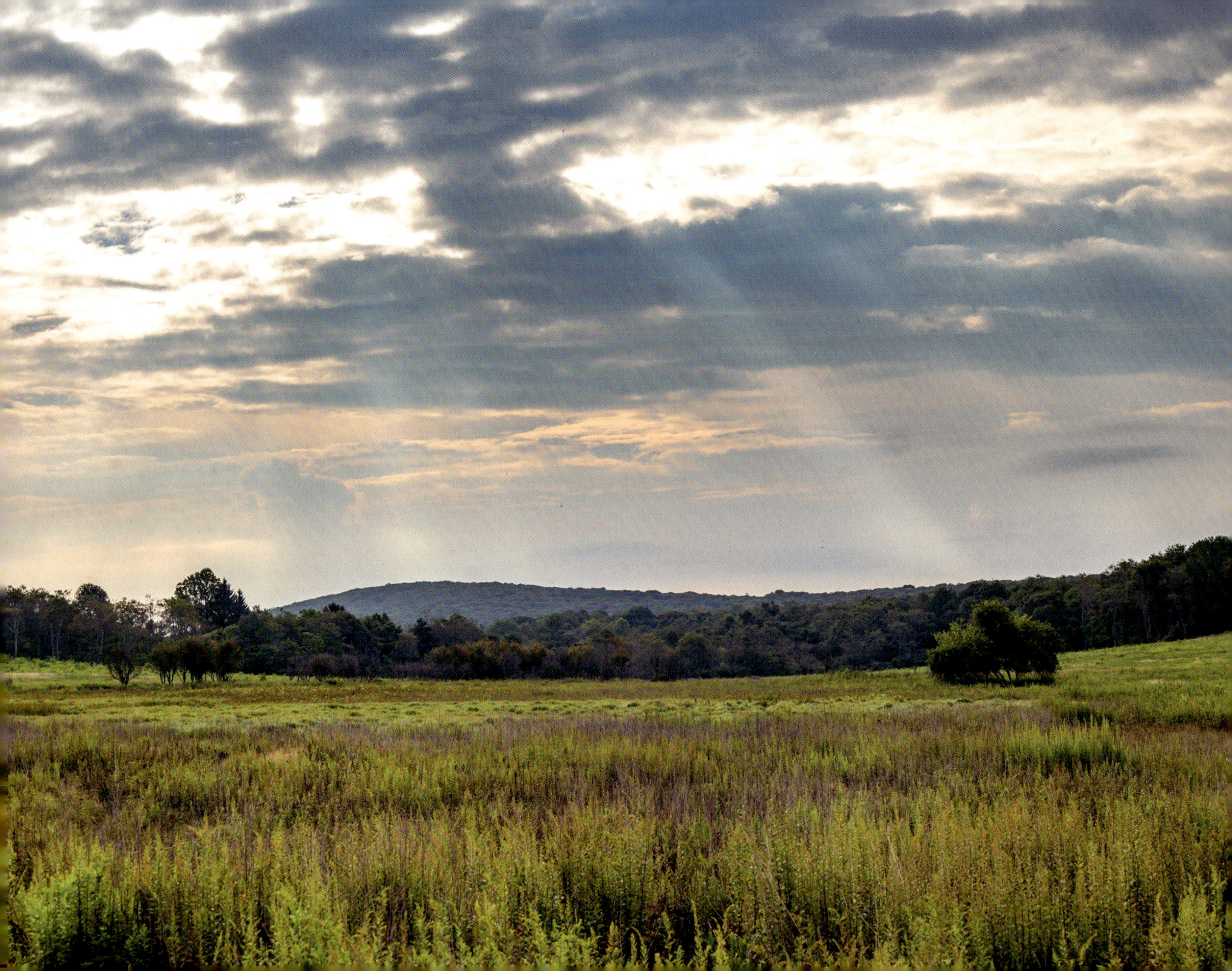

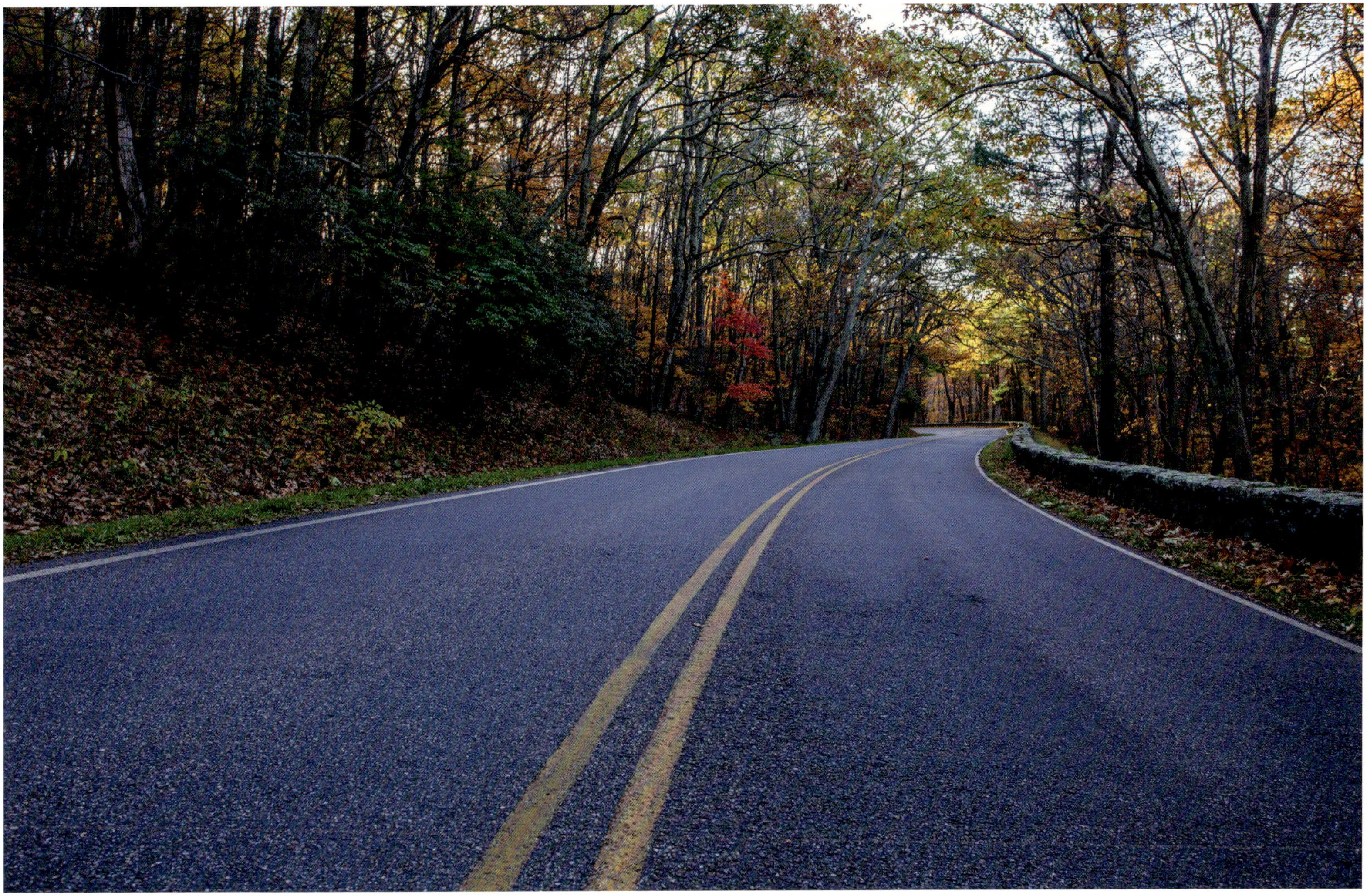

Northern Blue Ridge Mafic Fen

Skyline Drive, the 101-mile road the follows the crest of the Blue Ridge Mountains in Shenandoah National Park and connects to the Blue Ridge Parkway at its southern terminus

Ridge and Valley

A narrow belt running southwest to northeast, the Ridge-and-Valley Appalachians, as the name suggests, is characterized by long and sharply defined parallel ridges formed of sandstone and conglomerate rock. These rocks are much less resistant to erosion than the shale and limestone that formed the valleys.

The valleys between the ridges are straight. West of the Blue Ridge Mountains is the Great Valley (or Great Appalachian Valley), the largest valley of the province and its eastern border. The Great Valley has many regional names, such as the Shenandoah Valley in Virginia, the Hagerstown Valley in Maryland, and the Cumberland Valley and Lebanon Valley in Pennsylvania.

Most of the region's caves are found in this province due to its abundance of limestone, which is easily dissolved in water, forming caves. Also found here is perhaps the rarest ecotype of all: shale barrens. These dry, gritty areas are found on steep slopes and look like a piece of western North America gone astray. Most remarkable is that they are home to endemic plants found nowhere else.

West of the Great Valley, the ridges are narrowly spaced, with thinner valleys in between. The province is the parent range of the Allegheny Mountains, where some of the highest peaks in the watershed are found, including Spruce Knob in West Virginia and Dans Mountain in Maryland, and extends westward into the plateau region.

The Allegheny Front forms the transitional escarpment that separates the Ridge-and-Valley Appalachians from the Appalachian Plateau. The front is an important flyway for migratory birds.

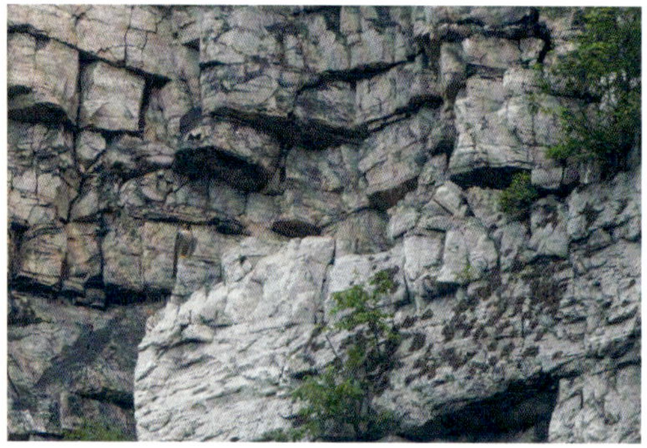

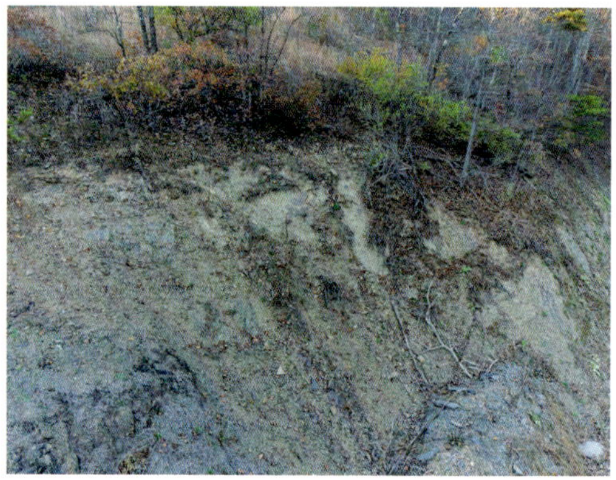

Exposed rock on Wills Mountain in Allegany County, Maryland

Shale barren in autumn

Appalachian Plateau

Rising up to the Allegheny Front, this rugged area with deep stream valleys is the westernmost province of the watershed. Unlike the sharper ridges in the Ridge and Valley, the plateau is characterized by wide, flat-topped highlands and much more rounded hills, though there are high ridges found here as well. It is made up of layers of sedimentary rocks such as sandstone, conglomerates, and shale. In this province are the rich beds of coal for which the plateau is most known. Part of the Chesapeake Bay watershed, this area is located in far western Maryland, western and north-central Pennsylvania, and West Virginia. The Allegheny mountain range (Alleghenies) is within the plateau.

Here the Eastern Continental Divide forms the watershed's eastern boundary, separating the rivers that flow eastward to the Atlantic Ocean (and Chesapeake Bay) from those that flow into the Gulf of Mexico.

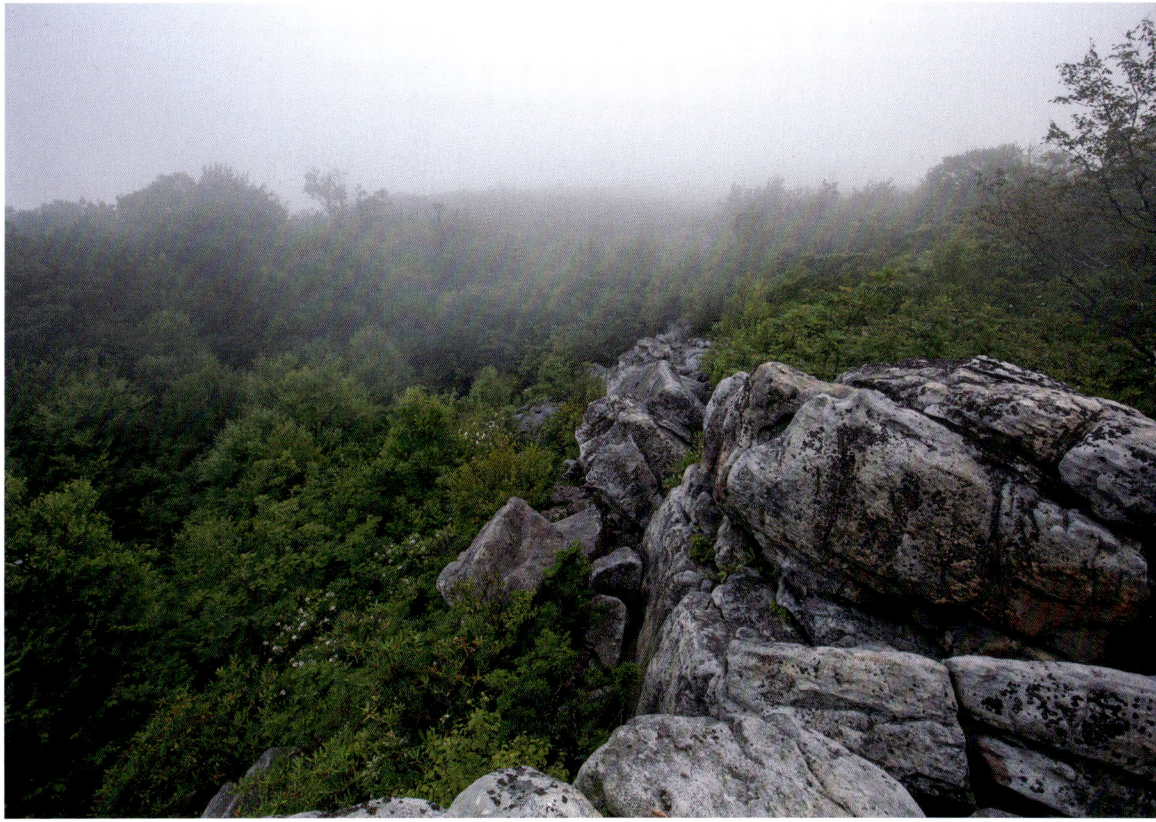

High Rock, Savage Mountain, Garrett County, Maryland

Today in the Mountains

Today there are numerous parks (national, state, and local) and preserves that protect portions of the Central Appalachians. However, with a warming climate, it is hypothesized that the higher elevations of the Central Appalachians will become a safe haven for species fleeing changing ecosystems. Additionally, there are issues with invasive species, with finding a balance between conservation and energy development, and with the restoration of forests. Because the watershed covers such a large land mass, what happens to an area hundreds of miles from the headwaters, or mouth, affects the Chesapeake Bay. Pollutants from development that flow into the Susquehanna River in Pennsylvania are detrimental to the subaquatic vegetation of Maryland and Virginia.

Author's note: Most of the species found in this book are found throughout the four states of the Central Appalachians: Pennsylvania, Maryland, West Virginia, and Virginia. Locations of many scenes and names of species are disclosed in the photos' captions, especially if the species is publicly disclosed as residing in that park or preserve. However, there are both protected and endangered species that may be extremely limited in their range and very sensitive to disturbance. For their protection, so as not to risk harassment or poaching of the species, the locations of these species are not disclosed.

Prior to spring's arrival, the gray birch is covered in hoarfrost during a winter storm.

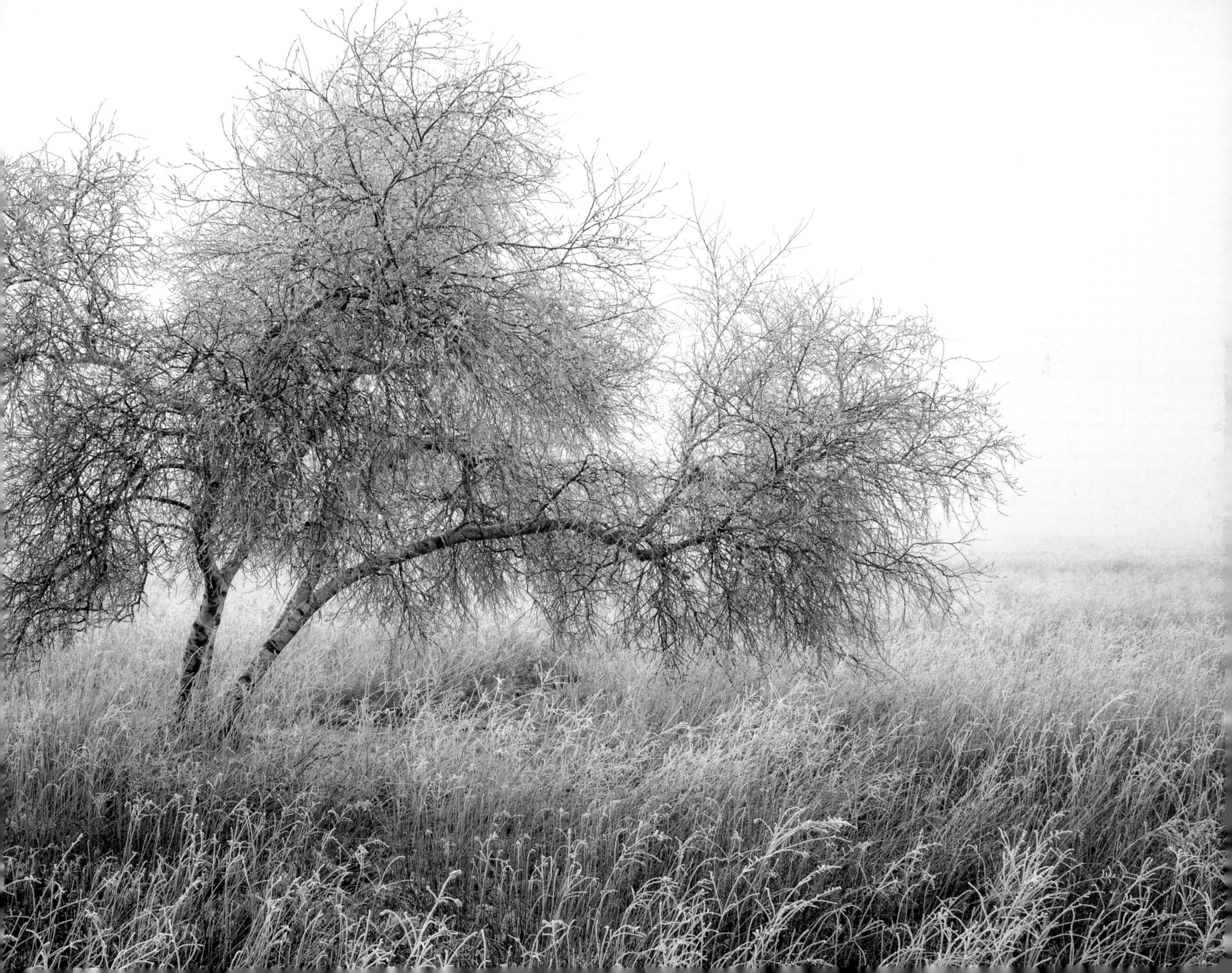

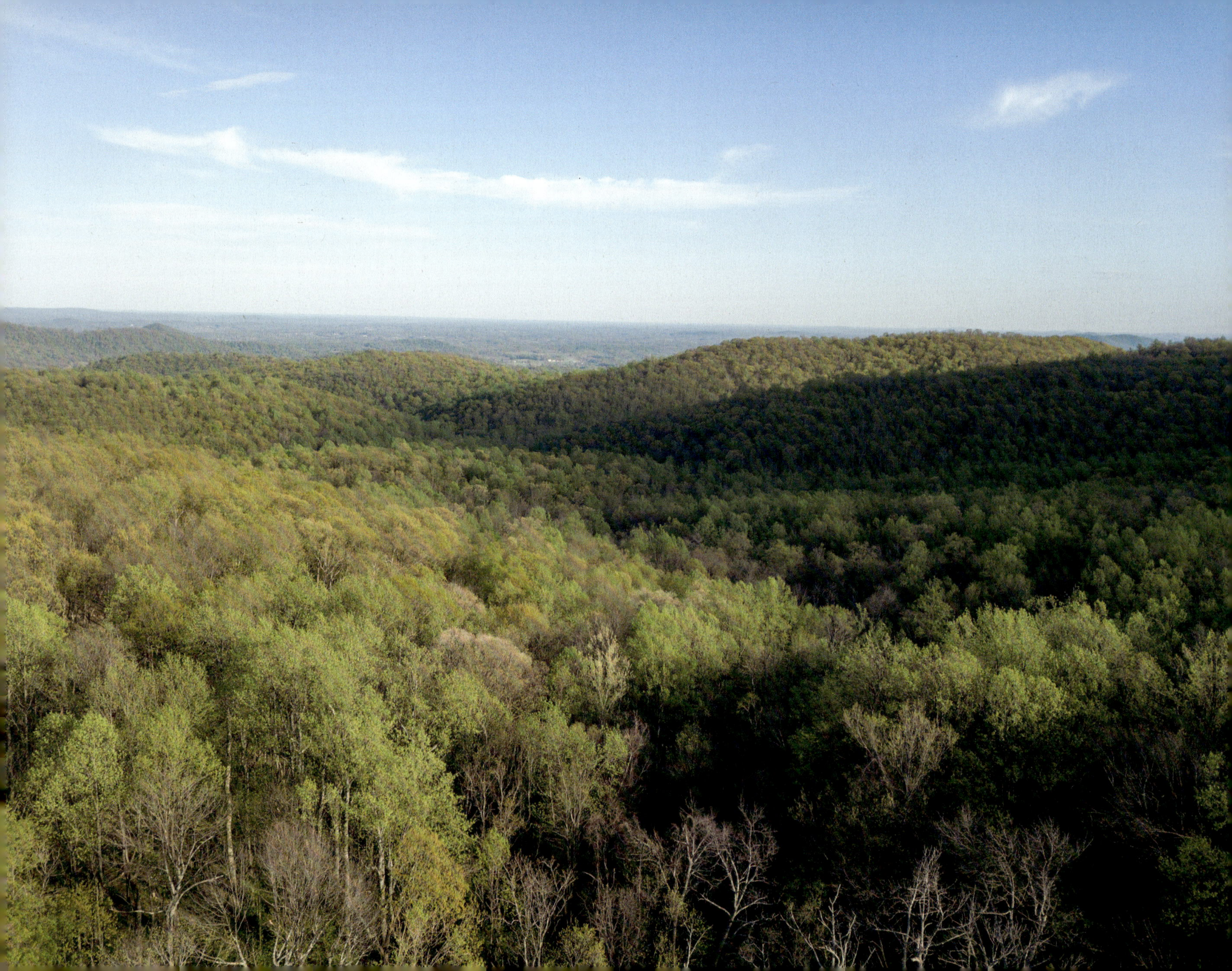

SPRING

The season of growth and renewal begins cool and muddy.
Meltwater flows from thawing ice before the first torrents of rain fall.

One of the earliest wildflowers to bloom is round-lobed hepatica, adding color to a still-brown forest floor. An early rite of the season occurs when American woodcocks return to their leks after sundown. Meltwater and early rain fill into depressions, forming vernal pools and creating mating hotspots for amphibians such as wood frogs and spotted salamanders. The gelatinous egg sacs will soon hatch tadpoles and larvae, beginning their metamorphosis in the waters of the pool.

Come April, spring ephemerals such as spring beauties and trilliums carpet the landscape, taking advantage of the sun's warmth shining through the bare limbs of the canopy. But soon the mountains will showcase shades of light greens and yellow, or simply "spring green." Complementing the greenery are the blooms of smaller trees such as flowering dogwoods and eastern redbuds, as well as azalea and rhododendron shrubs.

High in the Blue Ridge Mountains of Shenandoah National Park, a solitary gray birch begins to leaf out in the Big Meadows area. A spring storm approaches, covering the meadow in a haze. Less than a month ago, the tree was also surrounded by fog but covered in hoarfrost.

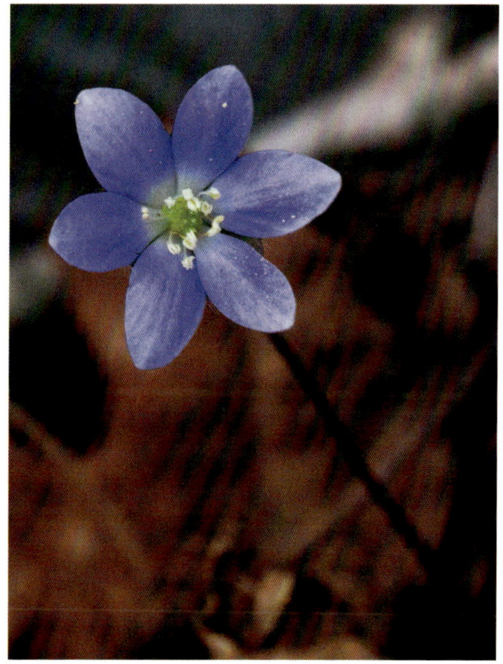

"Spring green" comes to the Blue Ridge Mountains near Chester's Gap, Virginia.

Round-loped hepatica. George Washington National Forest, Virginia.

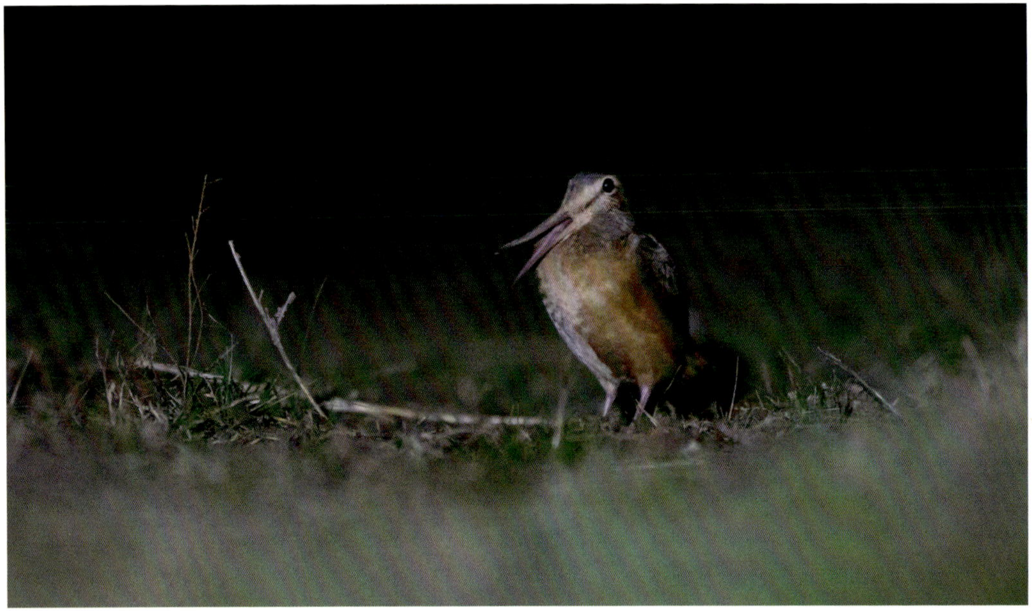

Life in the forest is animated and exciting. A black bear sow emerges from her winter den, and her cubs follow close behind. A cacophony of birdsong from returning neotropical migrants such as warblers and tanagers creates a natural symphony while the drumming of the ruffed grouse provides the percussion. In late May and into June, newborn white-tailed deer fawns will take their first steps. Their spots give much-needed camouflage during the precarious first weeks of existence, but they will soon be able to outrun any predator.

An American woodcock "peent" calling at a lek

White trillium in bloom in Shenandoah National Park, Virginia

False hellebore, Savage River State Forest, Maryland

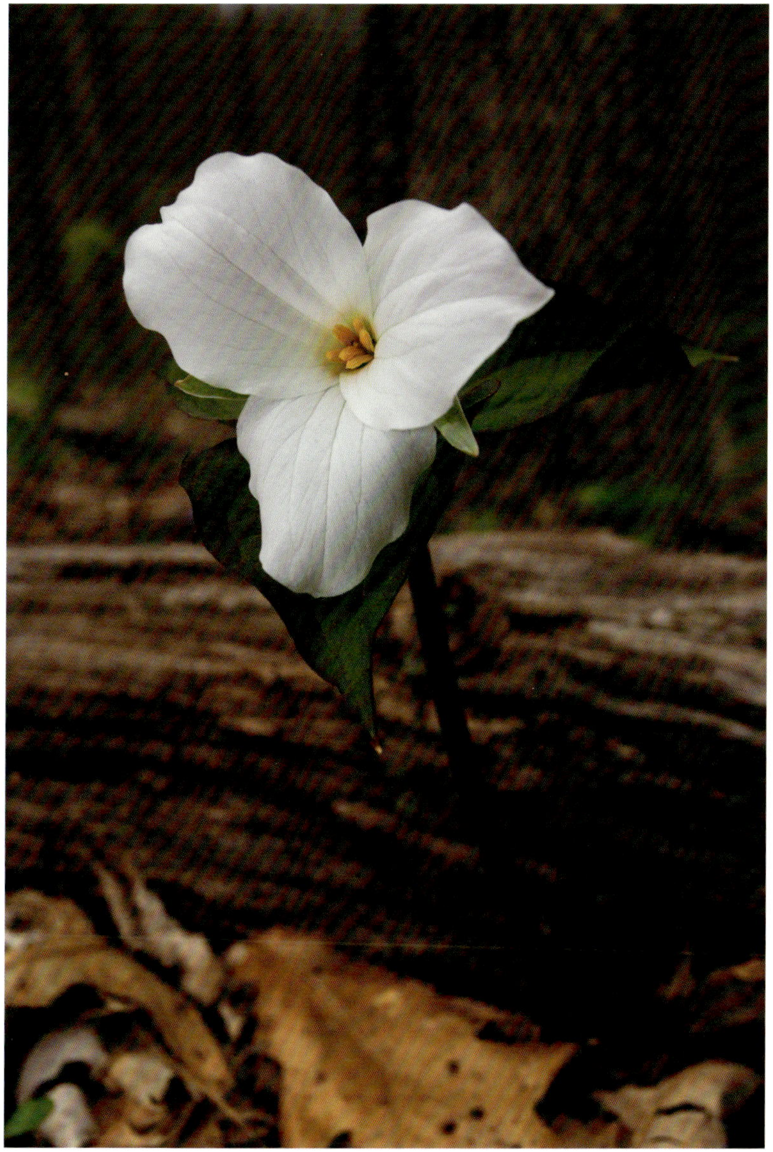

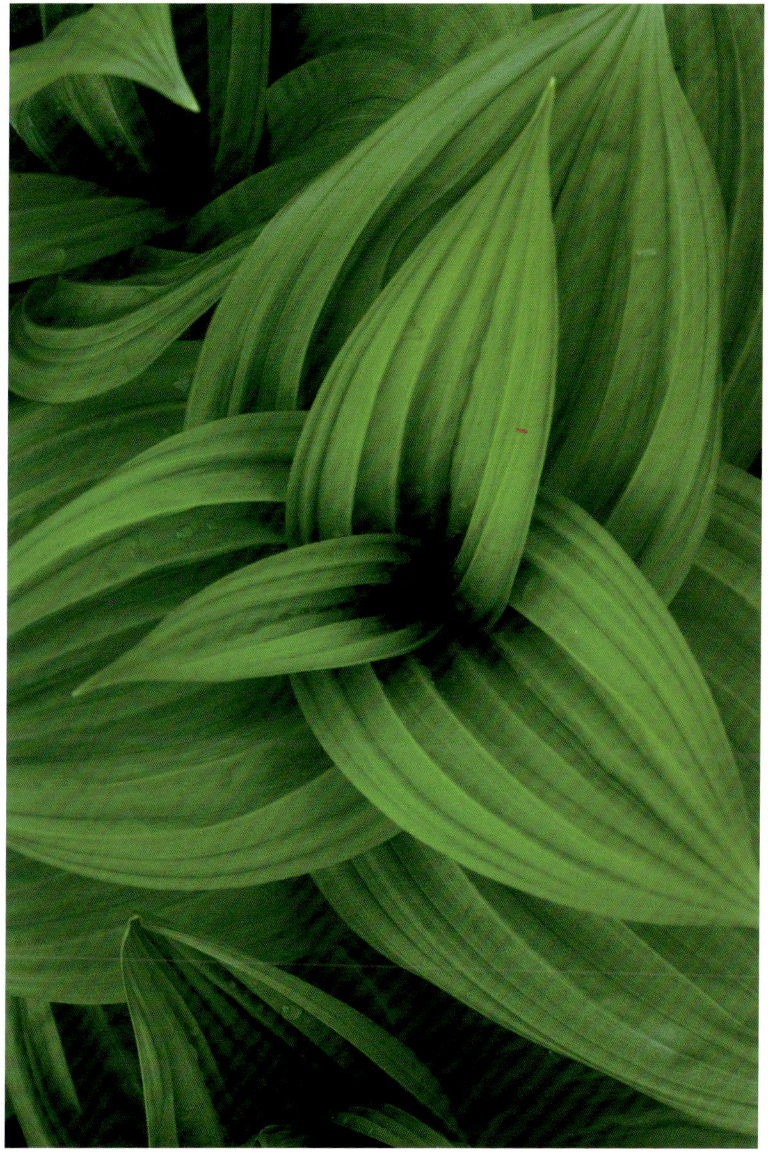

For the birch, the return of spring is wet and foggy, and soon the meadow will be full of new life. Shenandoah National Park, Virginia.

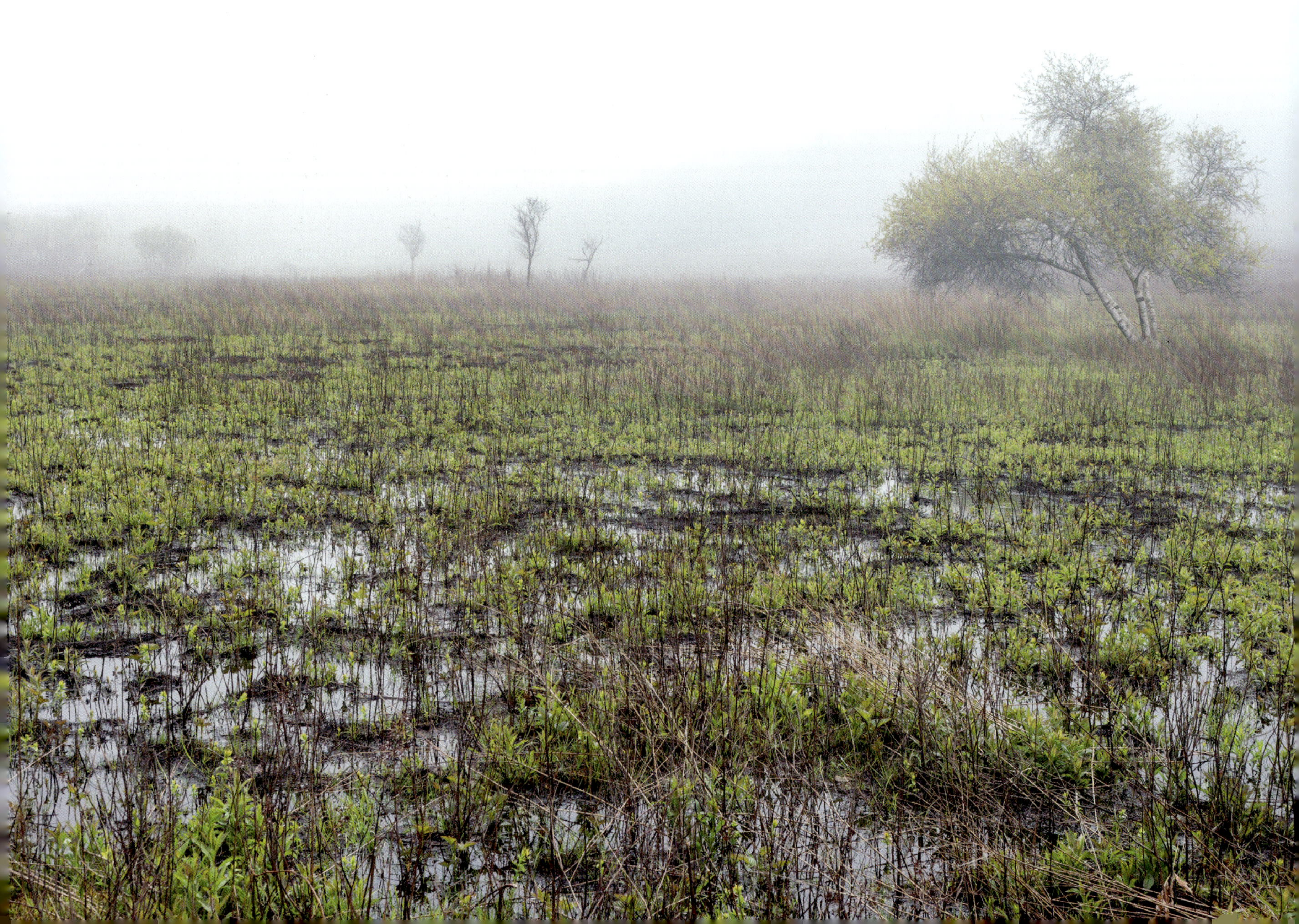

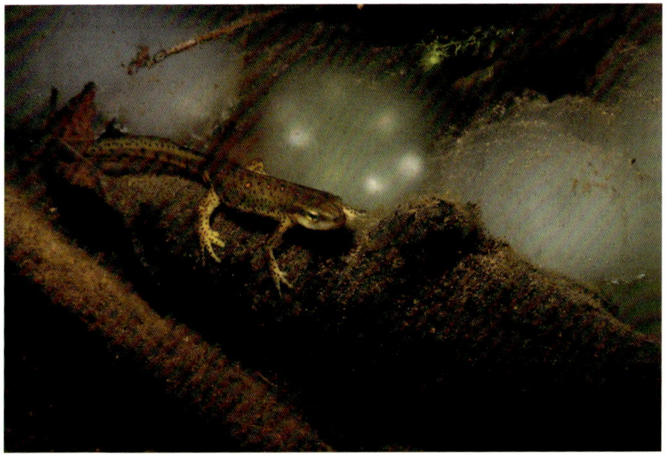

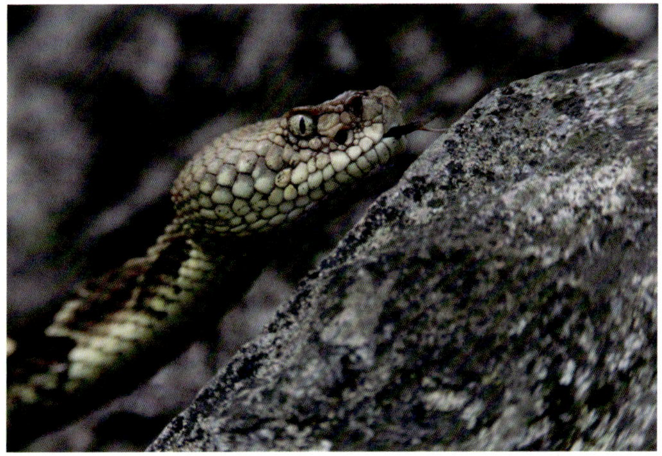

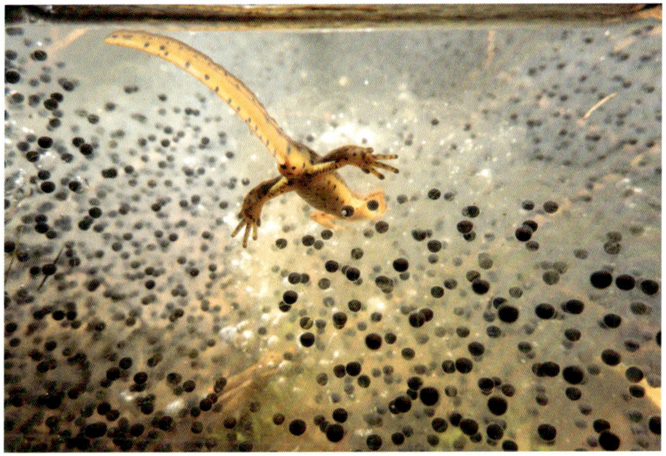

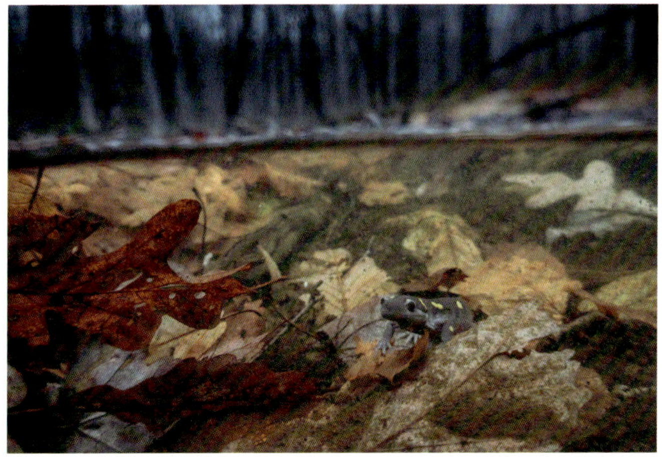

Warm spring rain and snowmelt form vernal pools, ephemeral habitat for amphibians such as the eastern newt to breed and a nursery ground for efts. Here, eastern newts use spotted salamander eggs as a food source. Frederick Municipal Watershed, Maryland, and Michaux State Forest, Pennsylvania.

A timber rattlesnake emerges from its winter den.

Warm spring rains signal the spotted salamander to return to the same pool of its birth, after living most of the year under the cover of leaves, logs, and soil. Frederick Municipal Watershed, Maryland.

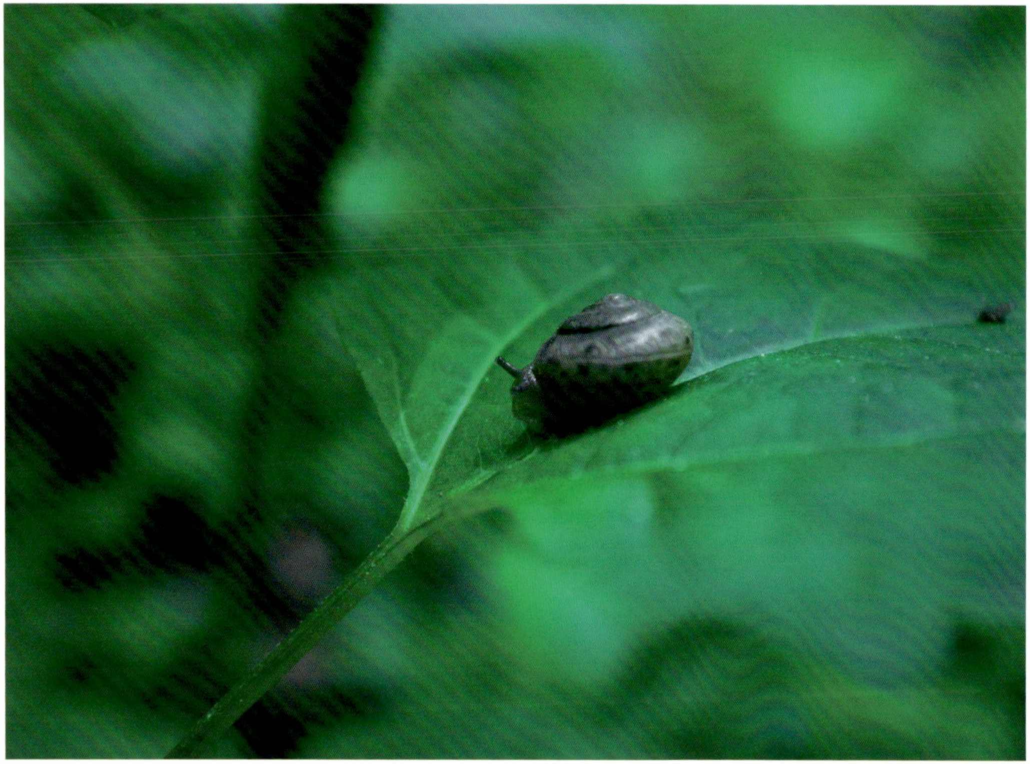

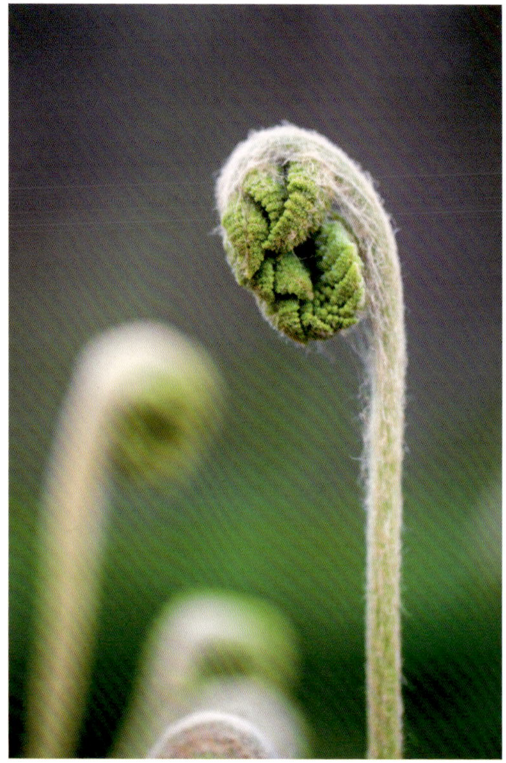

The Appalachian chain is home to over 260 species of snails, more than anywhere else in the world. Green Ridge State Forest, Maryland.

The exquisite spiral forms of fiddleheads (in this example, a cinnamon fern) are a welcome sight during spring. Finzel Swamp Preserve, Maryland.

Spring is the breeding season for the wild turkey, when mature toms will gobble and display to attract a hen. Shenandoah National Park, Virginia.

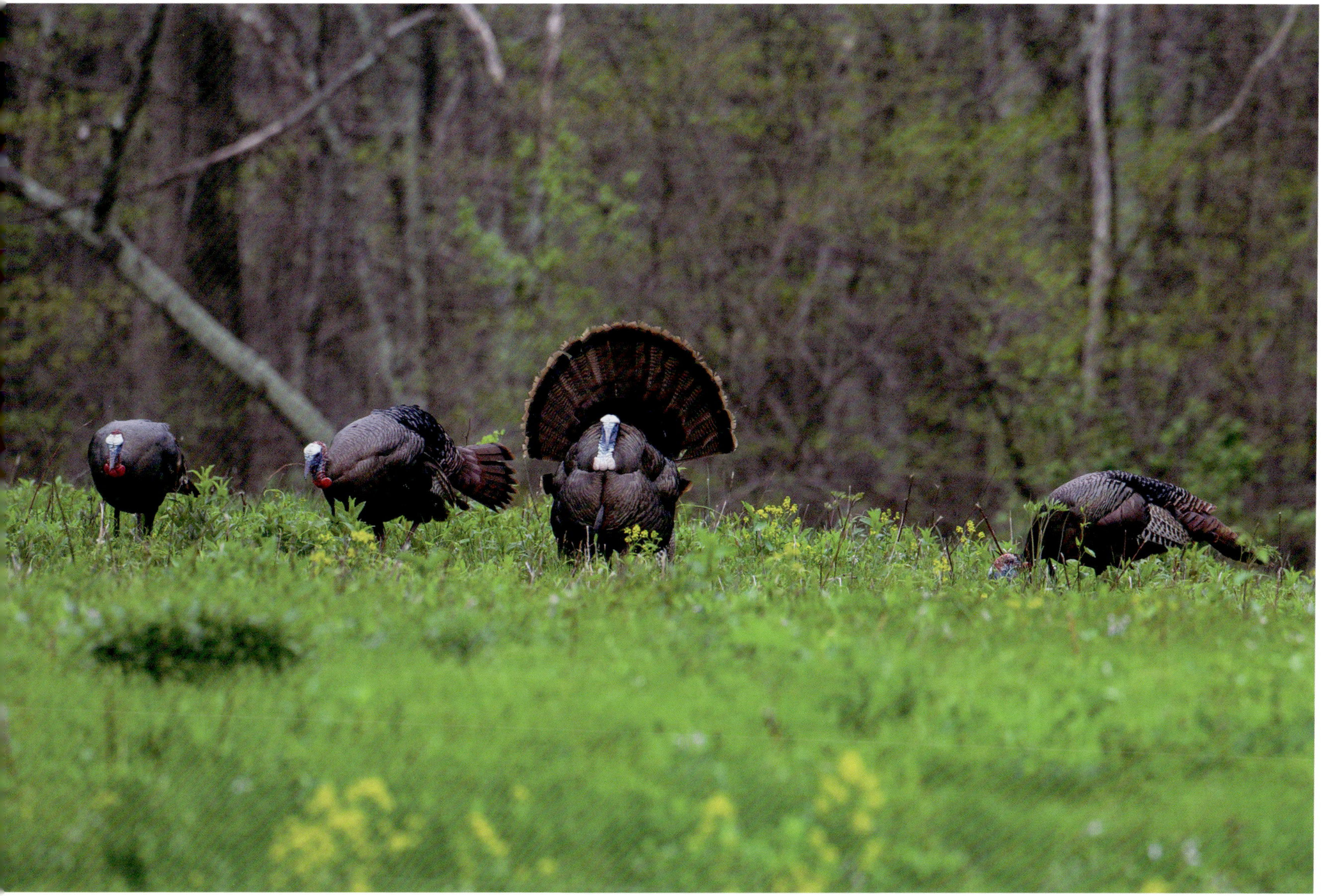

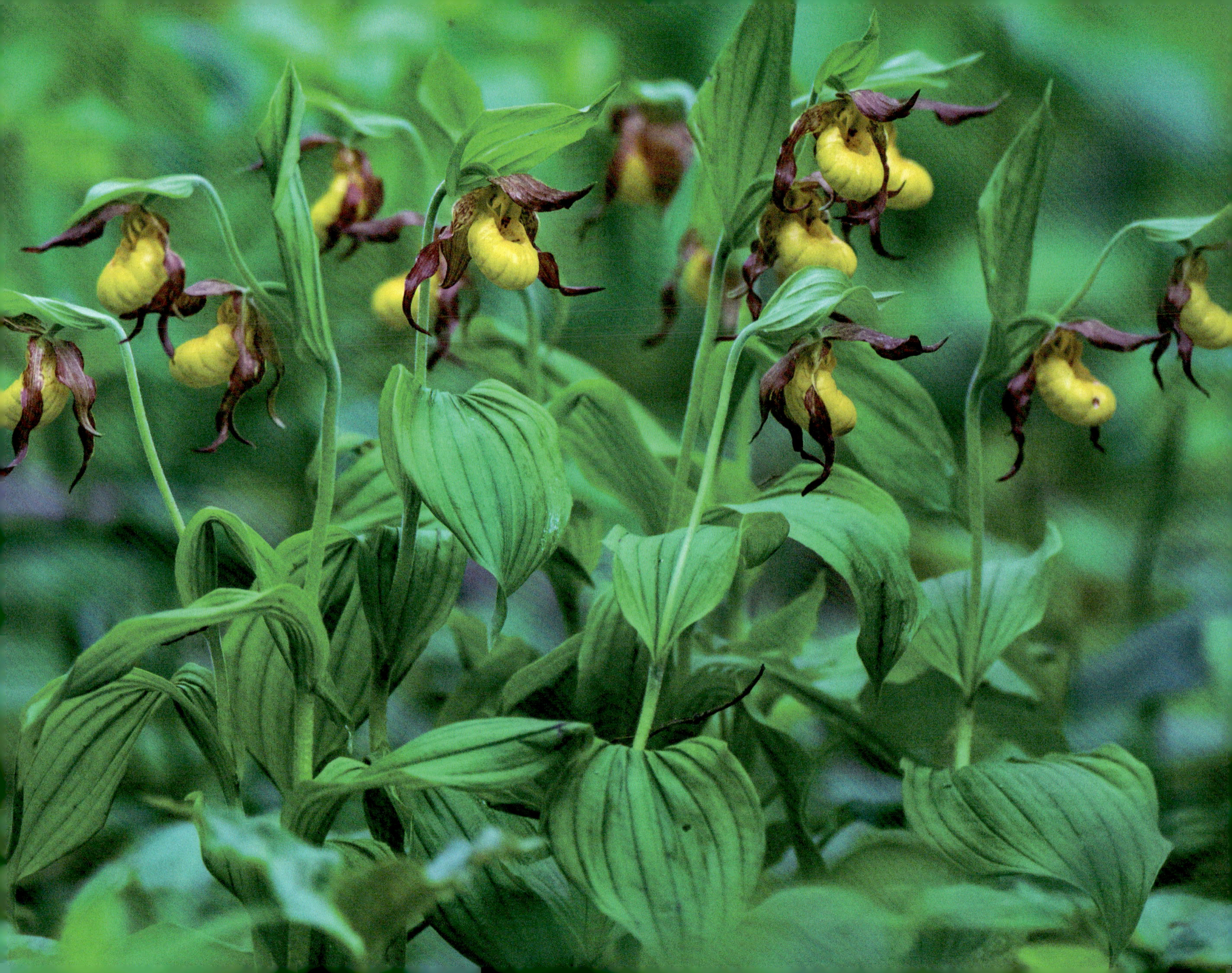

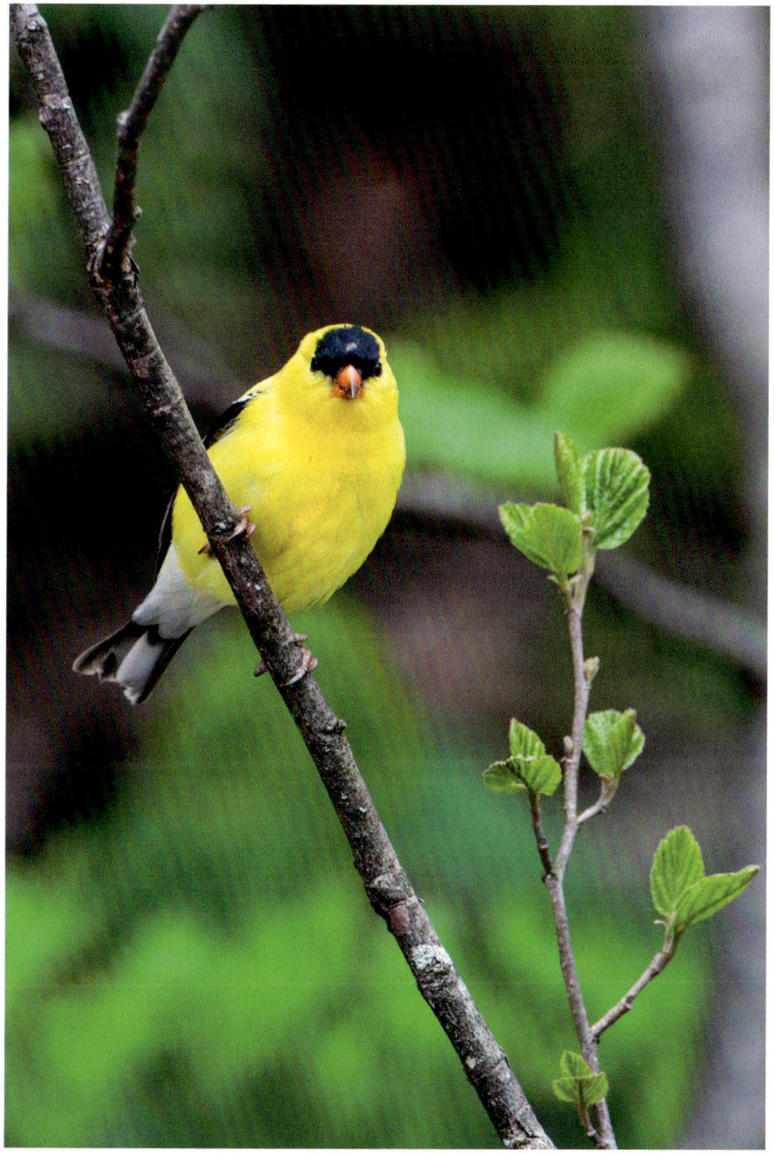

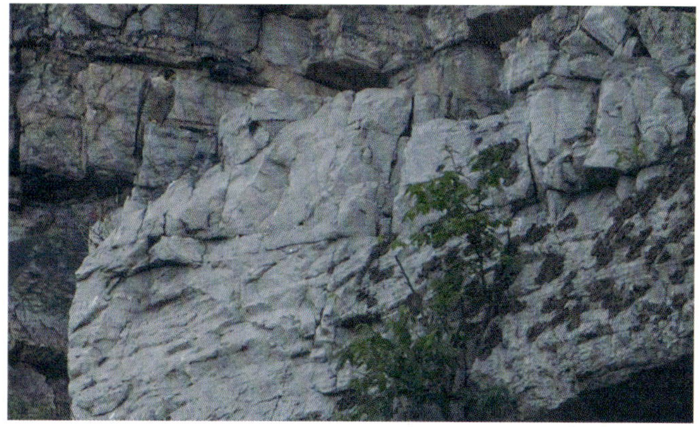

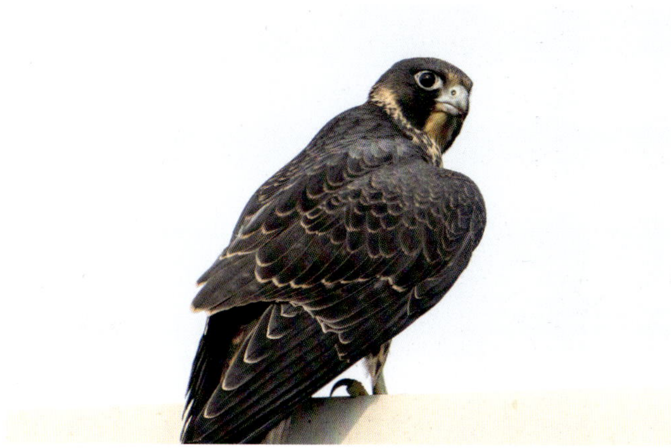

The striking yellow of the American goldfinch and a patch of small yellow lady slippers add splashes of color to the spring forest. Shenandoah National Park, Virginia.

Once decimated by the chemical DDT, peregrine falcons have returned to the mountains due to continuing reintroduction efforts. The falcons may sometimes be seen flying along steep cliffs where they nest.

The Morning with a Bobcat

One of my most memorable wildlife encounters was the morning I spent with a bobcat.

When compared to the other wild cats that inhabit North America, bobcats are fairly common throughout their range, though one would never know it because common does not equate to visible. These felines are incredibly elusive and well camouflaged, even when compared to other crepuscular animals.

I had originally planned to photograph newborn white-tailed deer fawns in Shenandoah National Park. It was late May and just the beginning of the birthing season. The fog, however, was so thick on that morning that I could see no more than 3 feet in front of me. I like fog. I find that it adds a bit of mystery to the environment. It also simplifies scenery when documenting the landscape, though it is not conducive to subject visibility when looking for fawns in a wet meadow.

Instead, I decided to capture some nearby waterfall scenes. I figured that water, some fog, and a hint of spring greenery could make for nice photographs in lieu of wildlife. At least, that was what I convinced myself of the trade-off. The trail was slippery and muddy, and visibility was very poor. Under normal circumstances, I would descend rapidly to the falls, but I was carrying a camera-mounted tripod while traversing over wet rocks, and my pace was that of a caffeinated snail. I am glad it was so, because what soon happened was the most serendipitous moment I ever had in the forest. About five minutes into the hike, a bobcat crossed the trail, so close that we almost collided. My heart thudded and my breath stopped. A bobcat, a real, wild bobcat—not one of those plush toys in the gift shop—was looking at me. This moment alone was incredible enough, but what happened next was sublime.

Now, if one is of good enough fortune to cross paths with a bobcat, it is almost guaranteed to be fleeting. They escape, vanishing into the underbrush or rocks. This was not what happened. We briefly gazed into each other's eyes, as if we were

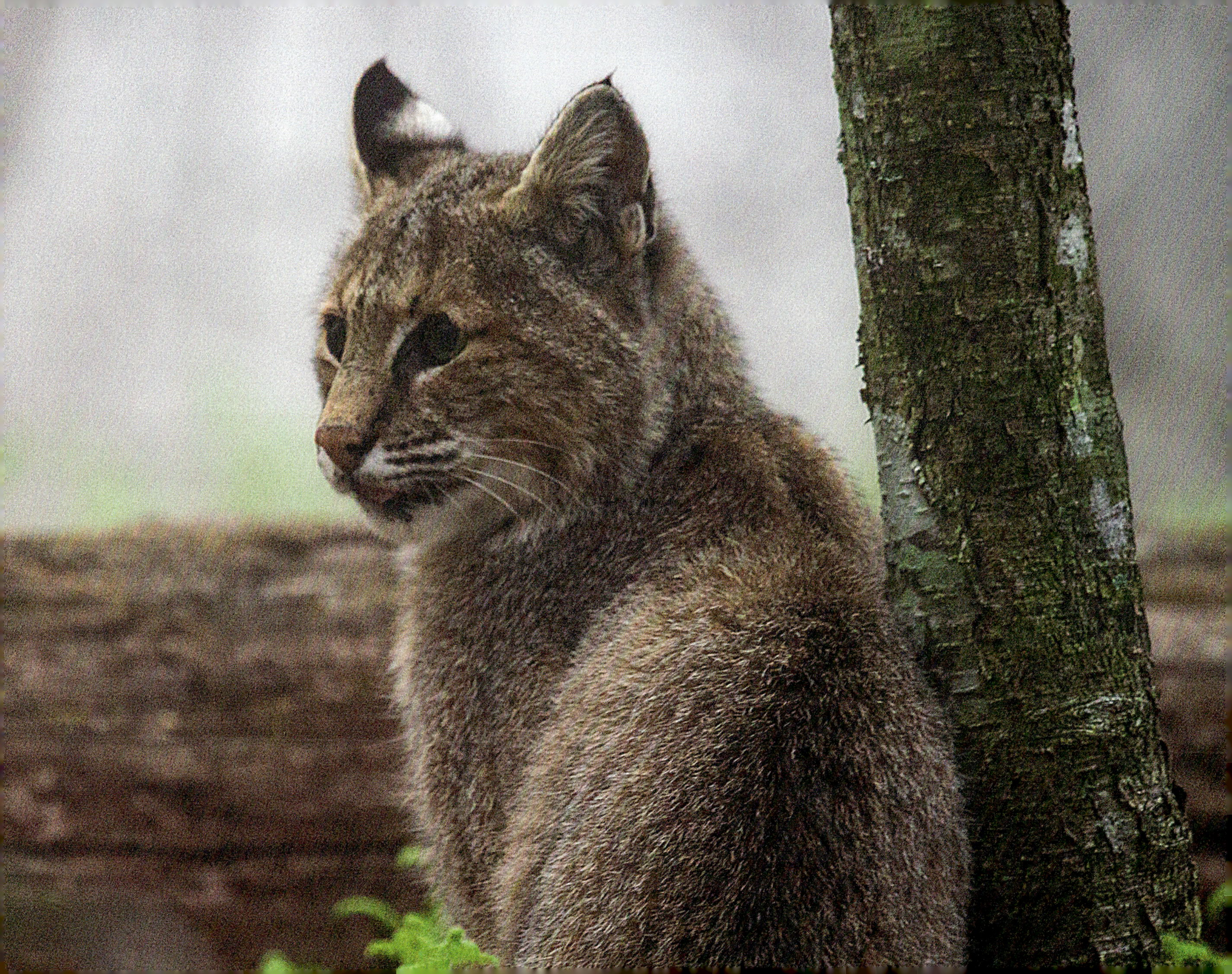

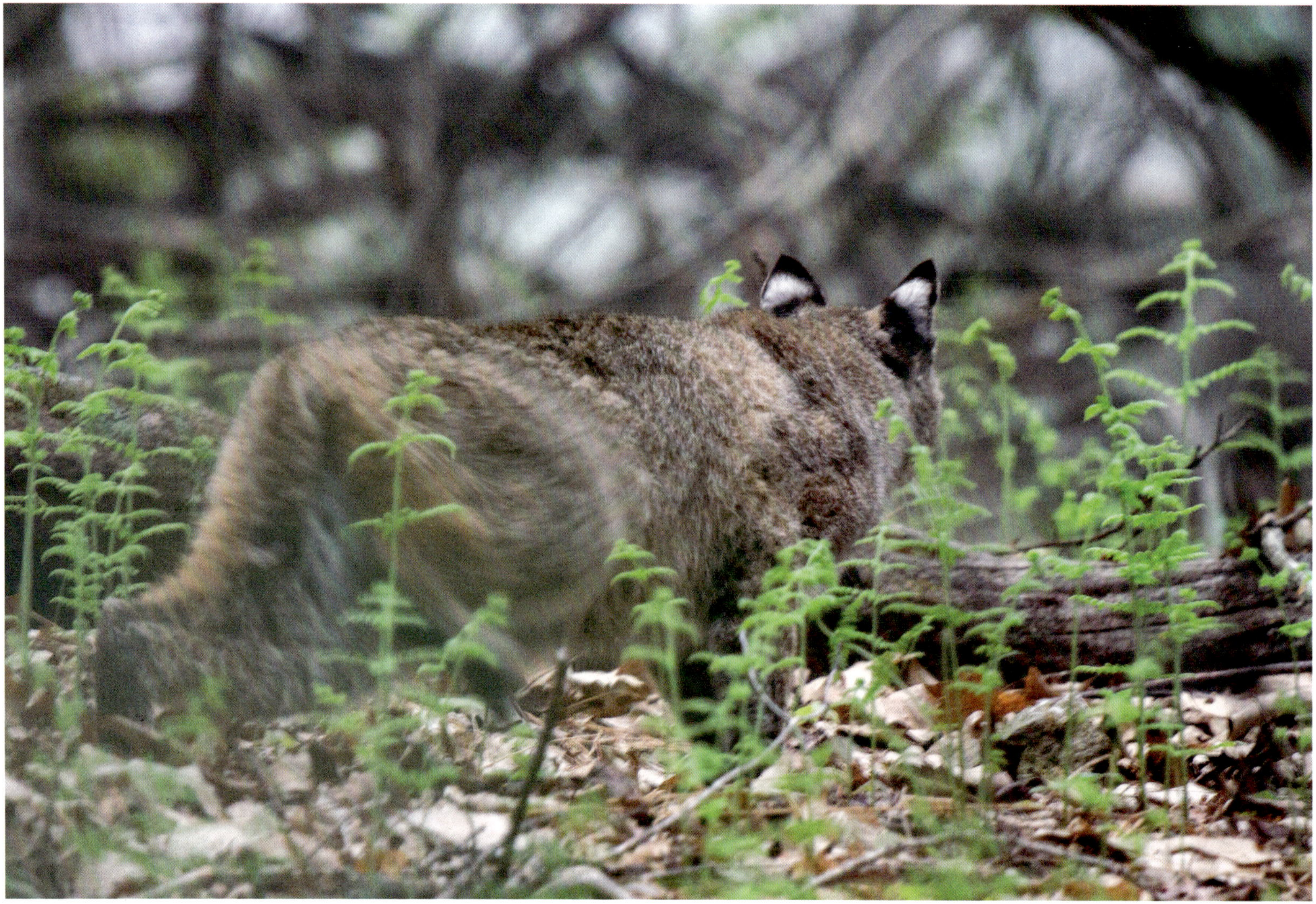

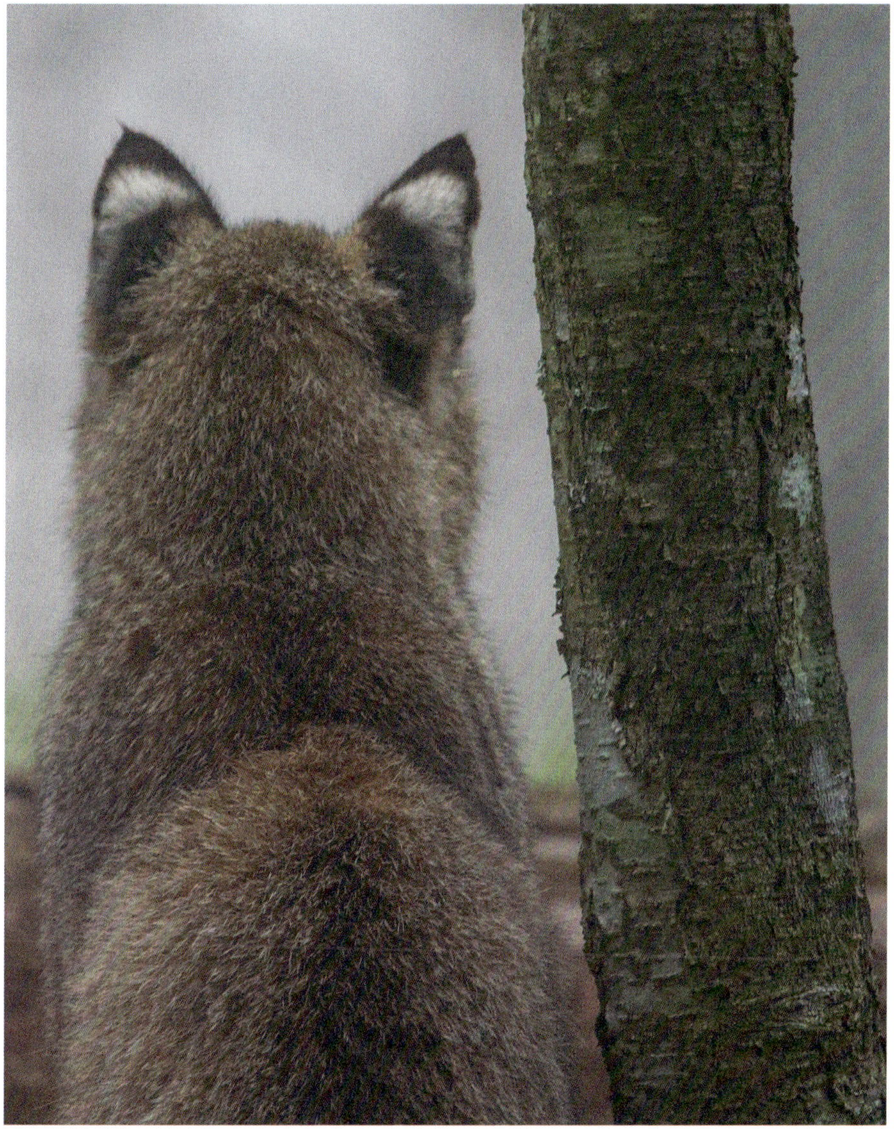

both caught in doing something we should not have been doing. For the bobcat, a chance encounter such as this is not supposed to happen. They rely on the element of surprise; they are not supposed to be surprised. Unbeknown to most, when you hike in bobcat habitat there is a good chance one is aware of your visit.

Next, the bobcat just continued on its way. Crouched and reticent, the well-camouflaged predator moved past the trees, almost disappearing into the forest floor. While my body was still frozen from excitement, my eyes continued to follow its movement. The bobcat continued to move in a crouched position, as if stalking prey. Instinctively I lowered to my hands and knees and began to follow it. Keeping a healthy distance, I moved when it moved, and I stopped when it stopped. On occasion it looked back at me, but the surprise in its eyes was gone. It was aware and then continued on the hunt.

It stopped at a fallen tree, where I was finally able to set down my tripod, not having considered that crawling with one would have been the least bit uncomfortable. I was following a bobcat, after all! Who has time to think about properly carrying your camera gear? I was even able to switch to a telephoto lens without it disappearing into the underbrush. This was definitely not your normal bobcat.

I photographed a few frames, though not as many as I would have liked. Slow and robotic I moved, since I was afraid that any strong movements would cause the cat to finally flee. Beyond the fallen tree, it was fixated on something. Now statuesque, it never turned to look at me again. In a sudden burst of action, the cat leapt over the log into some leaf clutter and caught a rodent-sized small mammal. It then disappeared like a bobcat does. An abrupt ending to an unlikely story.

That afternoon I called my father to tell him of the encounter, and he reacted in both shock and excitement, probably more so than any other field story of mine. I sent him a cell phone photo of the back of my camera's LCD screen, which only intensified his reaction.

In the years since, I have shared this story with many hikers, park rangers, biologists, and friends. What begins with disbelief turns to awe. I believe it is because bobcats are both predatory and mysterious. We are drawn to what we rarely see. Who spends a morning with a bobcat? Well, I do.

I have come across these wild felines only a handful of times since the encounter. It is always a brief glimpse of one running in the opposite direction. I do hope to have such a magical encounter again, though it is more likely I will win the lottery. But I recognize how fortunate I am to have experienced such a moment. Though, I must admit, whenever I learn that fog is in the forecast or I am near a waterfall, I cannot help but think it is time for the sequel.

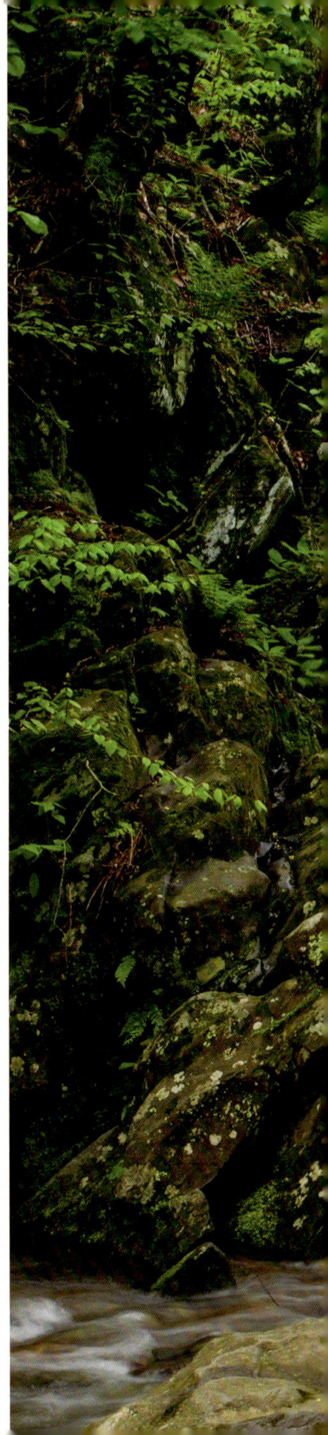

Dark Hollow Falls, Shenandoah National Park

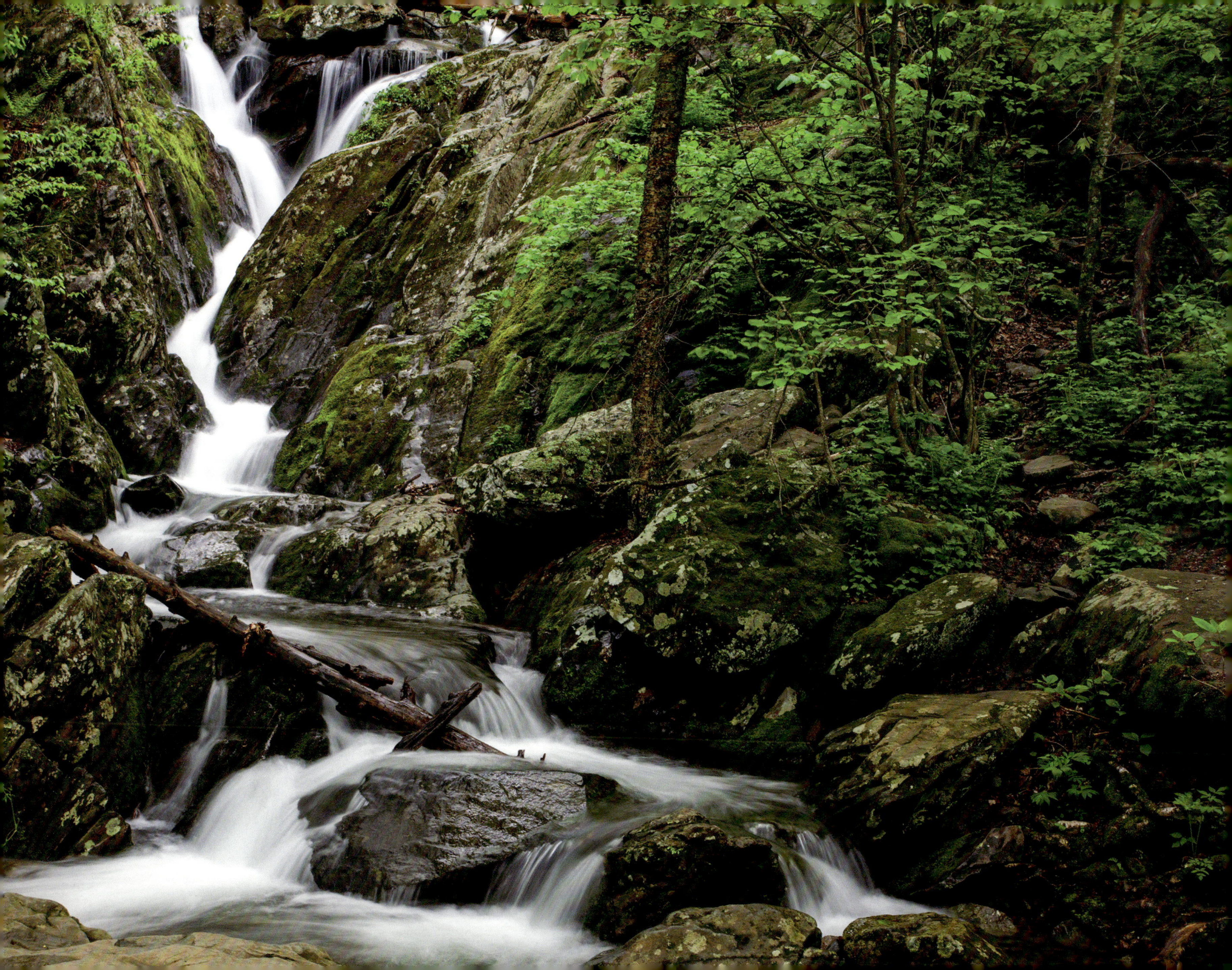

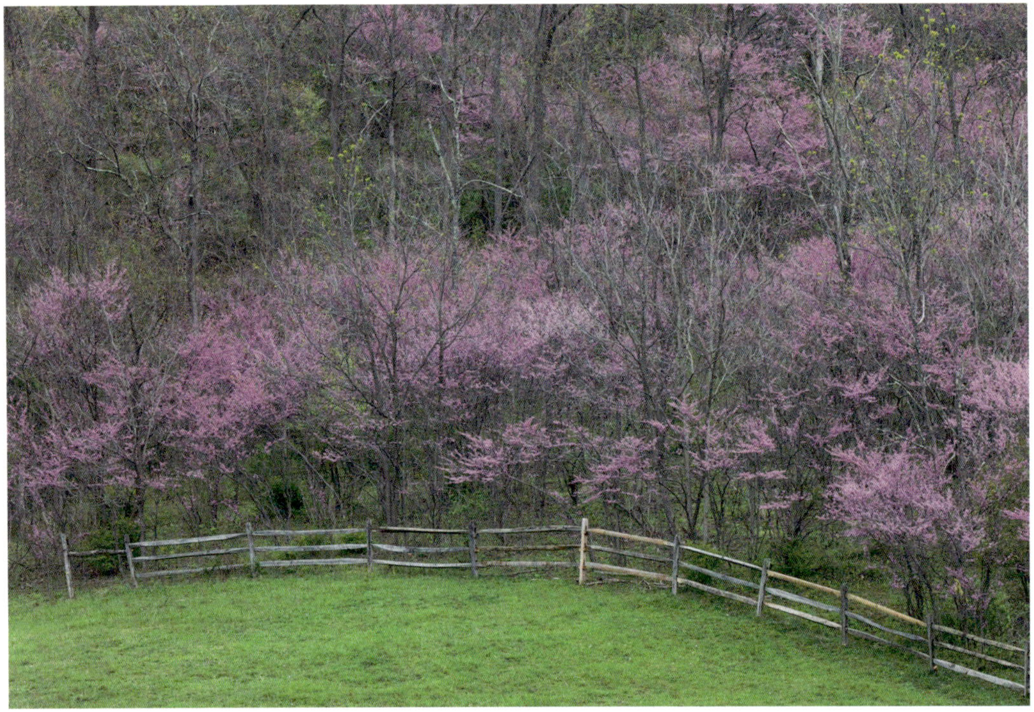

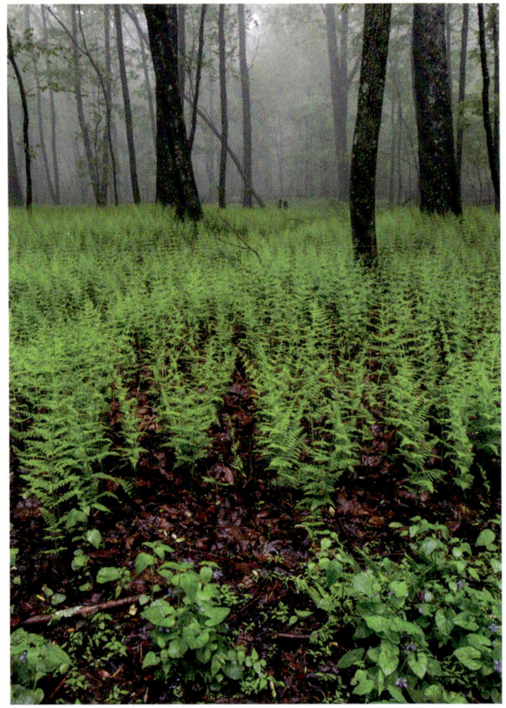

A rite of spring, the pink pastel blooms of the eastern redbud tree. Allegany County, Maryland.

By mid-May, ferns blanket much of the Appalachian forest. Appalachian Trail, Maryland

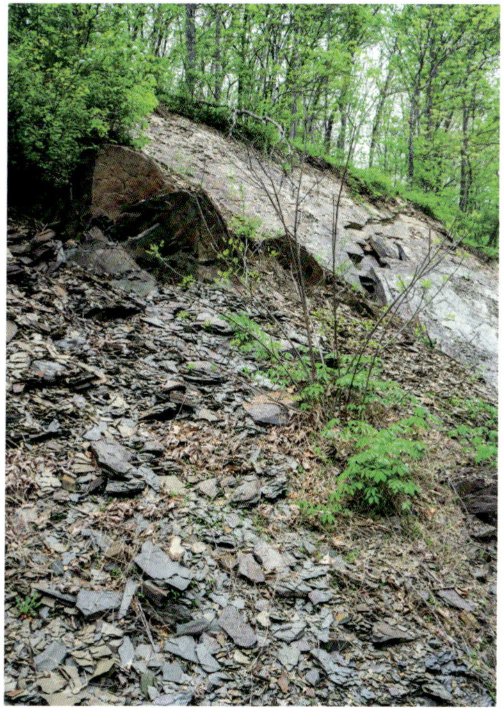

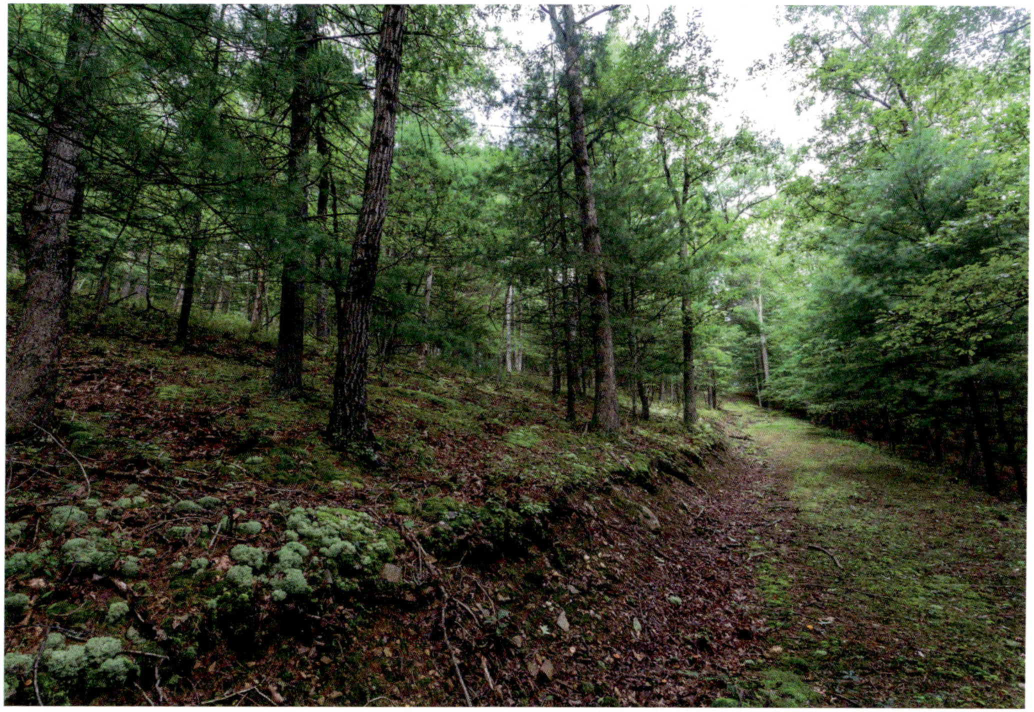

Shale barrens are rough, dry, gritty, and steep Central Appalachian habitat. Surprisingly, they provide a home to plants that are endemic or specialize in this harsh habitat. Green Ridge State Forest, Maryland.

Shale woodlands of oak and pine directly adjacent to the barrens. Sideling Hill, Pennsylvania.

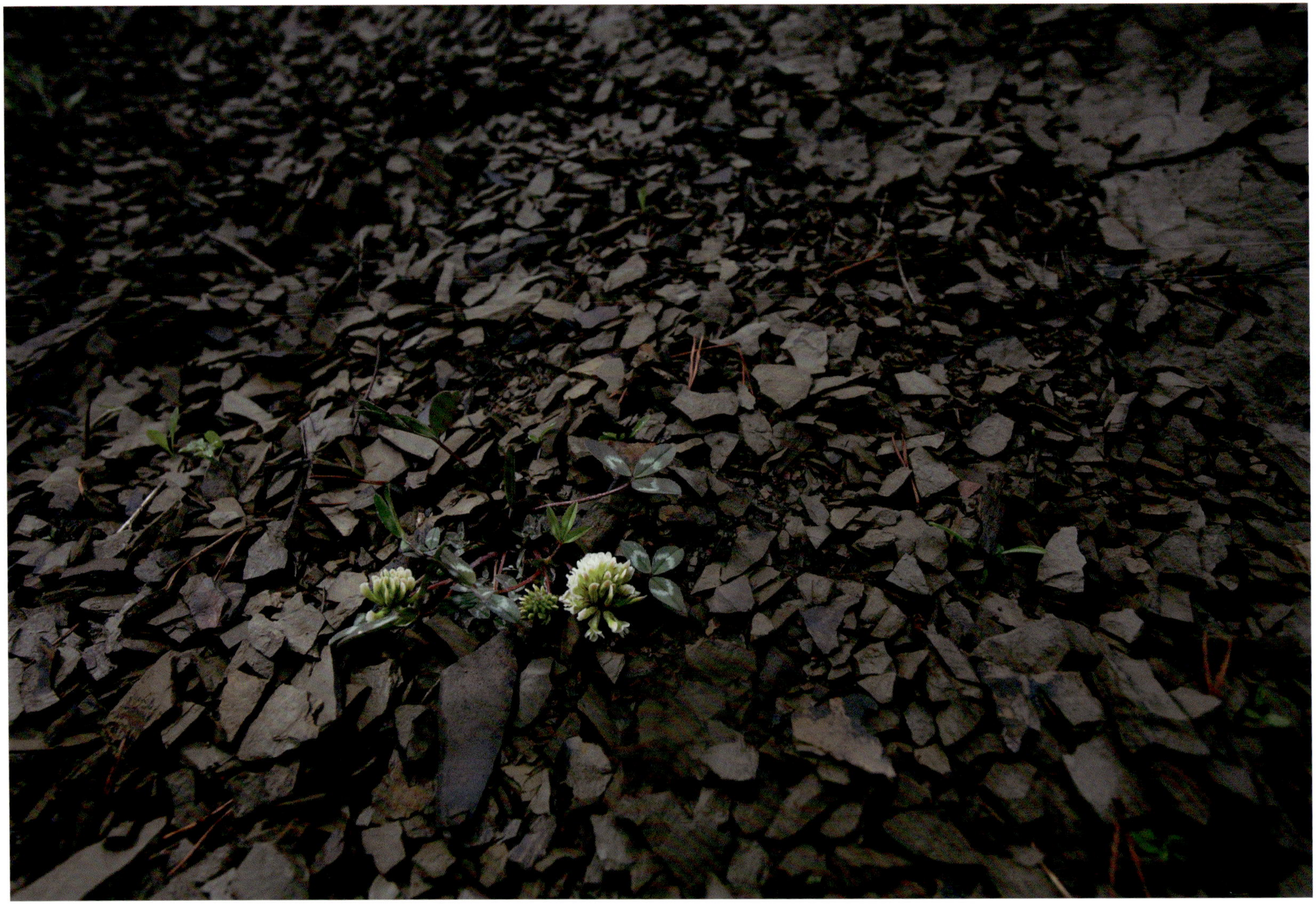

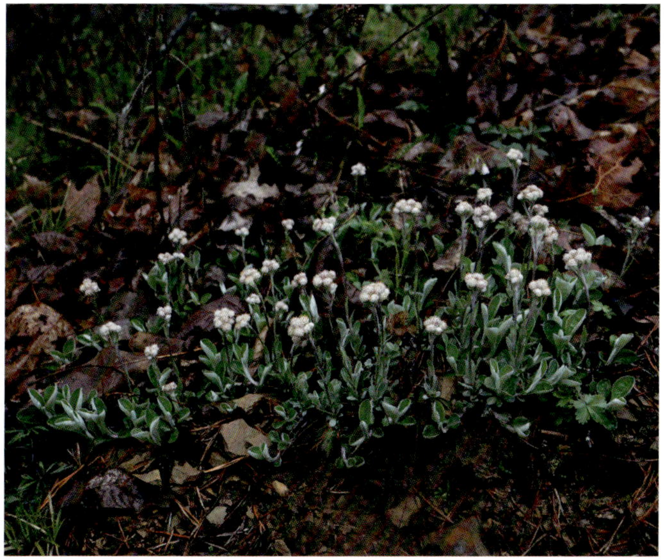

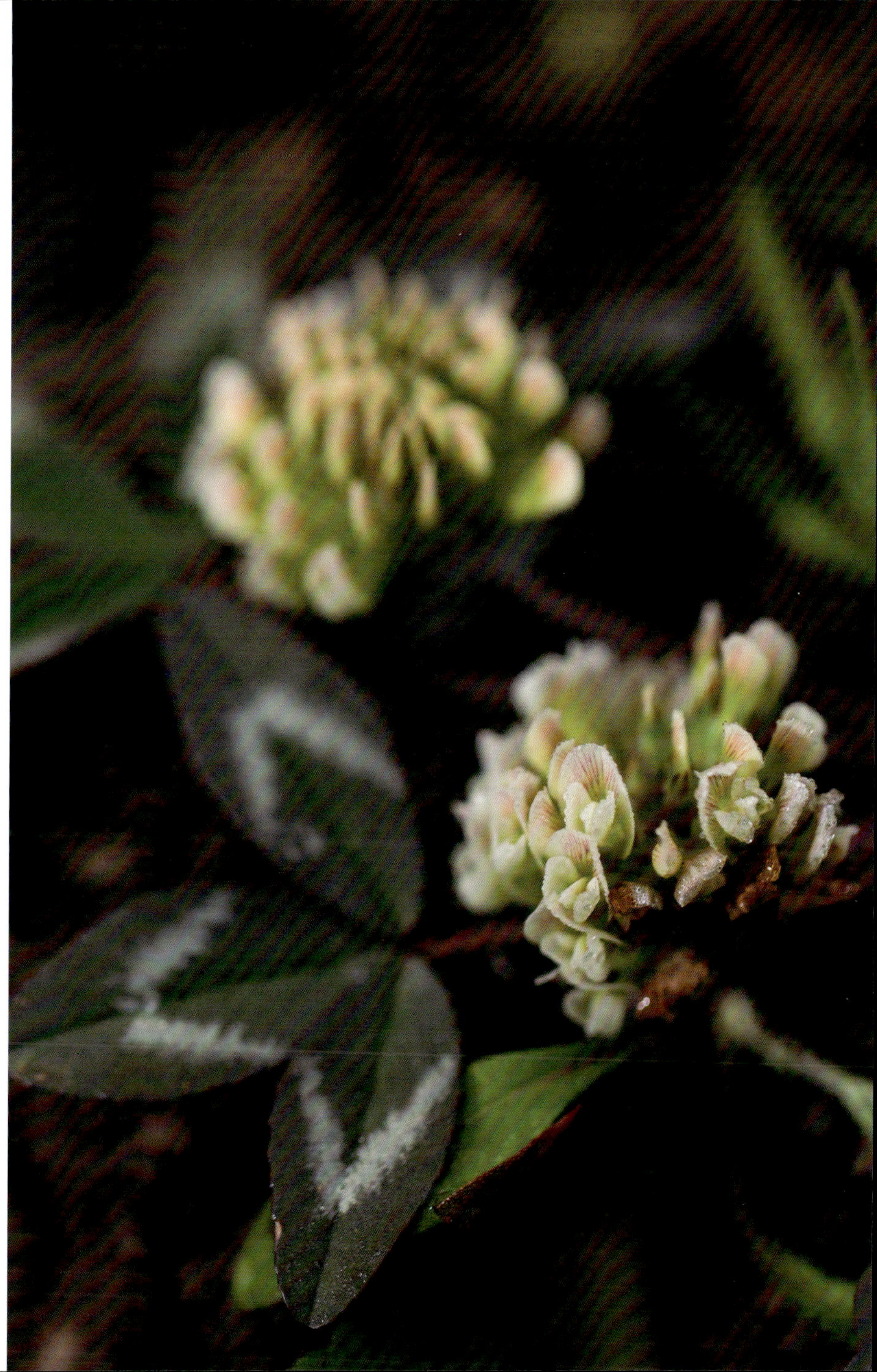

Kate's Mountain clover is a beautiful ephemeral that is a shale barren specialist.

Shale barren pussytoes, a species endemic to the shale barrens. Sideling Hill Creek Preserve, Maryland.

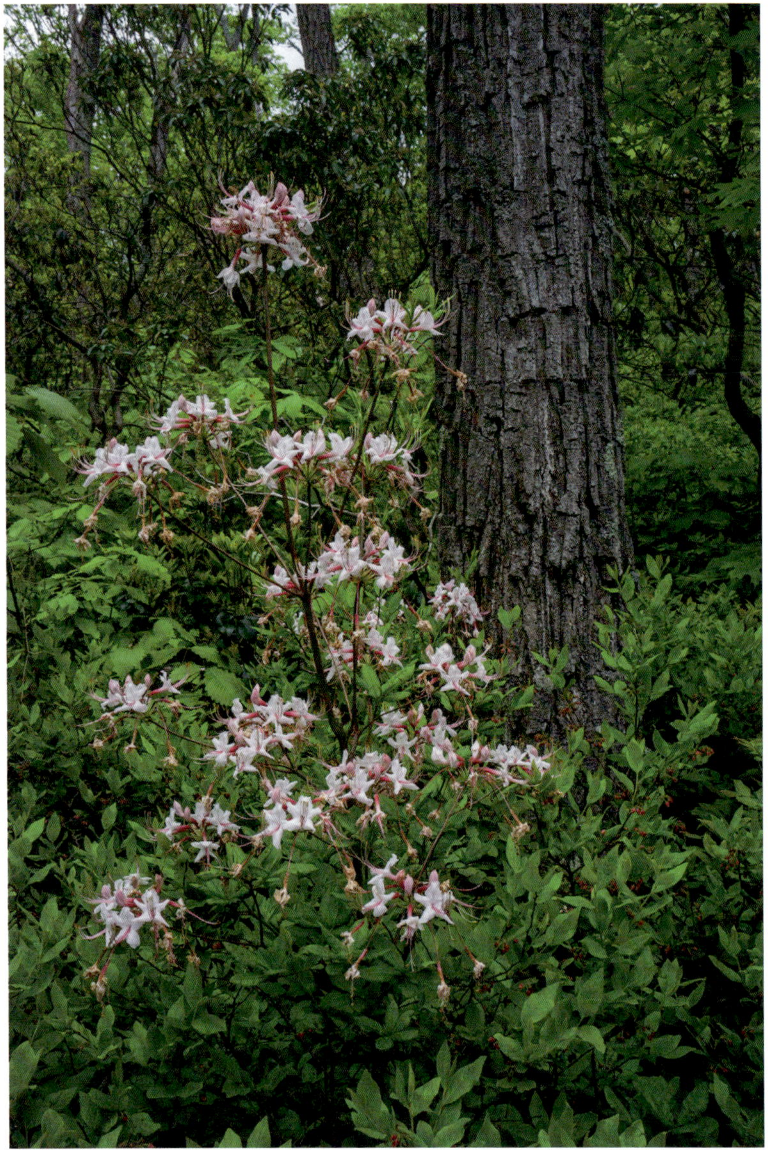

Pink azalea in oak-pine forest. The plant usually blooms in early May, but in the higher elevations the bloom is extended until late May. Gambrill State Park, Maryland.

A silver-spotted skipper feeds on the nectar of a pink azalea. Garrett County, Maryland.

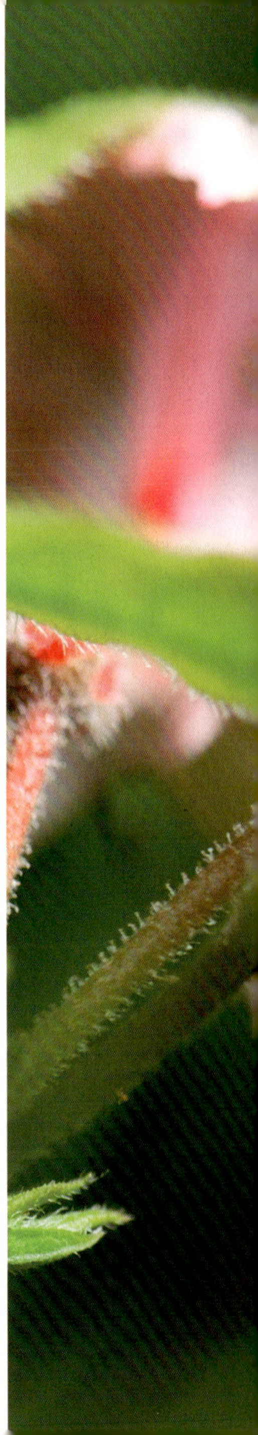

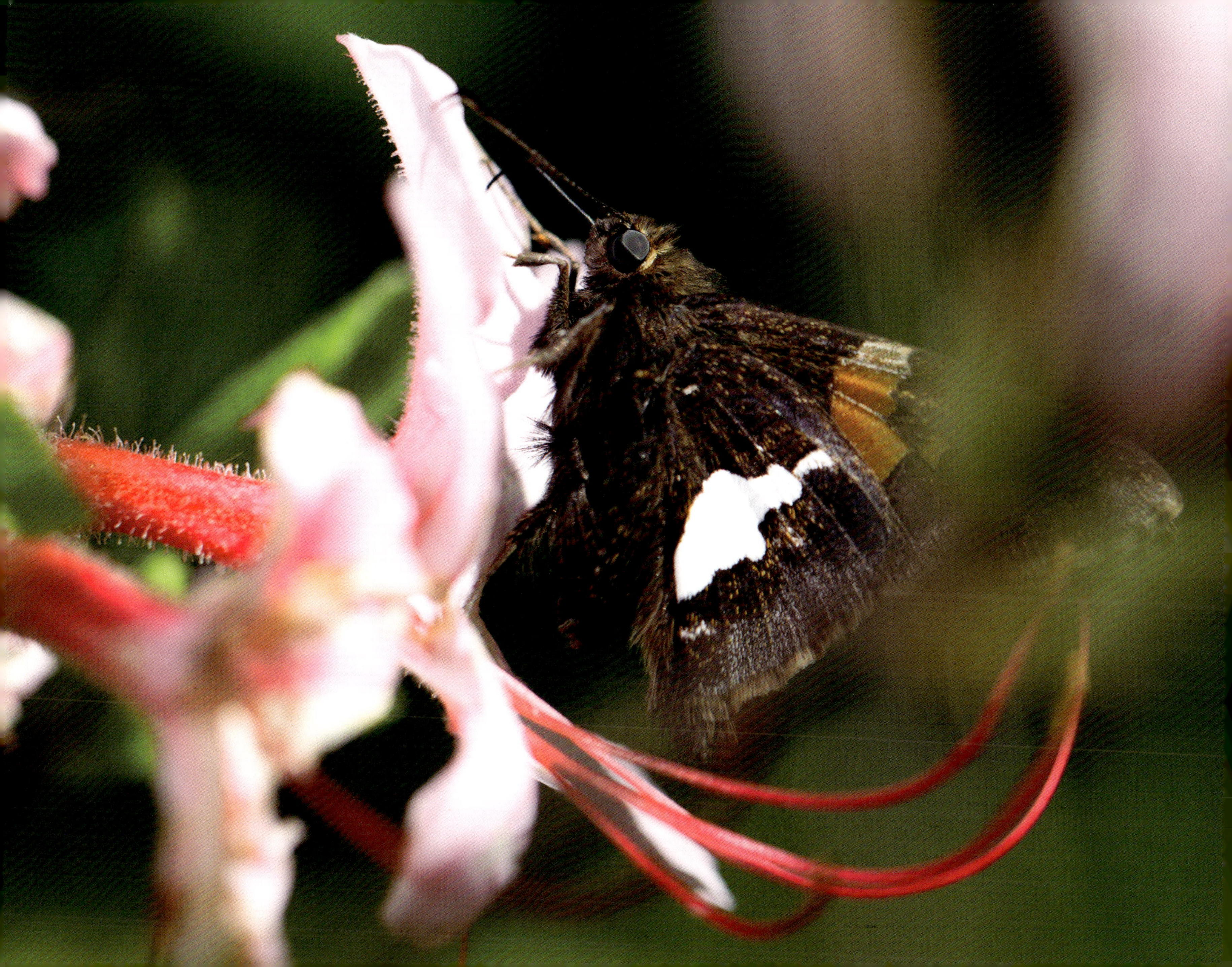

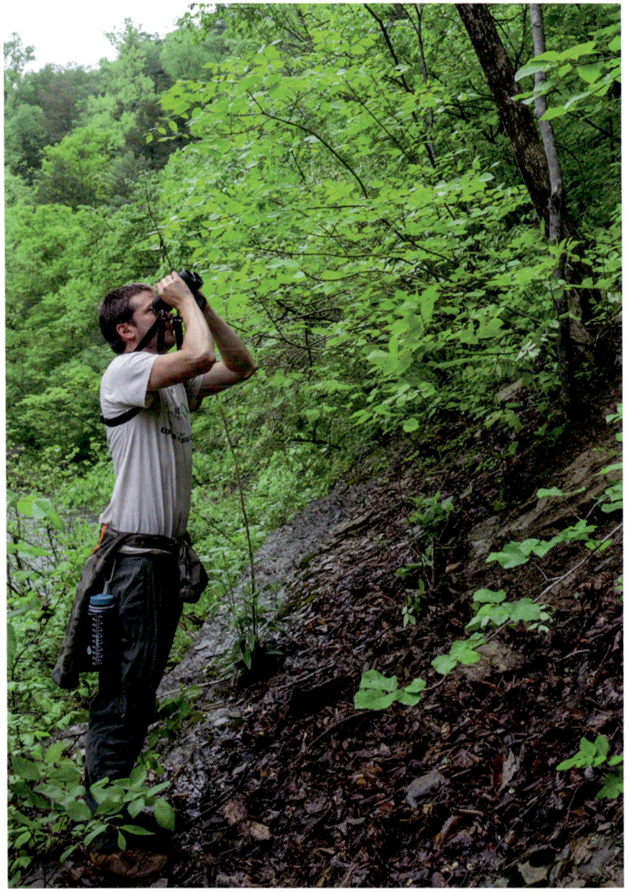

Biologist Robert Mulligan scans a ridge for a rare species of orchid.

The small white lady slipper, one of the rarest plants in the Central Appalachians, is highly sought after by poachers. It has been all but extirpated, except in a few small areas of the region.

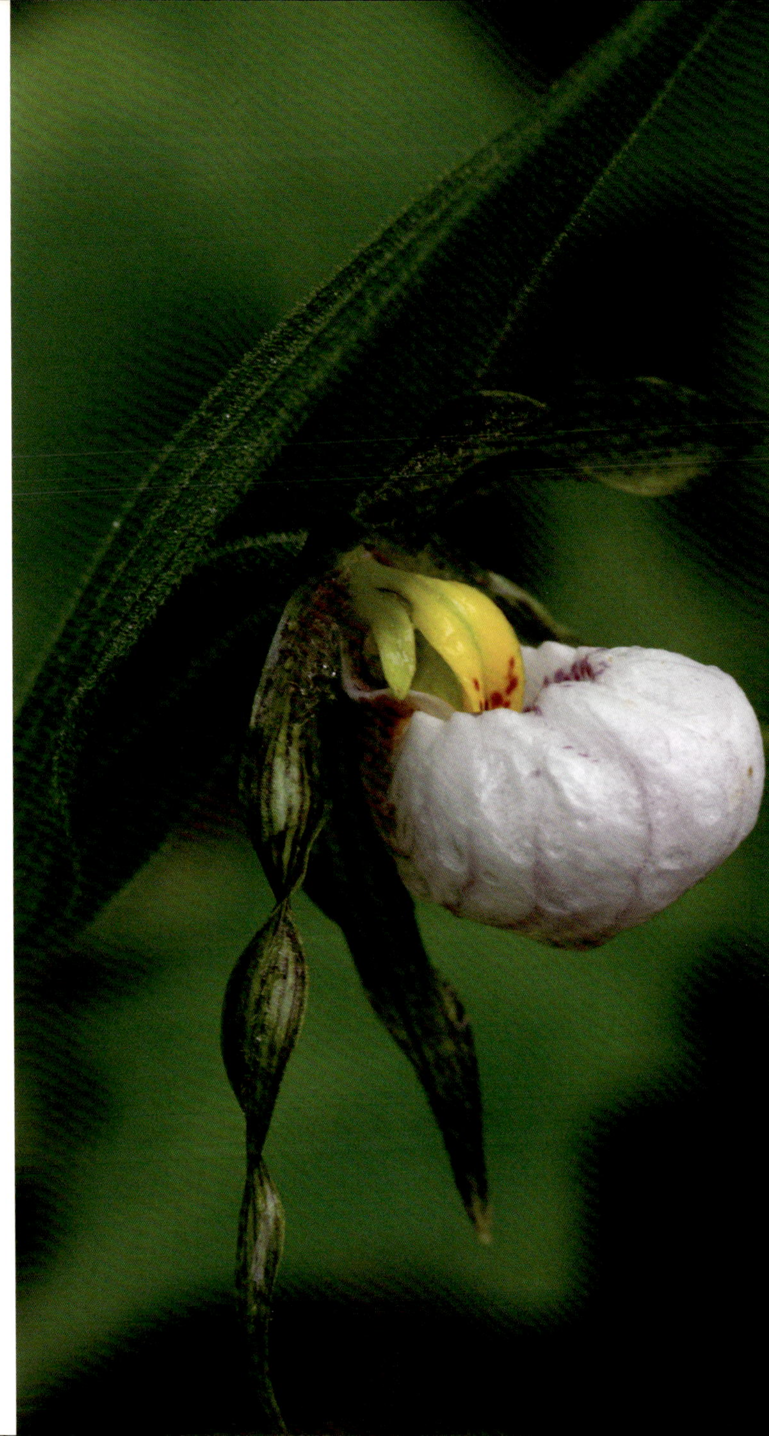

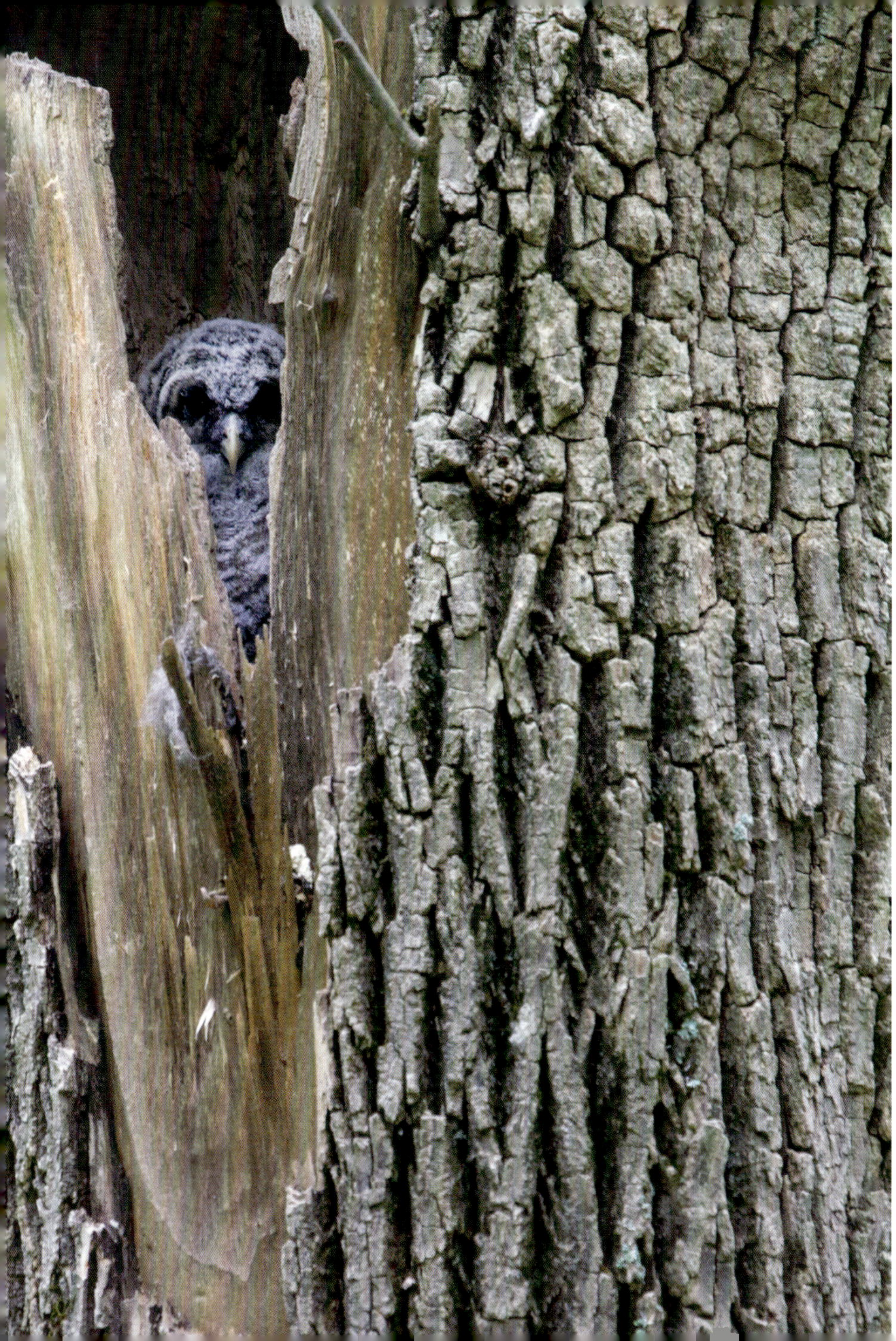

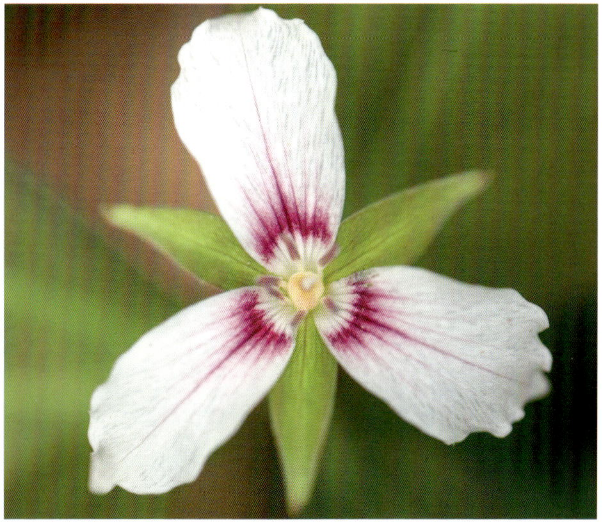

A barred owl chick watches from the protective hollow of a tree. Soon, it will become one of the mountains' most formidable predators. Shenandoah National Park, Virginia.

Color and new life symbolizes the season. But ephemerals such as this painted trillium never last long. Savage River State Forest, Maryland.

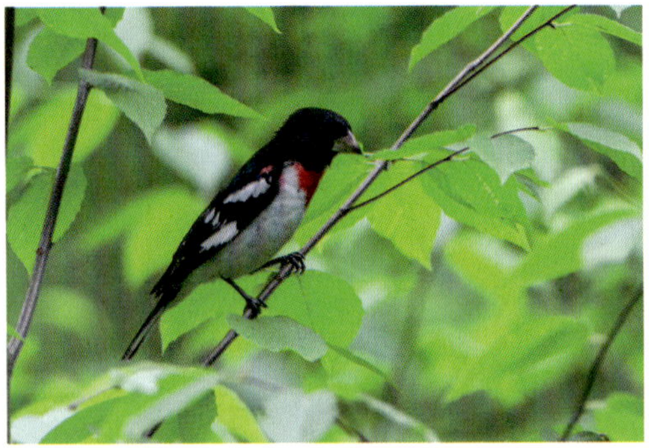

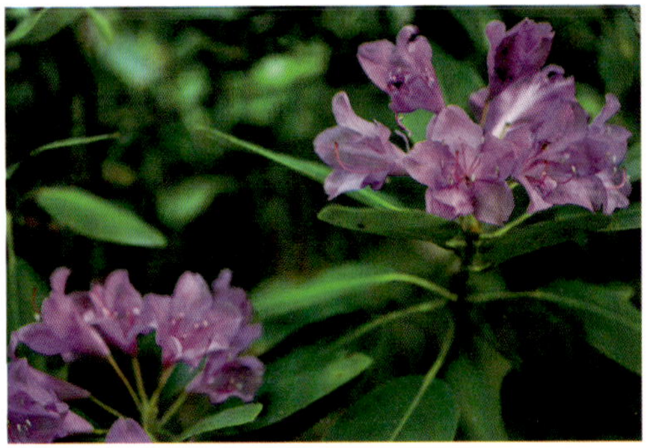

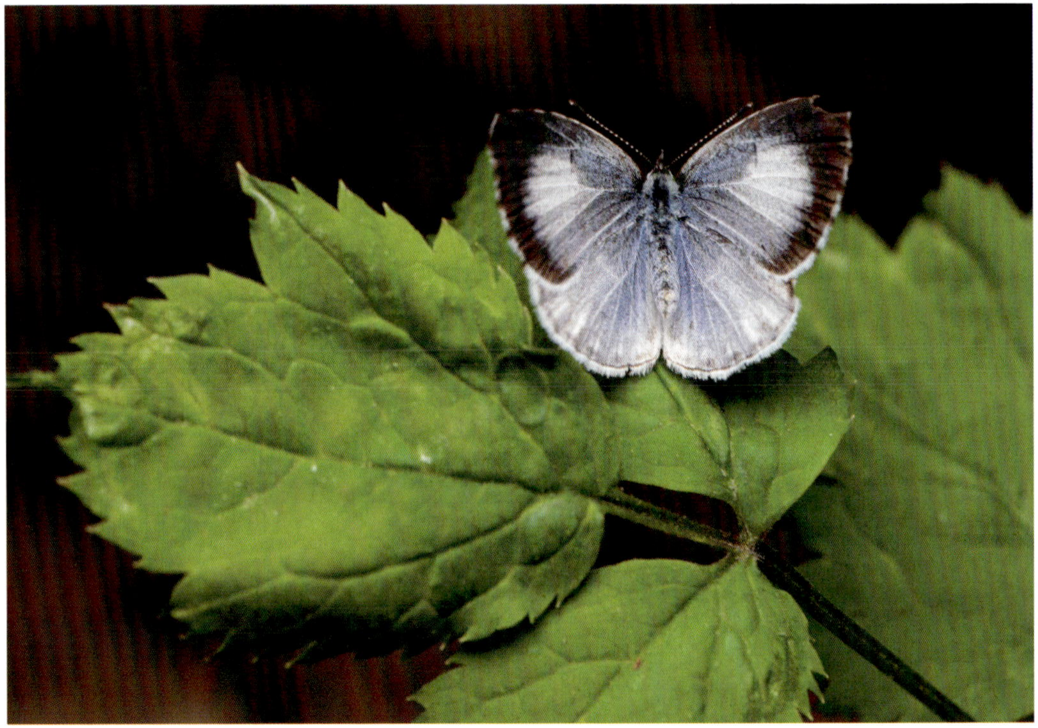

A rose-breasted grosbeak has returned to its breeding grounds in the mountains. Moshannon State Forest, Pennsylvania.

More commonly found along the high-elevation balds of the Southern Appalachians, the showy purple flowers of the Catawba rhododendron may be natively found in the Central Appalachians of Virginia and West Virginia. St. Mary's Wilderness, Virginia.

The Appalachian azure, a larger type of azure that is found in the Central Appalachians from late May into early June. The easiest ways to identify this azure are by flight time and by finding it on its host plant, black cohosh.

The cerulean warbler, one of the many warblers to nest in the region, is a beautiful blue warbler that returns to the Central Appalachians from the neotropics every spring to nest high in the forest canopy. Habitat loss both in wintering and breeding grounds has contributed to a 3 percent annual loss in its population since population surveys began in 1966. Eidolon Nature Preserve, West Virginia.

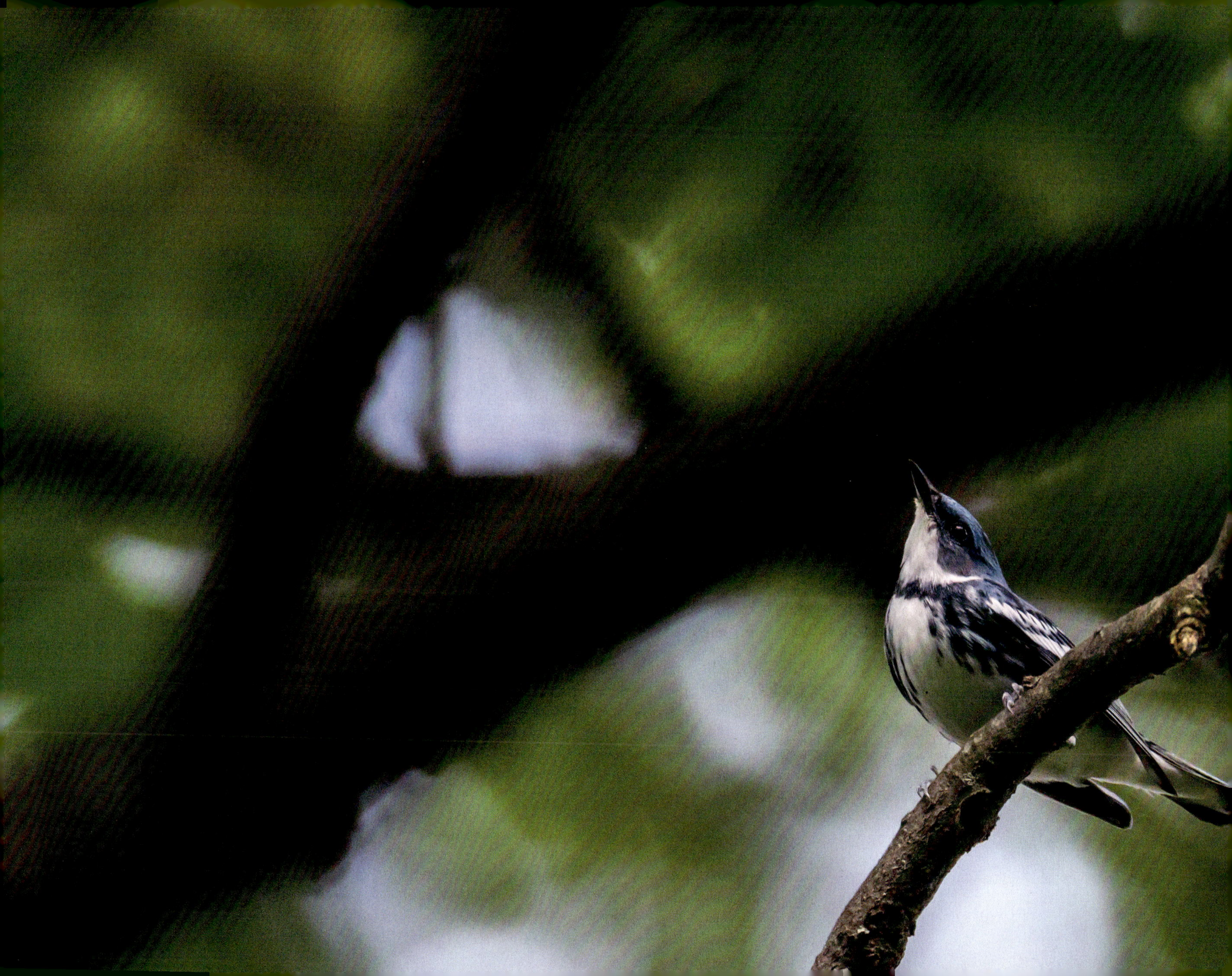

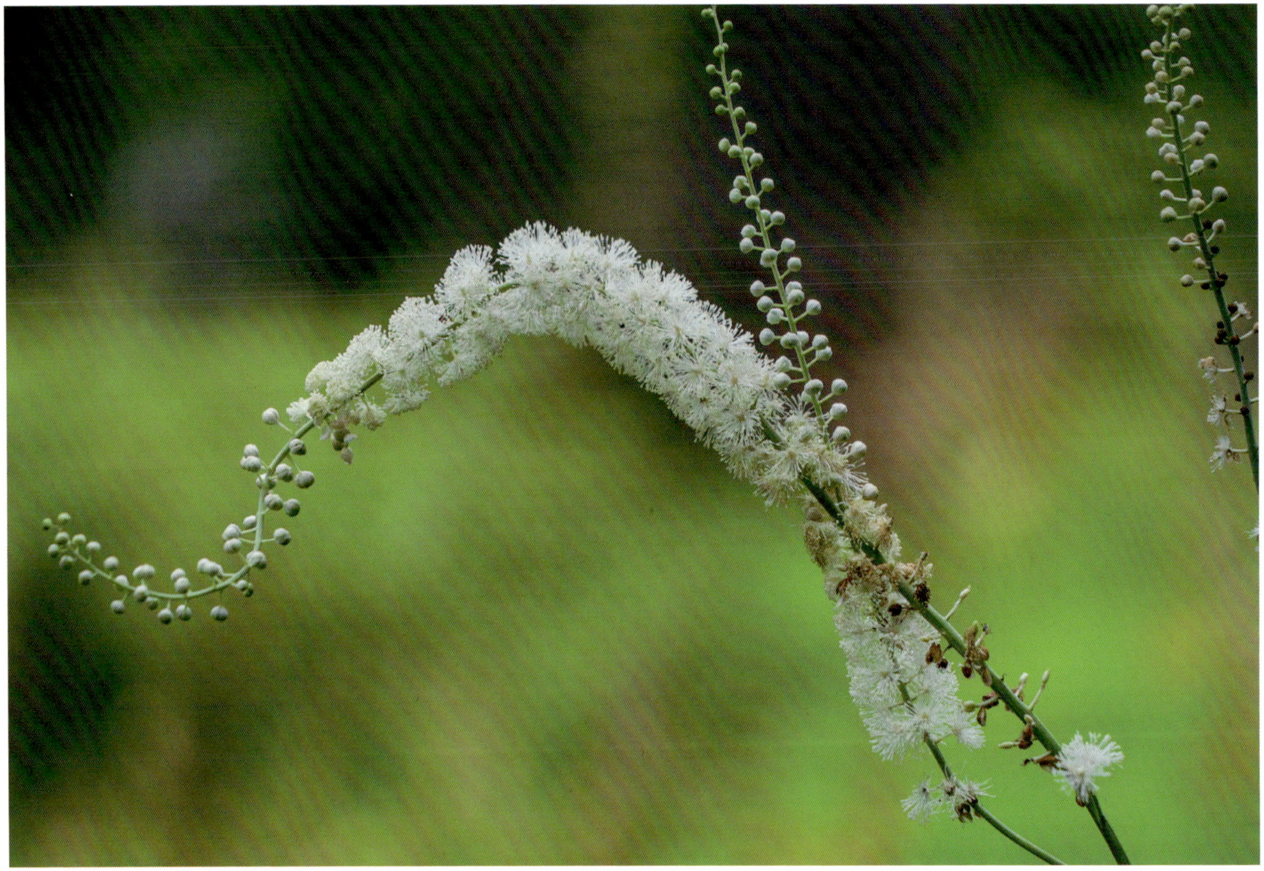

Black cohosh flower in summer

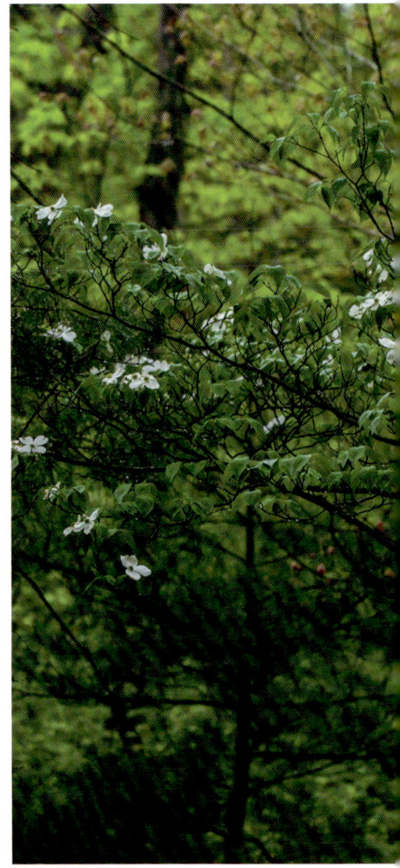

A flowering dogwood in bloom.
Catoctin Mountain Park, Maryland.

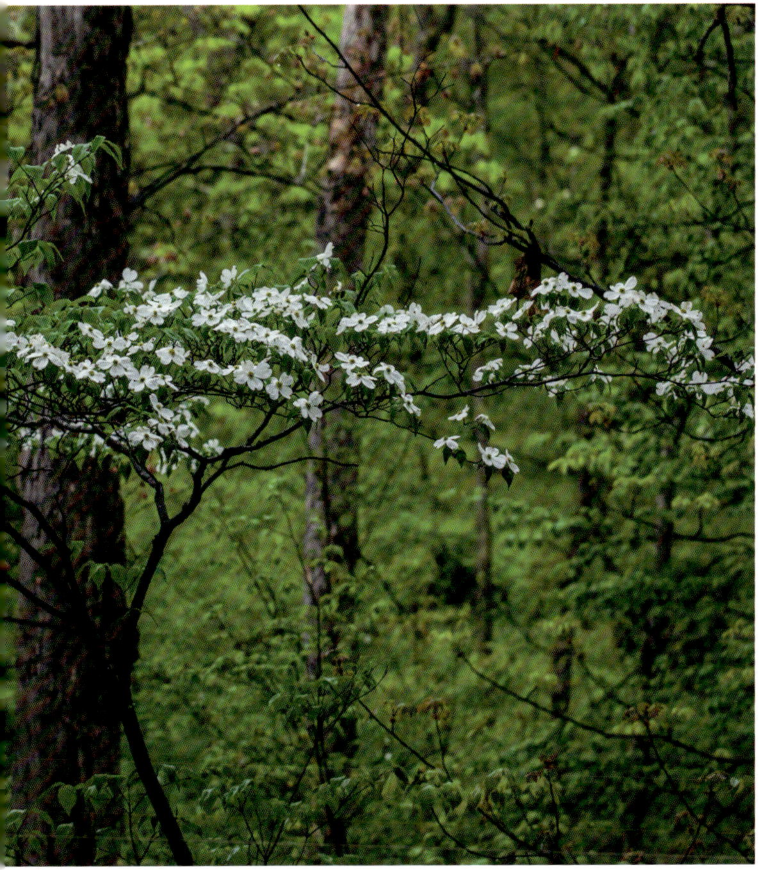

Reminiscent of snowflakes, the blooms of miterwort have an intricate, delicate beauty. Shenandoah National Park, Virginia.

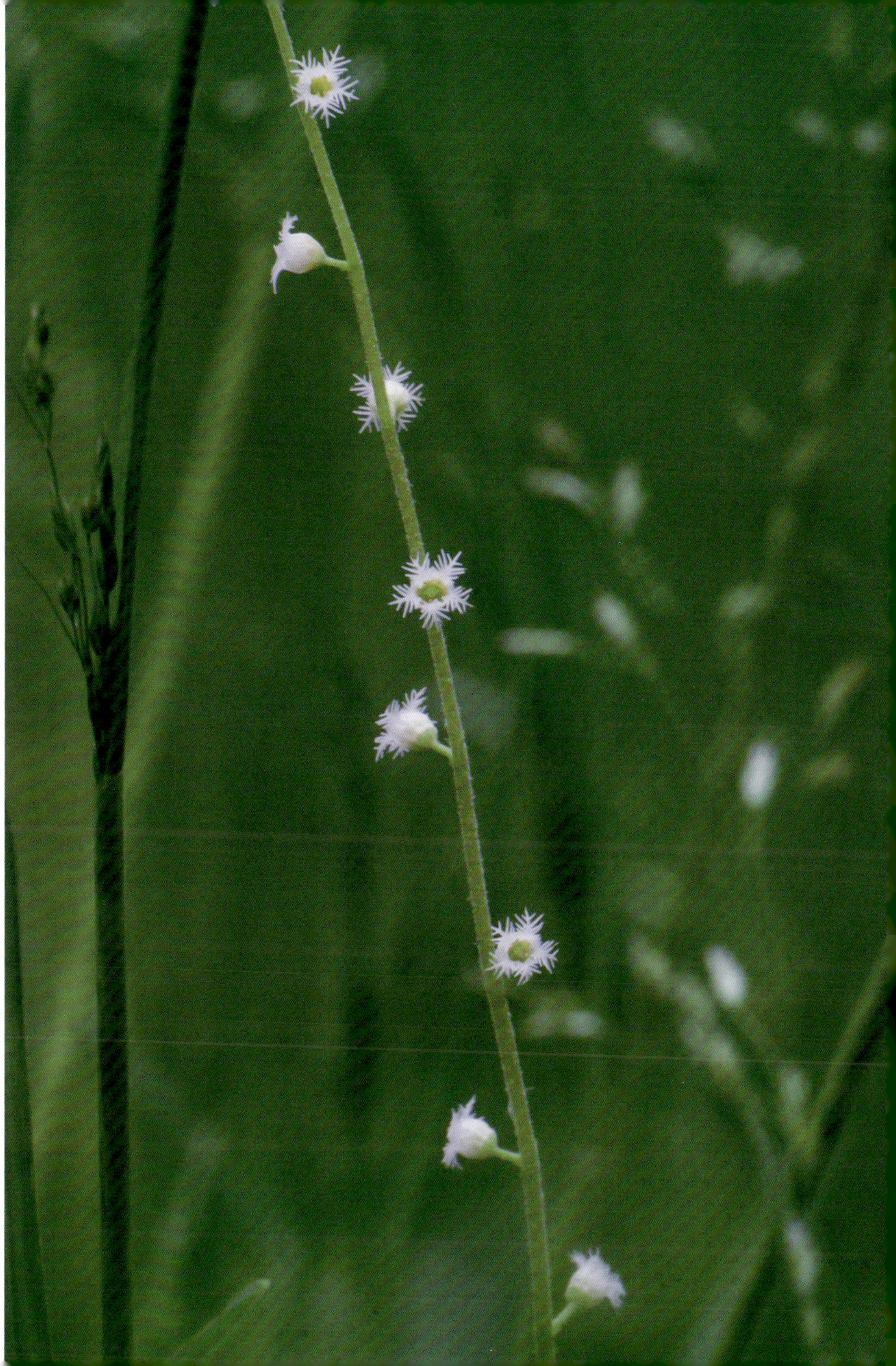

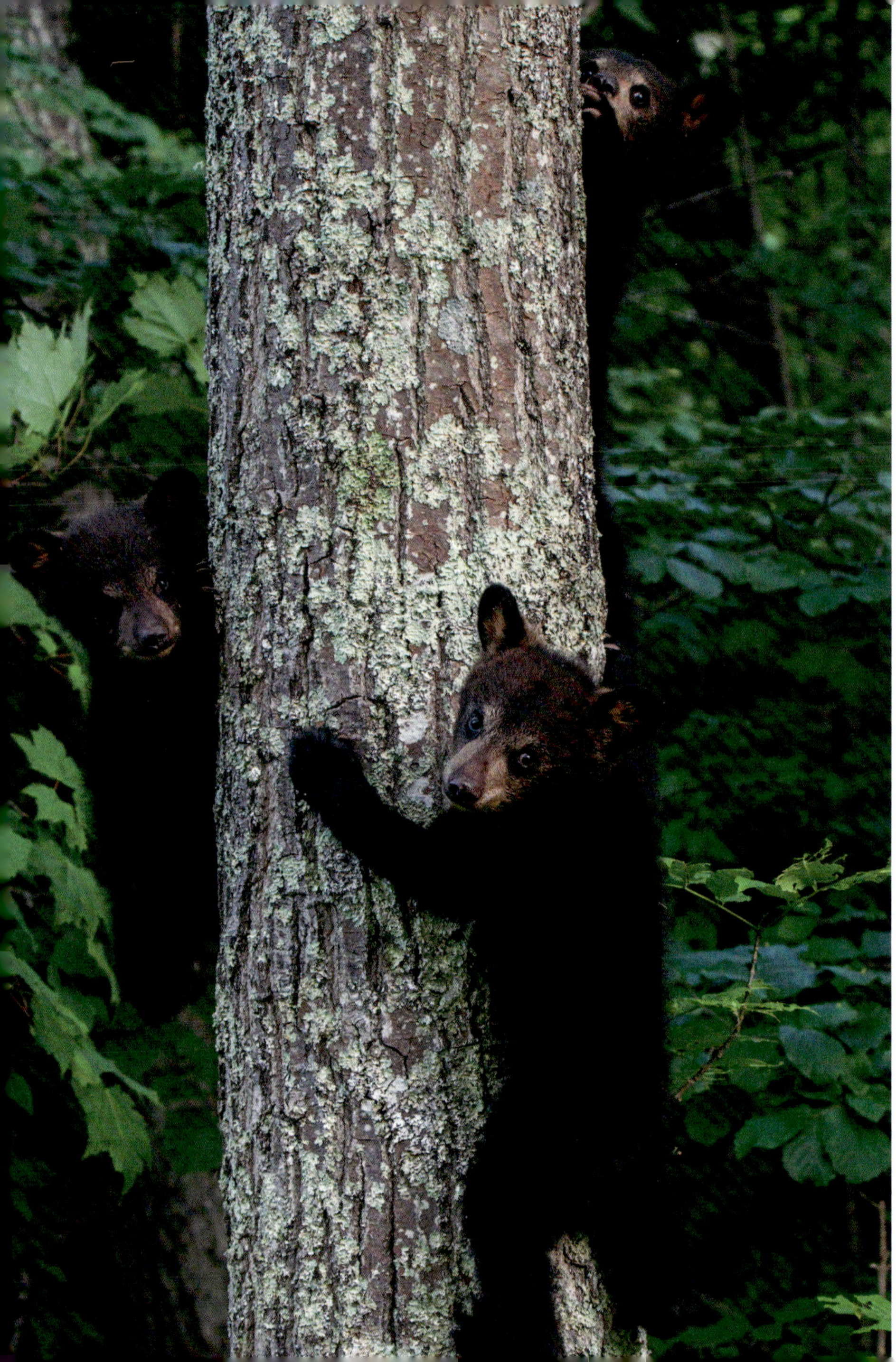

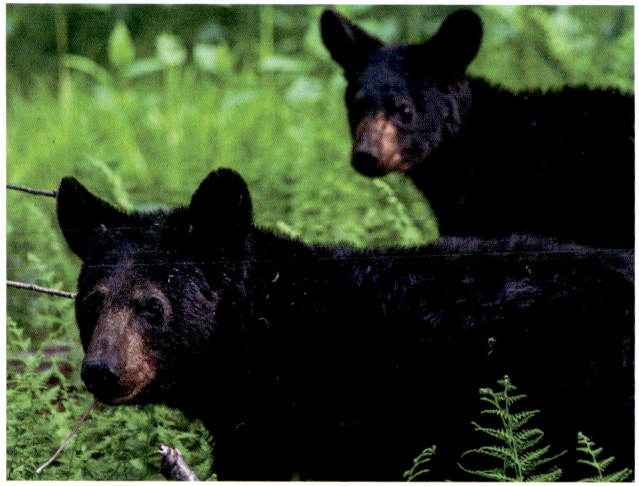

Black bear cubs are curious and playful with their siblings. Healthy populations of this charismatic mammal exist throughout the Central Appalachians. Shenandoah National Park and Blue Ridge Parkway, Virginia.

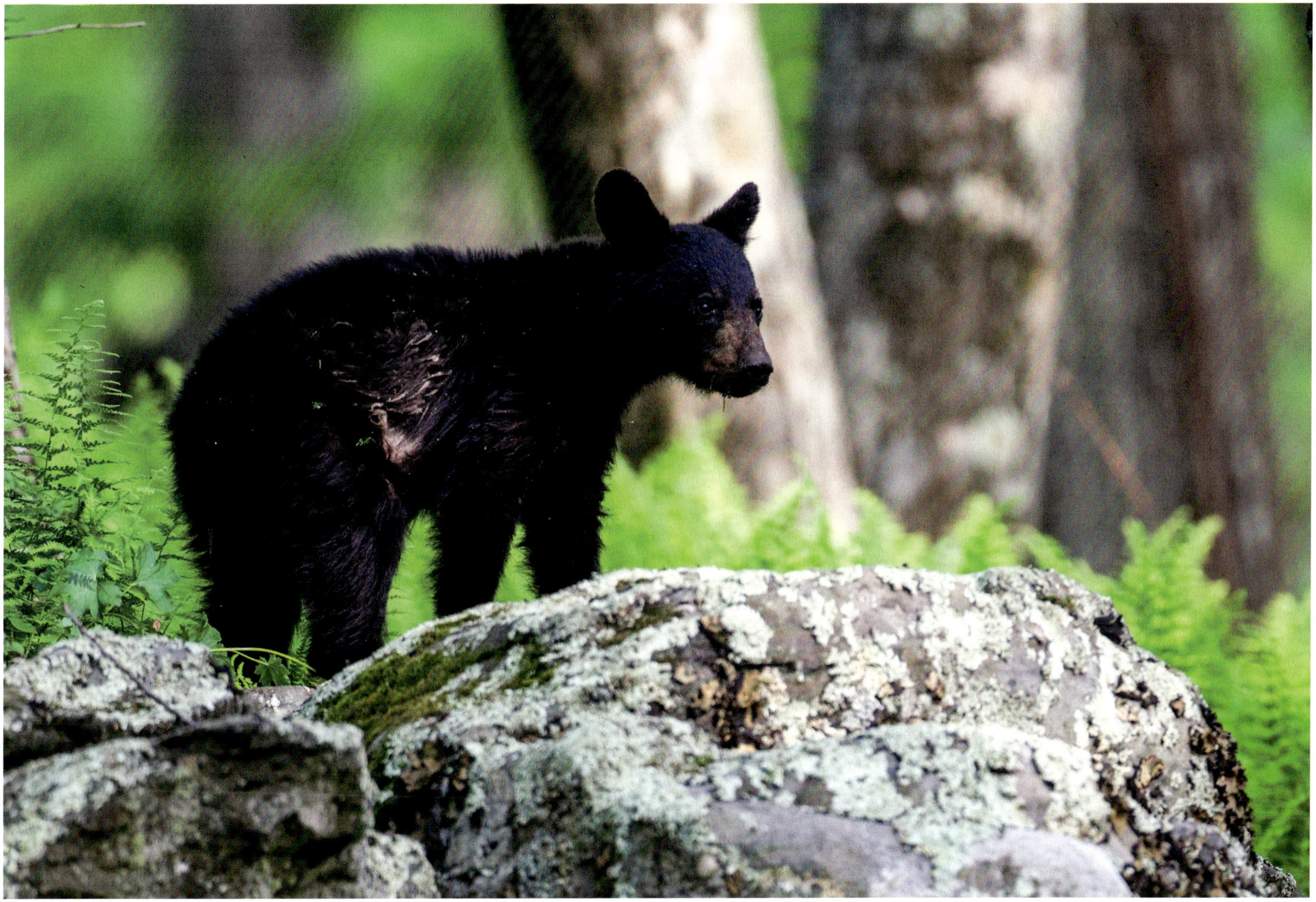

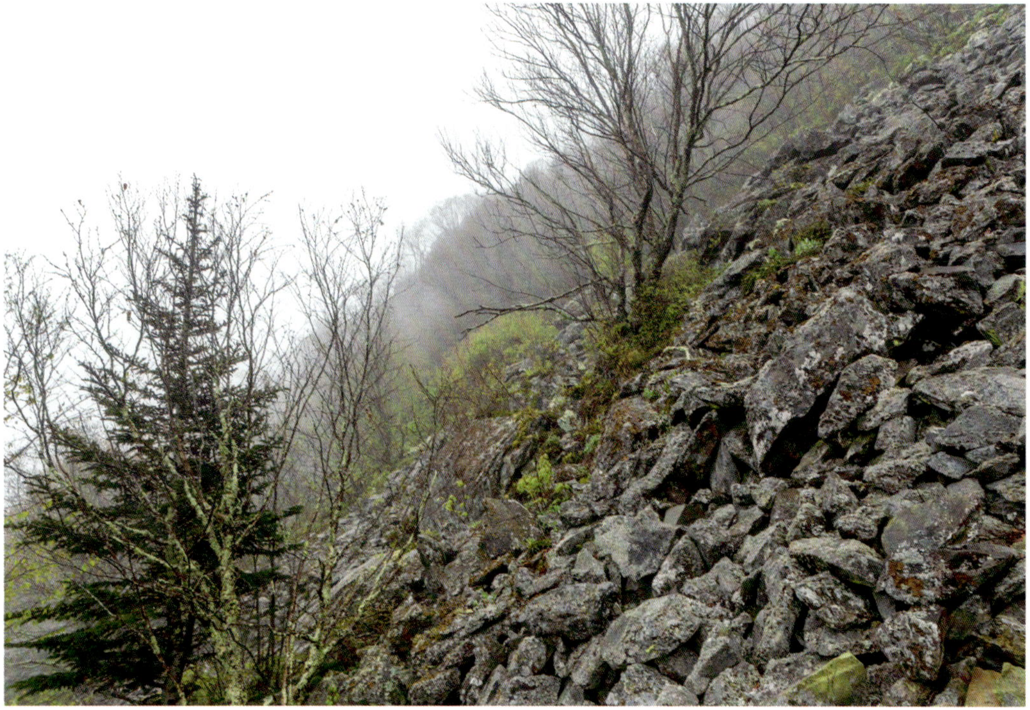

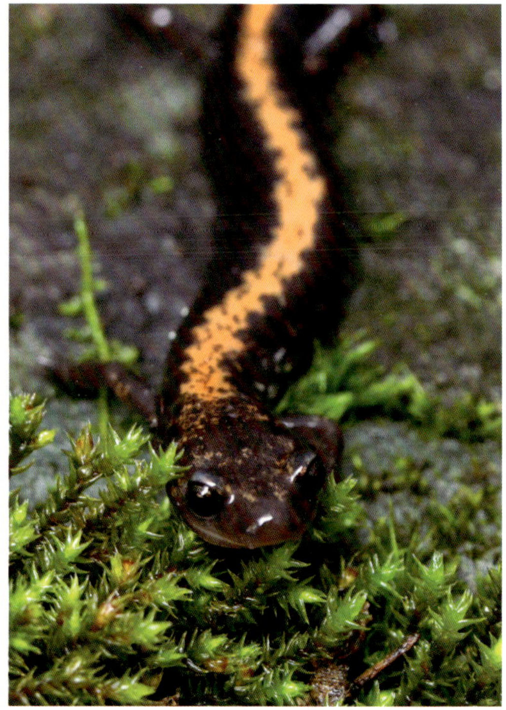

This cool, high-elevation talus slope in Shenandoah National Park provides habitat for once-plentiful species from the mountains' cooler epoch, such as balsam fir, and the incredibly rare Shenandoah salamander.

Existing only near the talus slope of three north-facing peaks, the Shenandoah salamander is a relic of the Pleistocene era and prefers the cool, dry habitat found here. The species may have the most restricted range of any tetrapod today. A warming climate is the species' biggest threat, along with competition from the common and expanding eastern red-backed salamander.

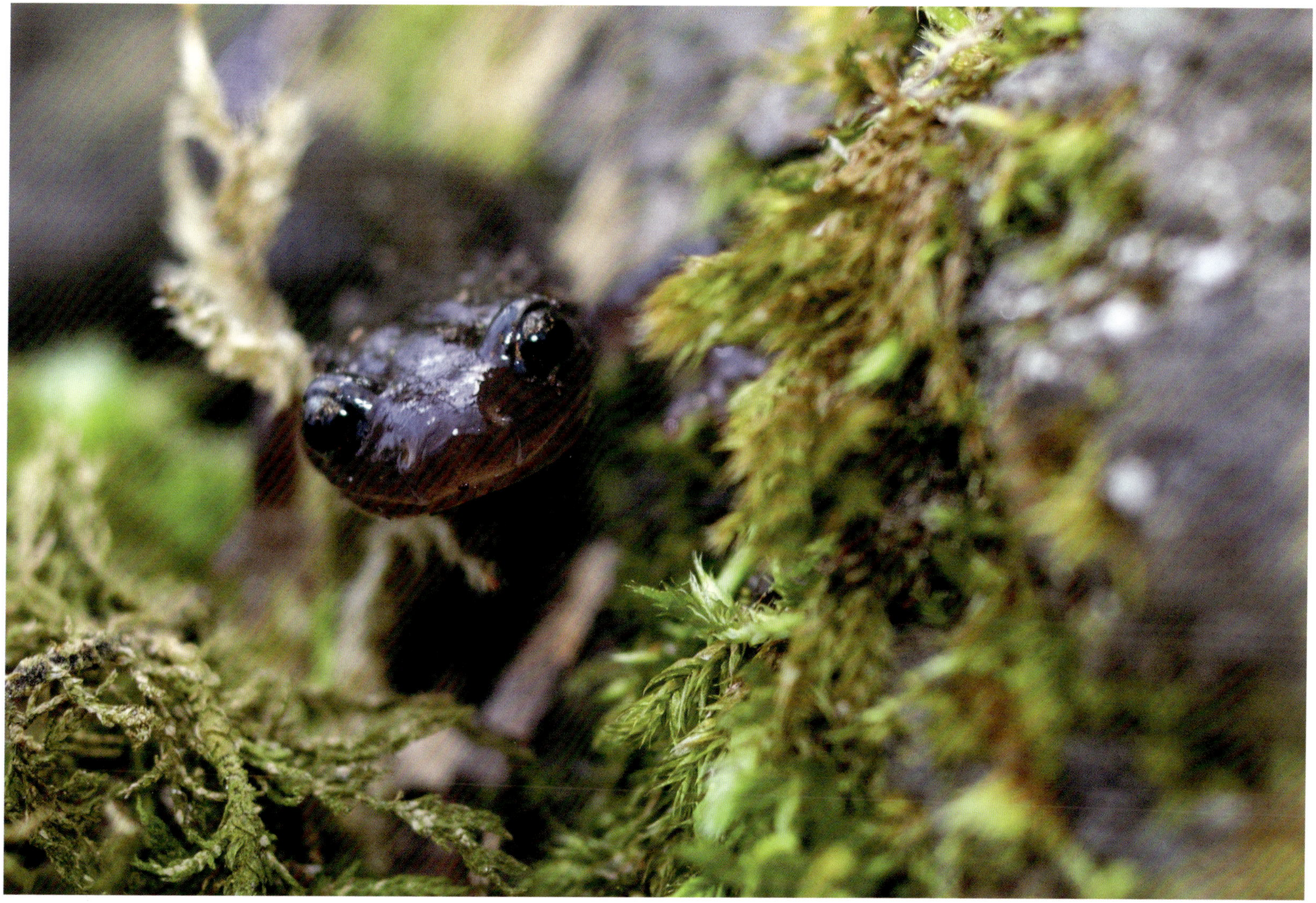

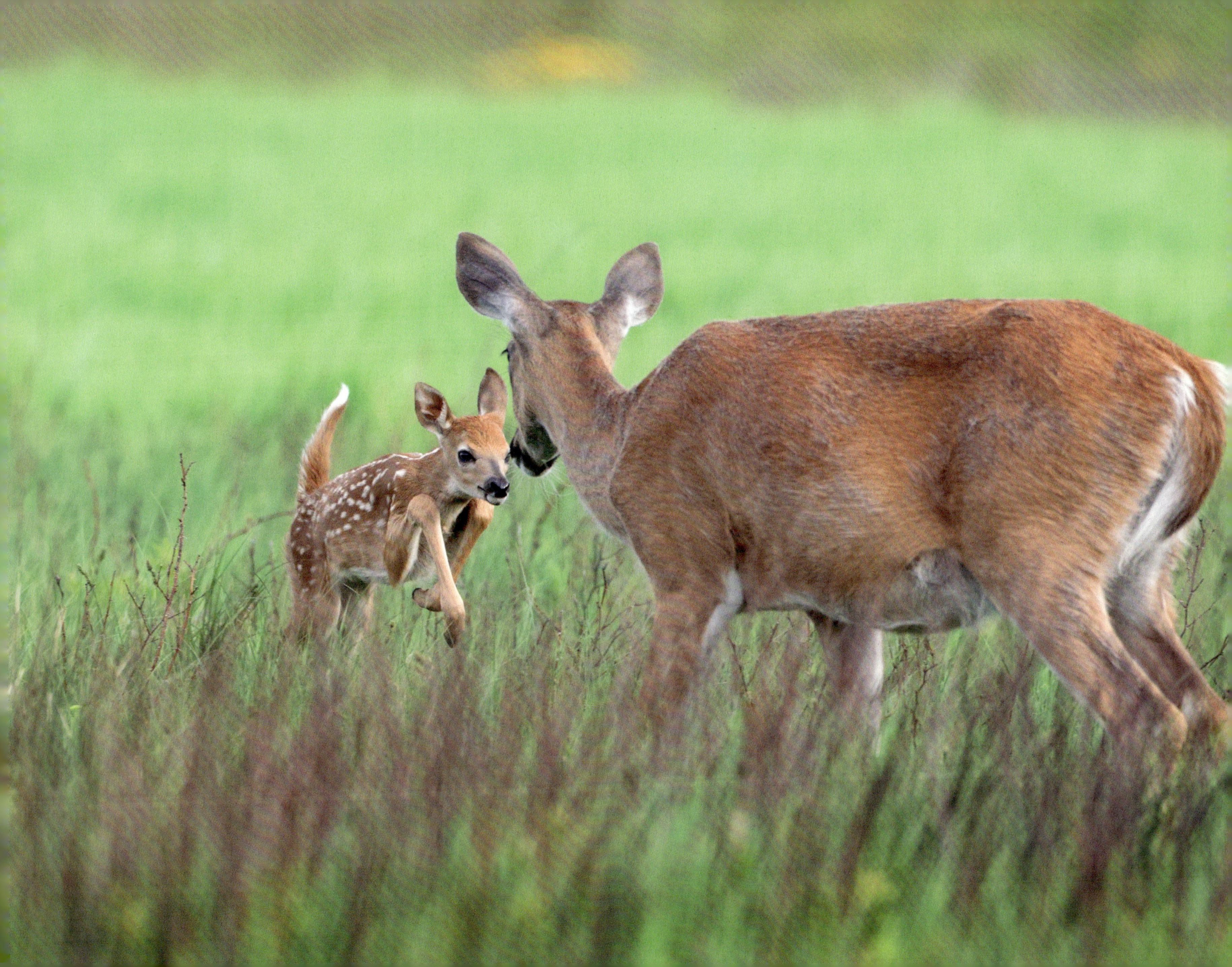

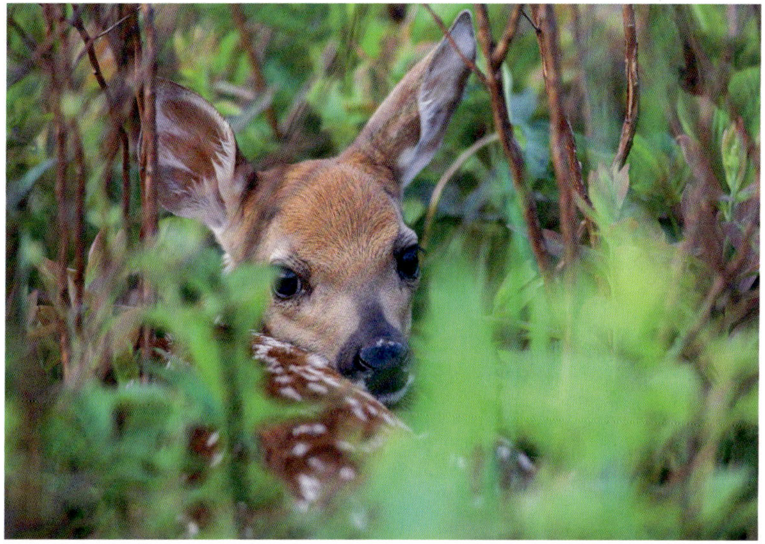

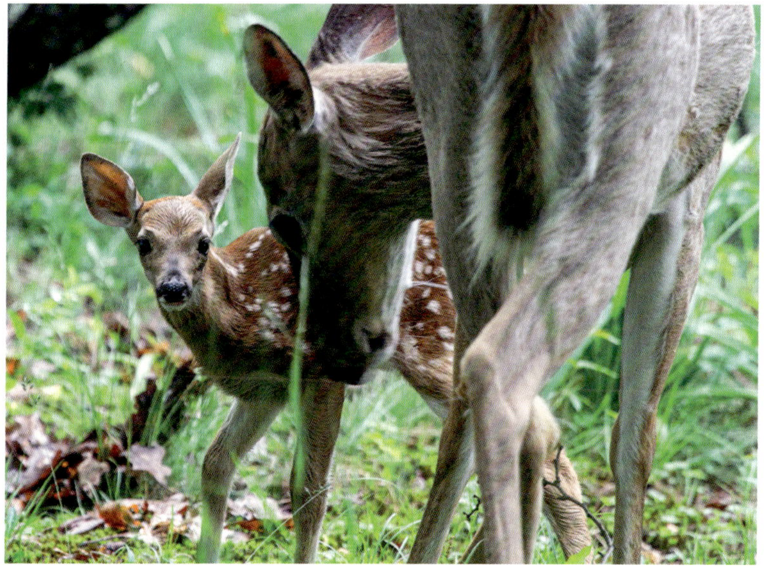

White-tailed deer are the mountains' most common large mammal. In late May and early June, spotted fawns are born in the meadows and surrounding forests. When mother leaves, the fawn will instinctually drop to the forest floor and is easily concealed among grasses, ferns, and shrubs. It is always a welcome sight to watch them strengthen their legs and gallop through the landscape.

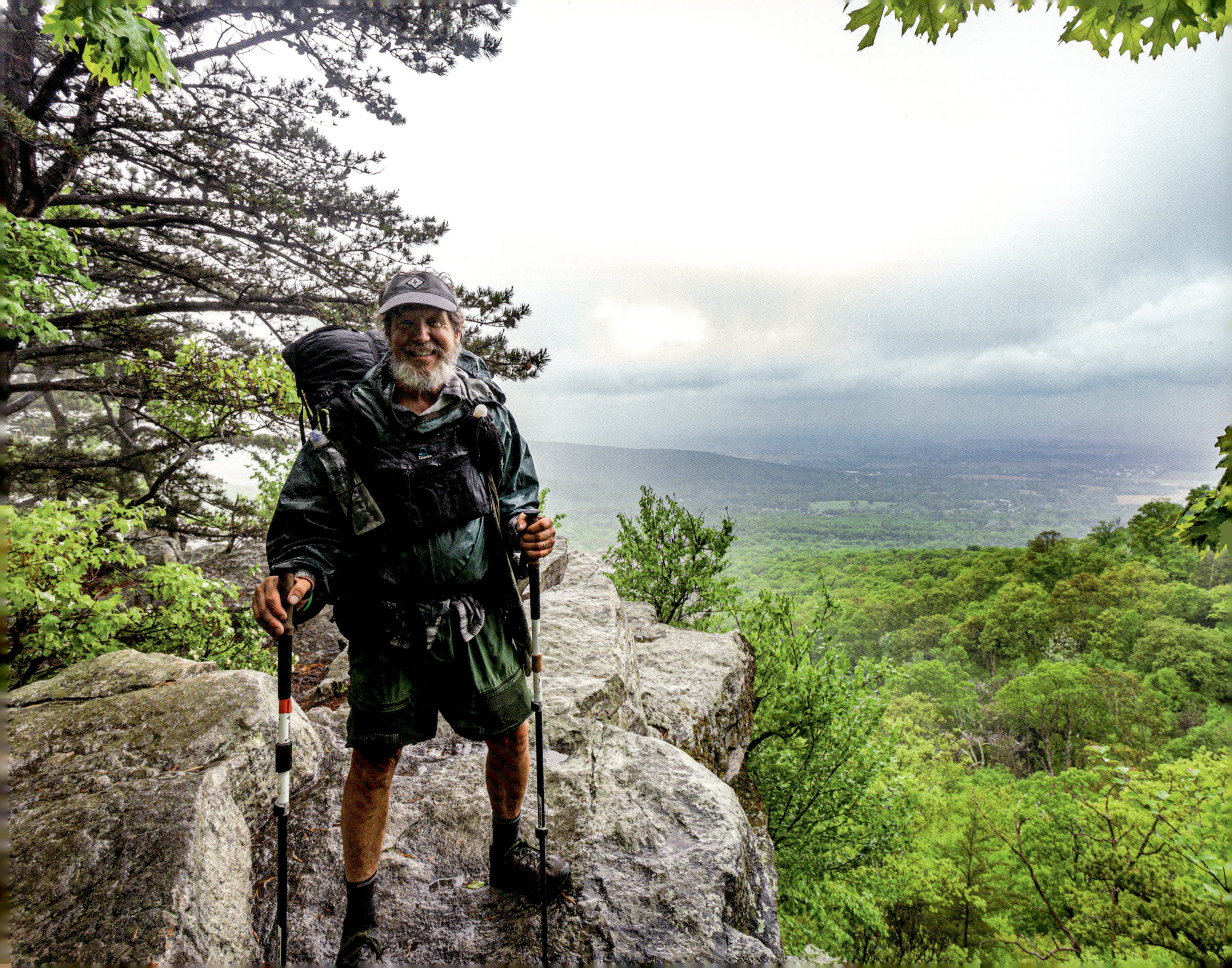

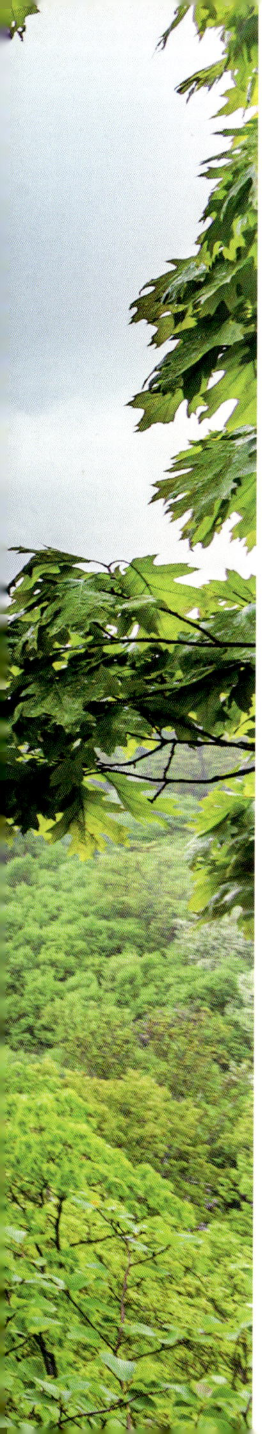

The Saga of Hops: An Appalachian Trail Thru-Hiker

It had been many years since I last saw Bill Minarik. We first met when I was a teenager employed at the National Aquarium in Baltimore, where Bill served as a director and senior staff member. We both shared a love of the great outdoors, playing guitar, and bluegrass music.

When I heard that Bill was going to hike the entire Appalachian Trail (the "AT"), I was immediately inspired. I have hiked portions of the AT in multiple states, and the idea of taking months of one's life to just hike is just very satisfying. I reached out to him to inquire about joining him on a part of his journey, and he was happy to oblige.

It was a damp and cloudy mid-May day when we reunited, since it had been raining in the days prior. As I made my way from the adjacent parking lot to the AT trailhead, it began to drizzle. Then, stepping out of the fog, like a butterfly emerging from its chrysalis, was the lean figure of a man with a very large backpack and even-larger smile. "Hey, Mark!" shouted the figure, with an animated wave. This was, and was not, the same Bill I knew so many years prior.

"You think a triathlon is hard?" he asks with a hearty laugh. "Try hiking the Appalachian Trail!" The damp hiking was easy compared to the snowstorms he had to endure since commencing his hike on March 4 at Springer Mountain in Georgia, the AT's southern terminus.

Sporting an unkempt beard attached to the figure of a marathon runner, his exterior was that of a mountain man—one who was well adapted to life on the world's most famous hiking trail. Prior to arriving in

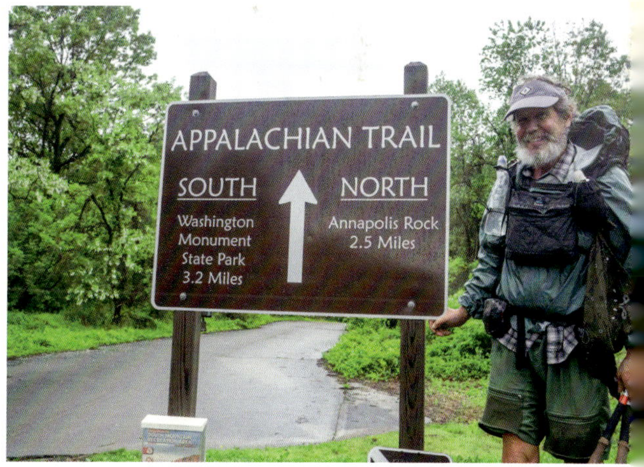

Harpers Ferry, West Virginia (which is the trail's mental halfway point; the geographical halfway point is at the Pine Grove Furnace in Pennsylvania), he had recently completed a section of the trail dubbed the "roller coaster:" a series of steep ascents and declines that last for over 13 miles. "Imagine going up and down, up and down. Then again with up and down, up and down," he explains. "These are tough trails!" he adds.

He even had a new moniker, "Hops." This, of course, was bestowed on him from fellow hikers for the way he "hopped" down the trail to join some fellow hikers at a campfire near the Easter holiday. It was also, in part, due to his passion for brewing beer. Thru-hikers, those who hike the entire AT in a season, have a culture in and of themselves. That culture dictates that they adopt what's known as a "trail name," a name that usually has some significance to the bearer, such as a location, personality trait, or inside joke. It also is simply a way to easily remember your fellow hikers and part of the camaraderie they share, as vagabonds who left the modern world for a time. And as Hops was quick to point out, it's not just other hikers that you share time with on the trail.

"I was settled in my tent and lying in my sleeping bag, and I noticed this rustling on the side," he says. The smile on his face now was childlike. "I look over and there's this rabbit that decided to come in and settle down. We made eye contact, and instead of running away it just snuggled down more! It was just do darn cute. I had a little roommate for the night!"

Besides rabbits, Hops shares anecdotes of the other wildlife he encountered on the trail. There was the owl that stalked him for a time. "A big barred owl," he adds. "It must've realized a scrawny hiker just wasn't worth it." Of the trail's abundant white-tailed deer? "They just look at me like almost another trail partner."

Fauna aside, what made him decide to embark on such a wild undertaking? The seed was first planted in his eleven-year-old mind when, as a boy scout, a scoutmaster shared the significance of this special hiking path. "He told us, 'If you go this way, you will go all the way to Maine, and if you go this way, you will go all the way to Georgia,'" he reminisces with that same fascination. Decades later, the seed had grown into a full-grown tree, a childhood dream realized.

We ate lunch and stopped in a laundromat so he could give his clothes a much-needed cleaning. While waiting, he shared more trail stories. Some of the most fascinating tales are of the "trail angels." These are people who leave food out for hikers at various destinations along the trail, hold cookouts, or offer to clean hikers' clothes. They expect nothing in return; they simply wish to help. It is part of that trail culture.

Our conversation turns to water and the watershed and how we both find it so fascinating that a drop of rain from the summit of Hawksbill Mountain in Shenandoah National Park or from Spruce Knob in West Virginia could eventually flow into the Chesapeake Bay. "I've had a chance to see lots of different rivers. I've kayaked 430 miles from Cooperstown, New York, through Pennsylvania to the bay. I've kayaked the Shenandoah and Potomac," he says, as his voice turns to a more reverent tone. "Nothing is more special than when

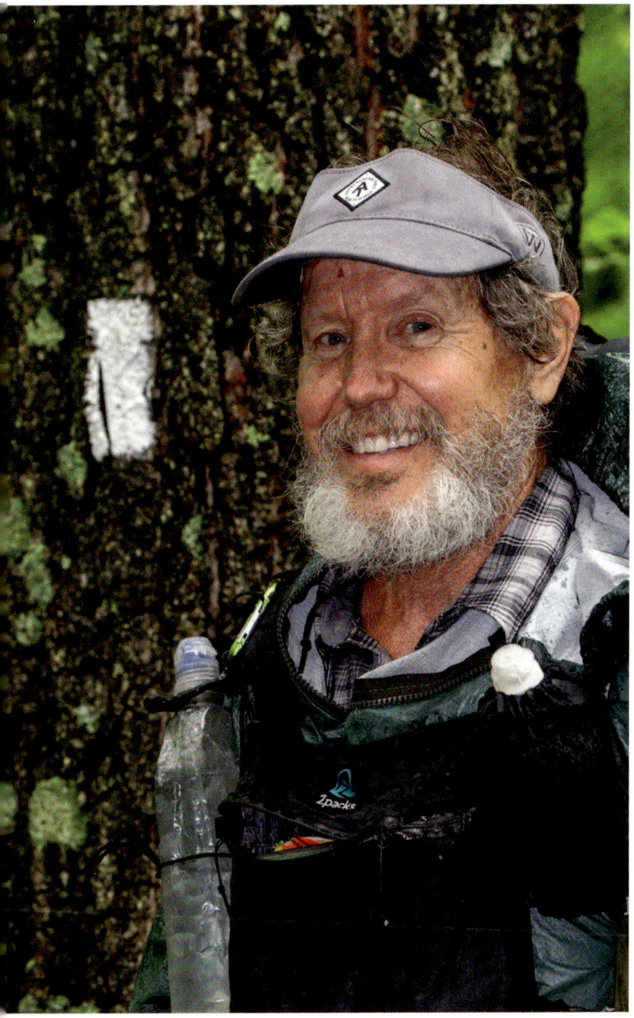

you see the water bubbling up from the ground, from these great rivers, the same water that eventually moves to the bay. And the ocean. After a hot day, when your water supply is diminished to nothing, feeling how blessed it is to have free, pure water. That the fish, the animals, and your fellow humans get to have that sacred water."

The simple desire to share in the substance of life after spending so much time alone is admirable. Hops is ecumenical in his desire to see more people experience the Appalachian Trail. "You don't need to do it for six months; you can do it for a weekend, and you'll have both a mental and physical change of mind. You can do an easy section, 3 miles in and 3 miles out, and see wildlife. With younger family members you can share the fresh air and the natural world. Whether you want a little or a lot, the Appalachian Trail gives it to you."

That's the gift of these mountains, and why it is so important to embrace outdoor recreation. The changes Hops alluded to are very much a re-creation of the human psyche to what it must have been during a more primitive time. Urban metropolises such as Baltimore; Washington, DC; Philadelphia; and New York City are only a short drive from the AT and most other areas in the Central Appalachians. If more people made the natural world a part of their lives, however small, it would alleviate some of the stress that stems from modern existence.

"When you step into these great eastern forests, which are so green and so lush, you realize immediately that you're not in the city anymore," he says. "If you turn over a log you might find an orange newt. If you listen, you will hear the birds chirping to one another. Mentally, you are very quickly able to get into that frame of mind that helps you to re-create."

Back on the trail, a gentle spring rain began to fall. Off the trail we saw a number of pink lady slipper orchids. Water droplets clung onto their labella, much like the droplets that fell from our raincoats.

We paused and admired the delicate form of a trio of orchids, which contrasted nicely to the multiple hues of green in the springtime forest. When other hikers emerged from the mist, Hops was quick

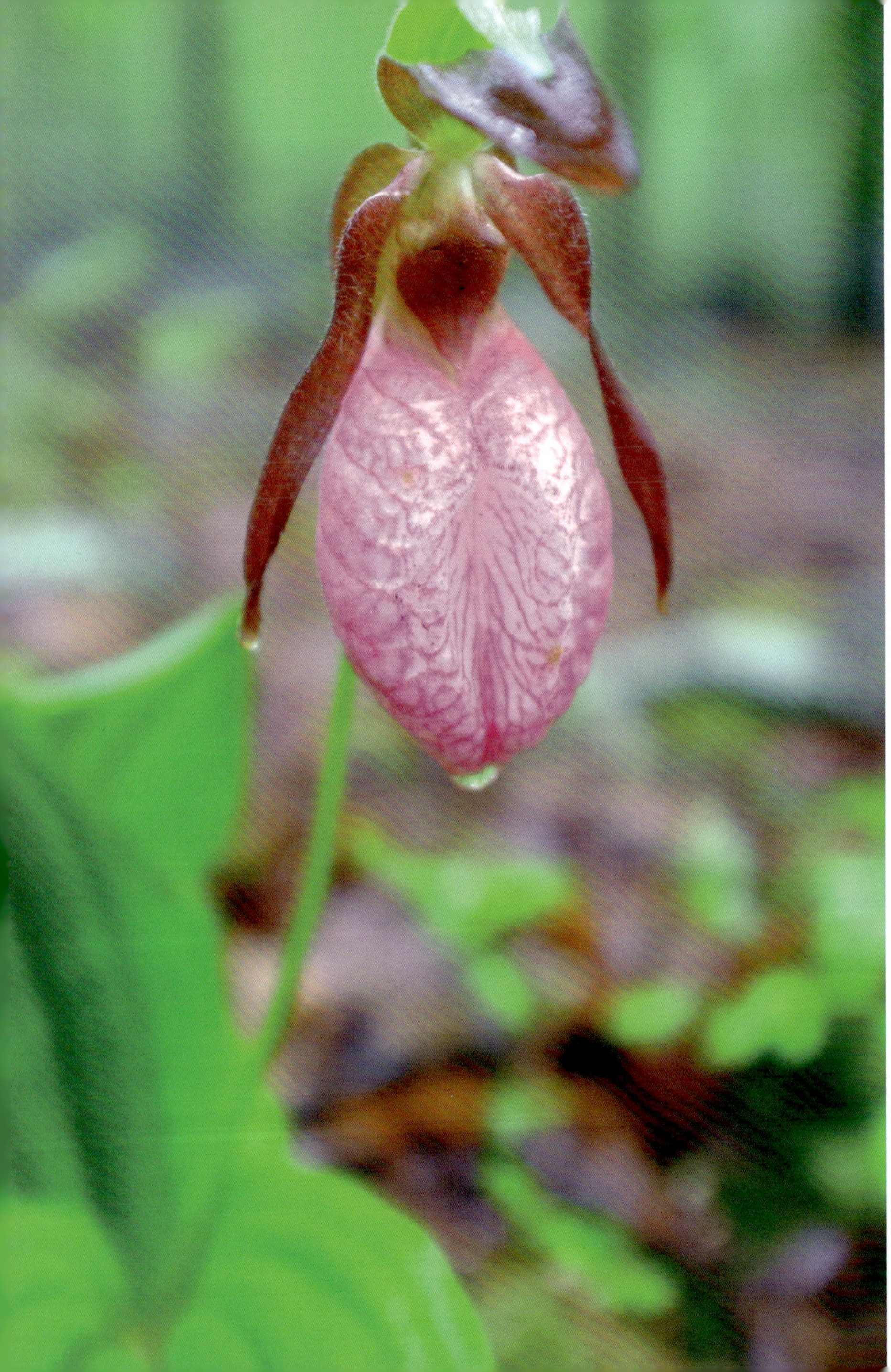

Pink lady slipper

to share the beautiful scene with them. We stopped at one of the life-giving springs to fill our water bottles and talk about our families. At the Annapolis Rocks Overlook, our brief reunion came to an end, but I felt privileged to have shared in a small part of his journey. On August 16, 2018, after five and a half months, traversing fourteen states and hiking 2,190.9 miles, Hops arrived at the summit of Maine's Mt. Katahdin and completed his journey on the Appalachian Trail. The experience and lessons from the trail remain and have had a lasting influence on him.

"I don't worry about things, because I know a 'bubbling spring' is coming up soon that will refresh me," he says with a chuckle.

Of the relationships he formed with other hikers on the trail, he said: "I know what making a friend is when they hike over the same bouldered mountains and muddy paths with you for hundreds of miles, only to laugh at the adventure."

And this is not the first book that features Hops! He appears numerous times in Ben Crawford's excellent work, *2000 Miles Together: The Story of the Largest Family to Hike the Appalachian Trail.*

Today, when not spending time with his grandchildren or tending to his historic farmhouse, Hops speaks and lectures on the trail, still adorned with his trademark mountain-man beard (though he now trims it). After those many months of living a wayfarer existence in the wilderness, he never took a break from hiking. "Hikers hike," he says, "and I do almost every day."

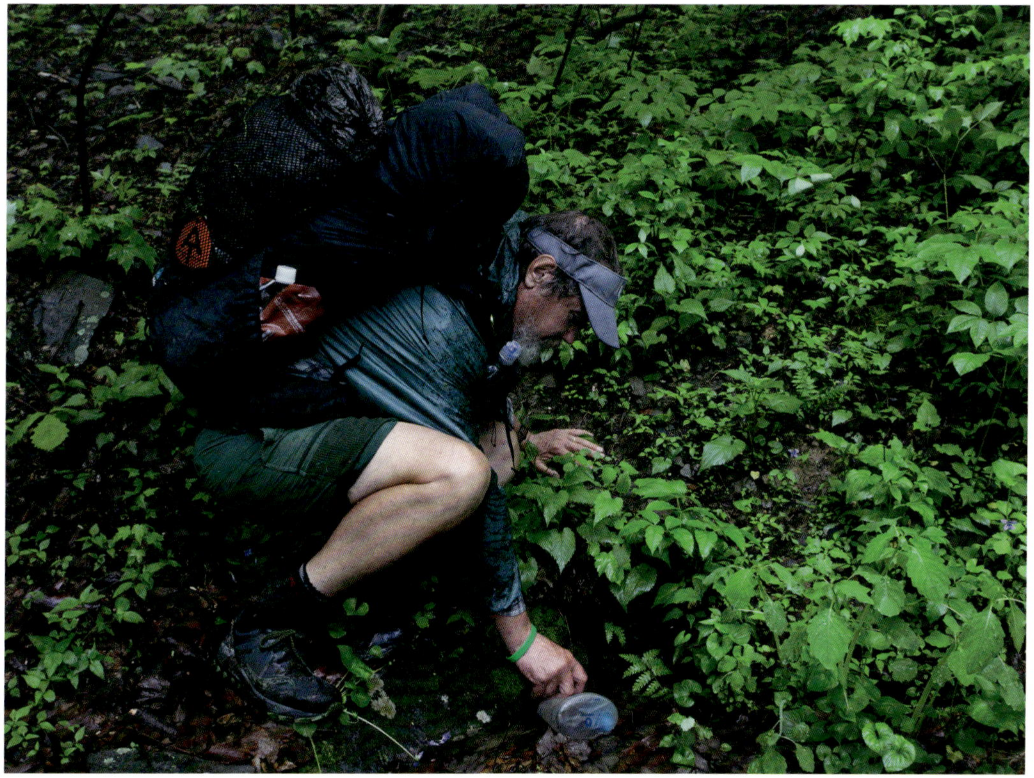

SUMMER

As the summer solstice approaches in mid-June, the delicate blooms of mountain laurel line the many trails.

Lack of light pollution allows for excellent celestial viewing, and the core of the Milky Way is best visible during this time. In the dark, a pack of coyotes are heard in the distance. What begins as a few high-pitched, ghostly howls quickly transforms into a crescendo. Closer still, a barred owl calls to its mate. By nightfall, these predators become very loquacious. Their signature mnemonic, "Who cooks for you? Who cooks for you all?," replays again and again.

Days are humid and afternoon thunderstorms are not uncommon. The cooler temperatures of the mountains provide relief to animal and human alike. Rushing waterfalls flow into the mountains' many tributaries. Butterfly activity is high, each species having a dedicated flight time. The dashing white flowers of great rhododendron are now in bloom. In July, blueberries and huckleberries provide a tasty treat to various mammals, birds, and hikers. Reptiles, such as timber rattlesnakes, sun themselves along the rock outcrops. As evening transitions into twilight, the meadows and riparian forests sparkle in the

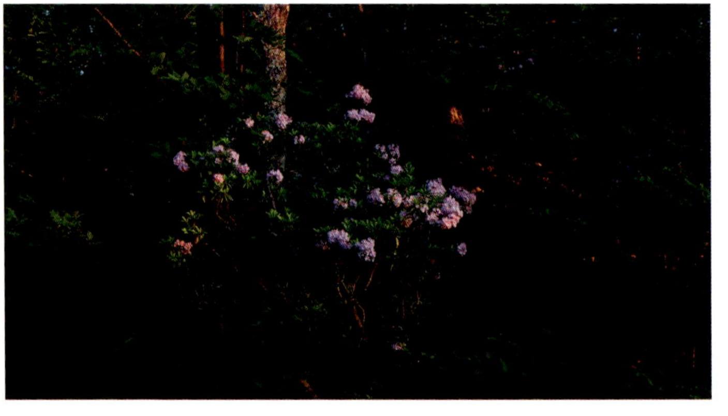

Mountain laurel in bloom. Michaux State Forest, Pennsylvania.

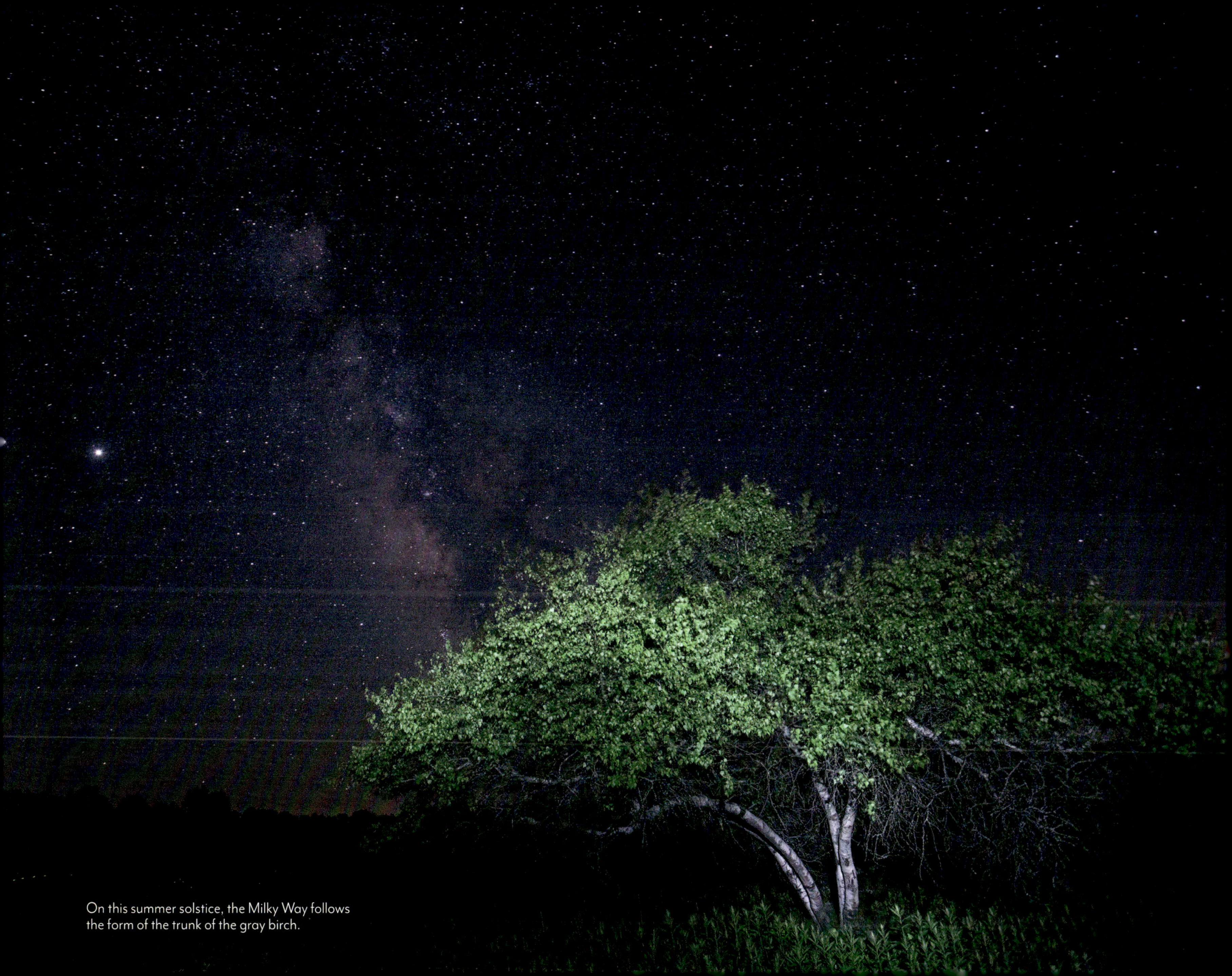

On this summer solstice, the Milky Way follows the form of the trunk of the gray birch.

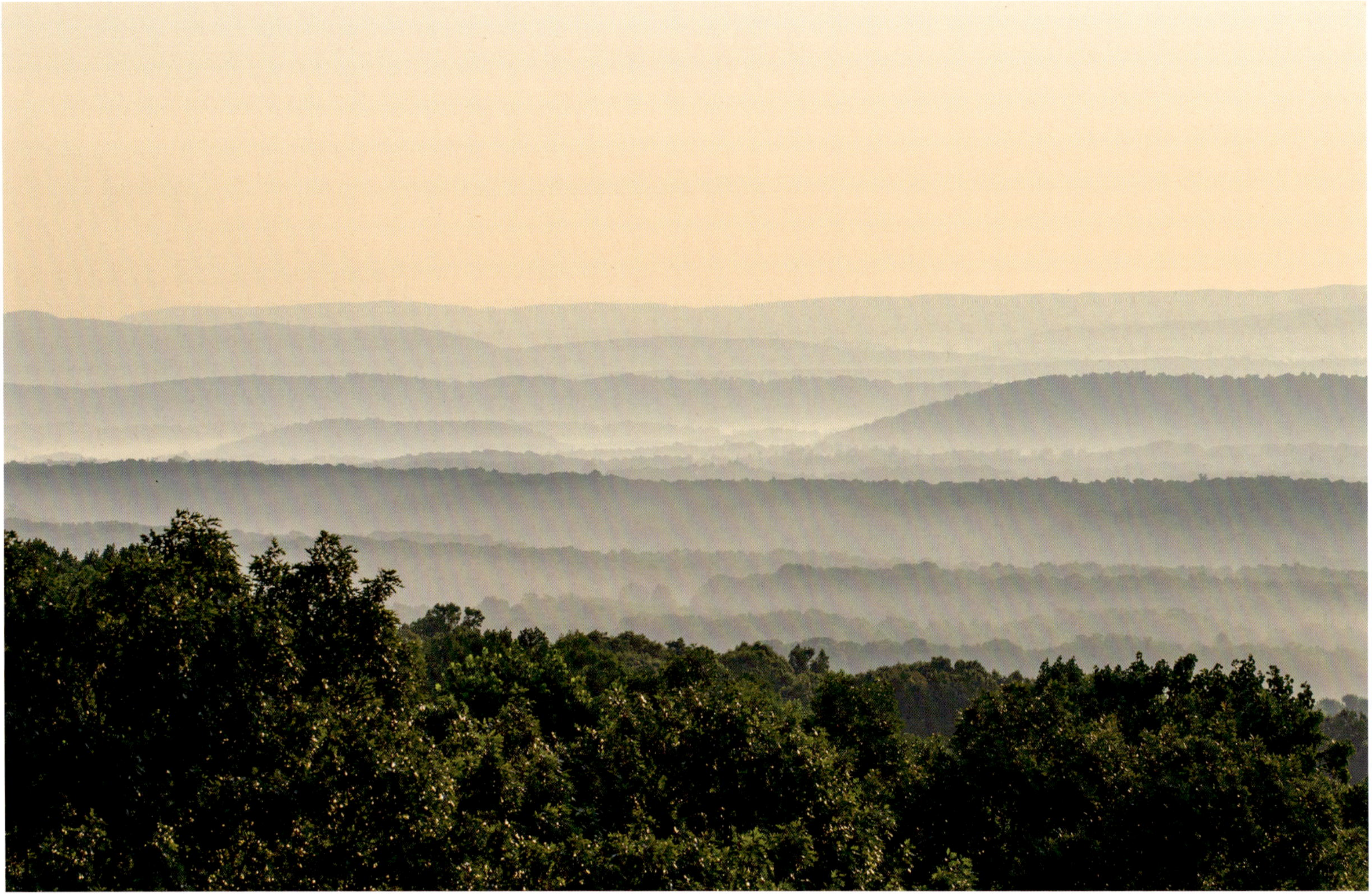

The warmth of summer is felt through the fertile valleys and the mountains of the Ridge and Valley Province. Washington and Allegany Counties, Maryland.

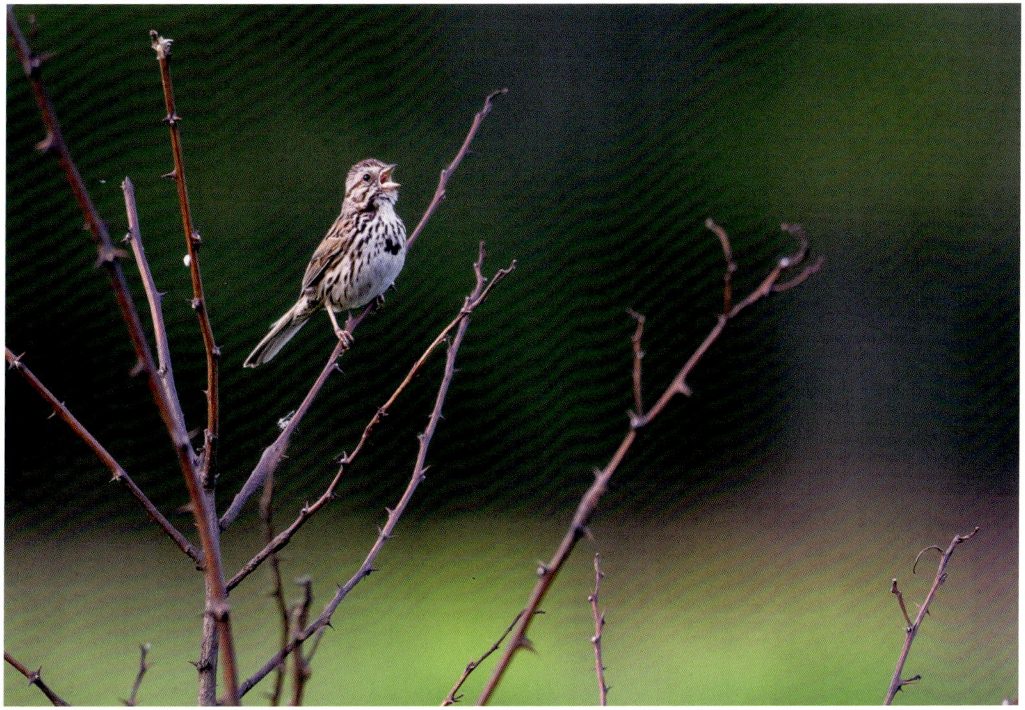

yellow flickers of fireflies—not flies at all, but a type of beetle. Their bioluminescent communication is a summer rite of passage for many.

By August, birds such as cerulean warblers and alder flycatchers begin to migrate south. The red berries that adorn the mountain ash have fruited, as have as the first apple trees, and both will continue into the autumn. Into September, blooming goldenrod provides energy to migrating monarch butterflies.

As the mountain rivers become cool on September nights, the rare eastern hellbender begins its breeding season. Once a creature of mountain lore, the hellbender is one of the mountains' most distinctive inhabitants.

Mountain laurel and lush ferns are a typical June scene in the Blue Ridge Mountains. Shenandoah National Park, Virginia.

The song of the chipping sparrow is heard throughout the valleys and meadows on summer mornings. Sky Meadows State Park, Virginia.

A black bear yearling prepares to climb a tree to escape the heat of summer. Shenandoah National Park, Virginia.

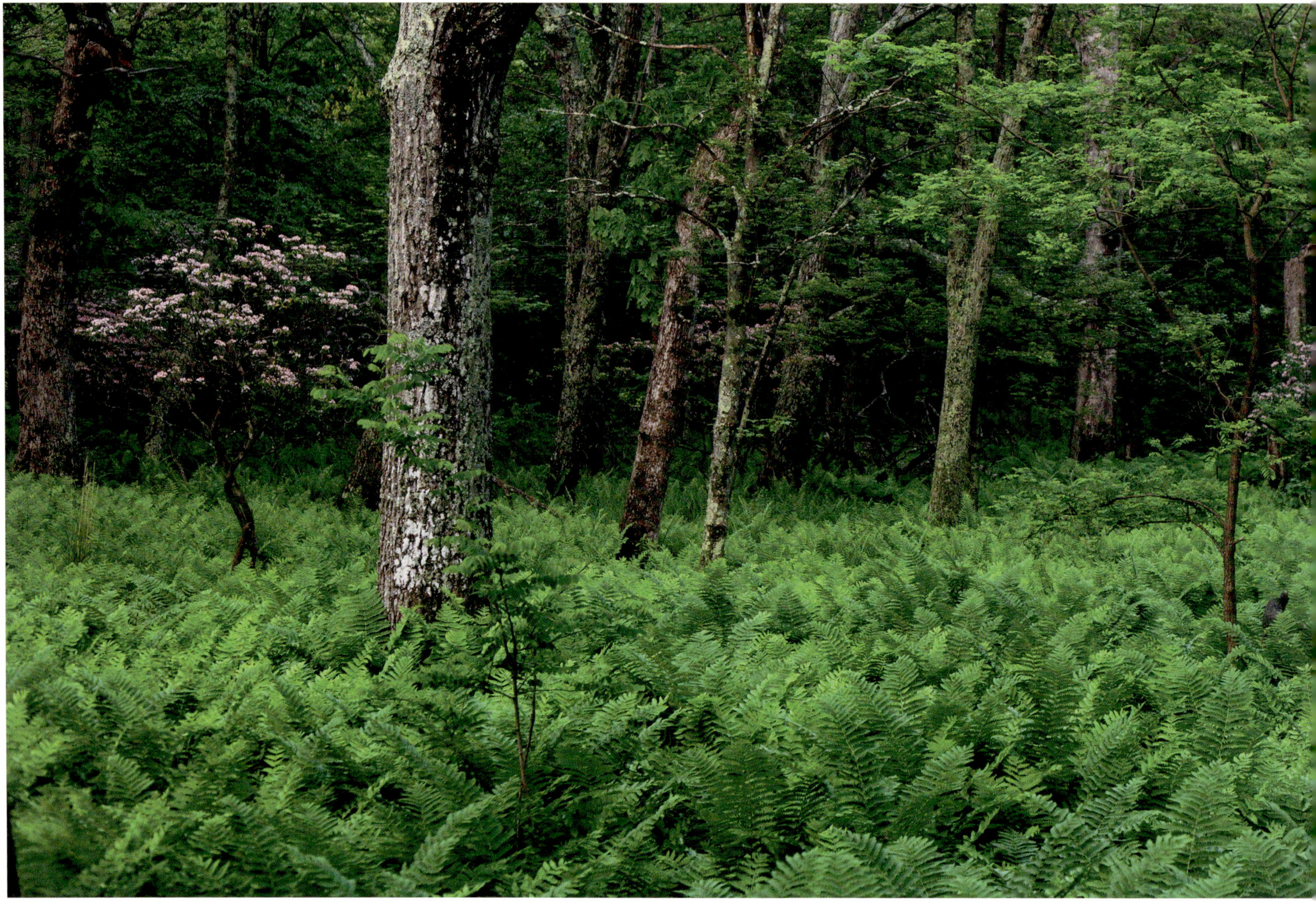

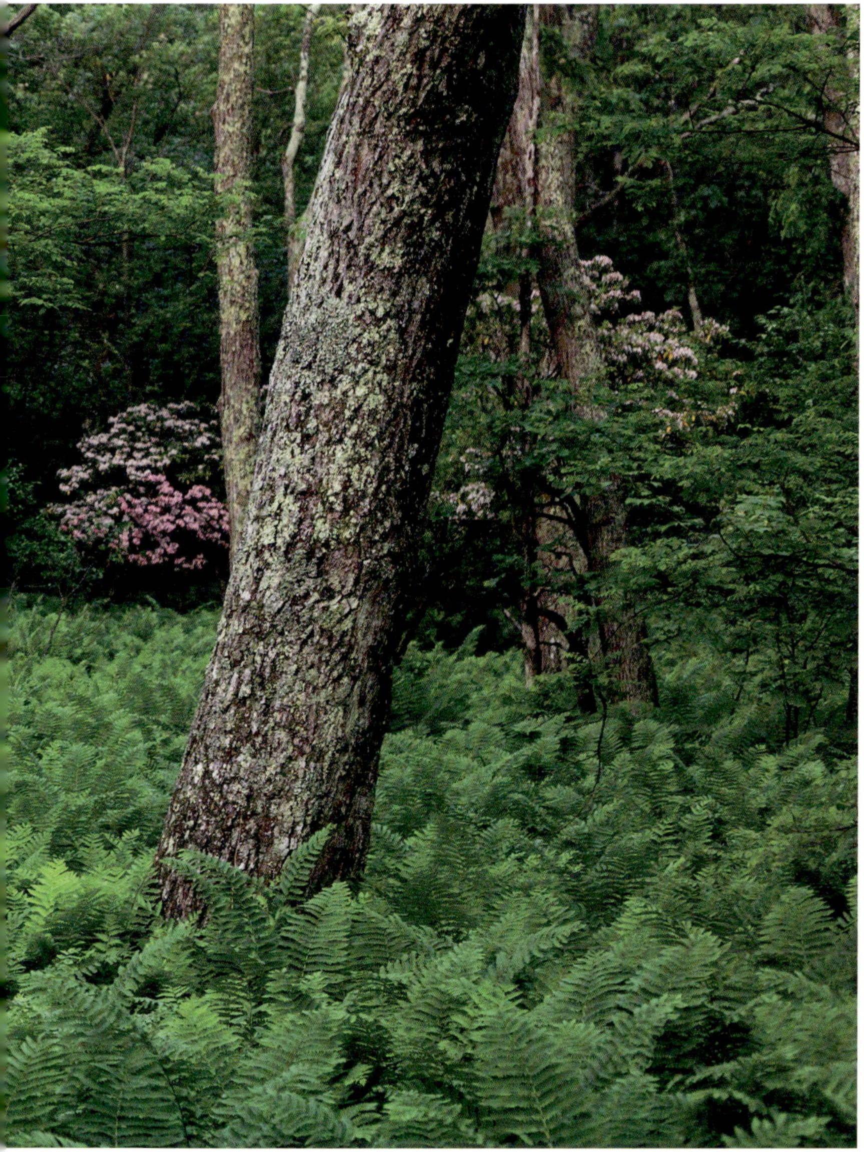

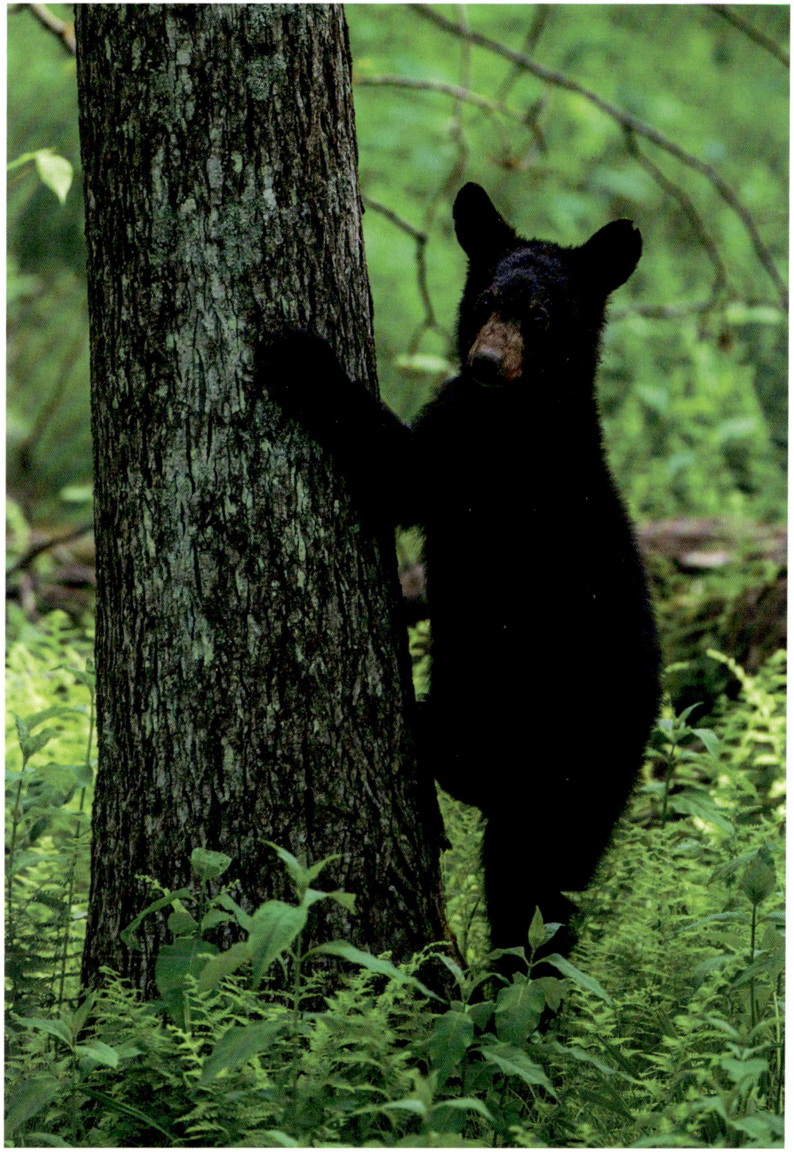

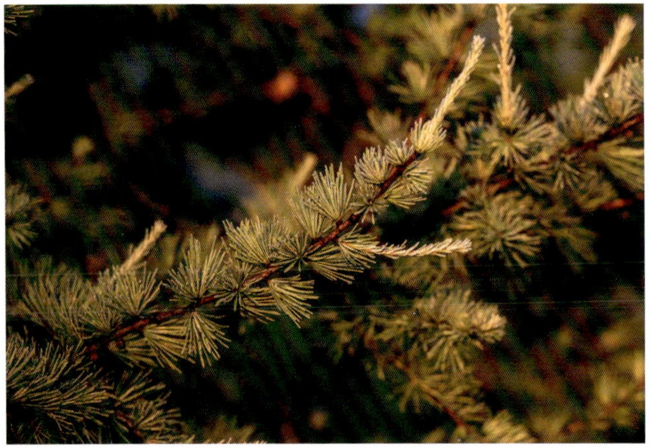

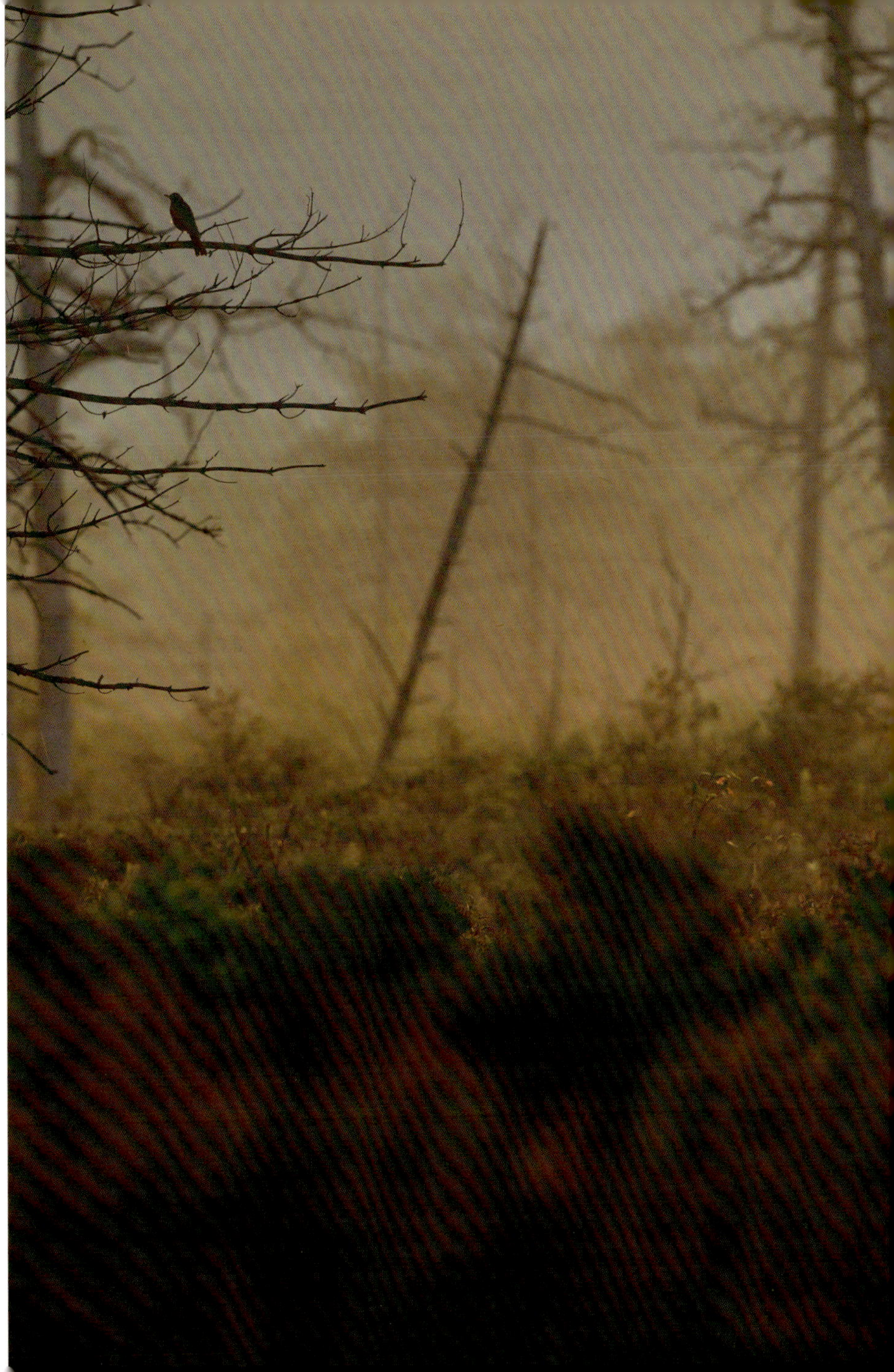

A robin is perched on the dead snag of a tree in the remnant boreal fen that is the Finzel Swamp Preserve, Maryland.

The needles of a larch, also known as the tamarack tree, are illuminated by the rising summer sun. Finzel Swamp Preserve, Maryland.

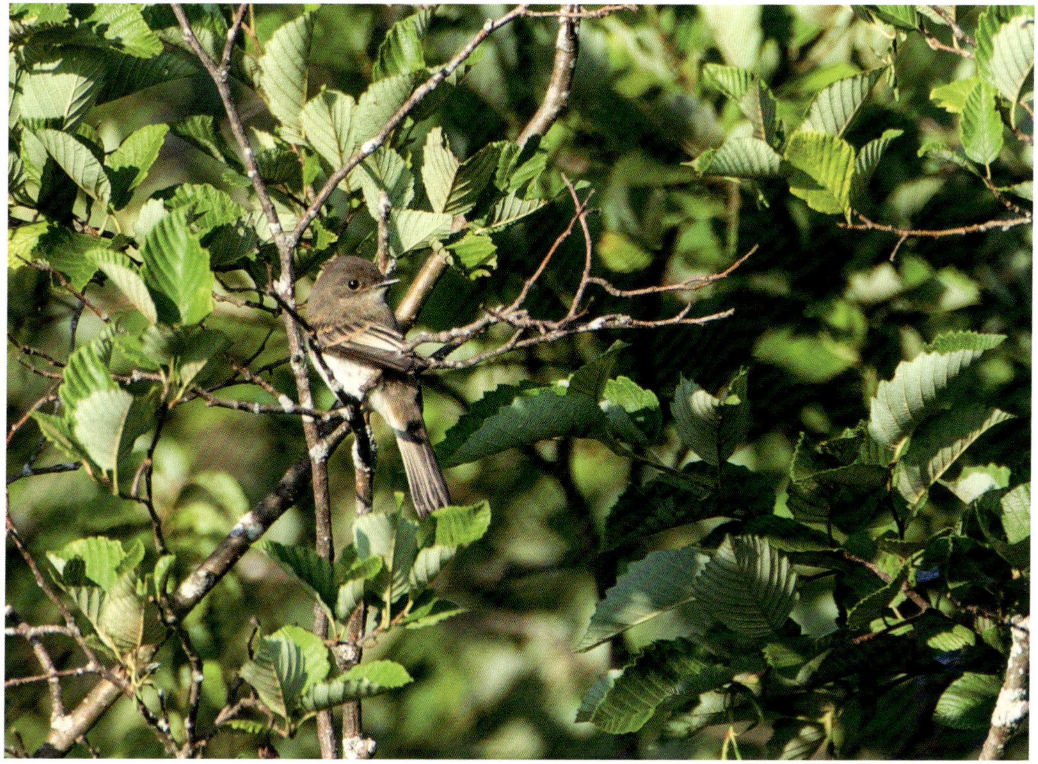

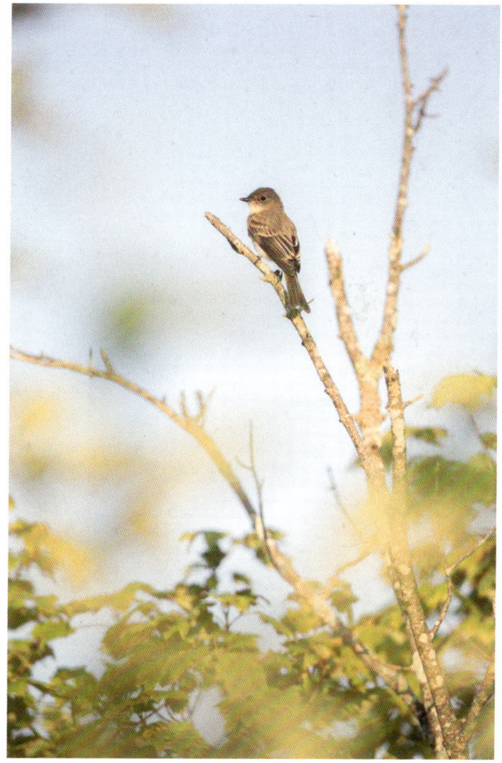

The alder flycatcher is a rare breeding bird for the region and prefers high-elevation bogs and shrubby wetlands. As its name suggests, it may be encountered in speckled alder thickets. Finzel Swamp Preserve, Maryland.

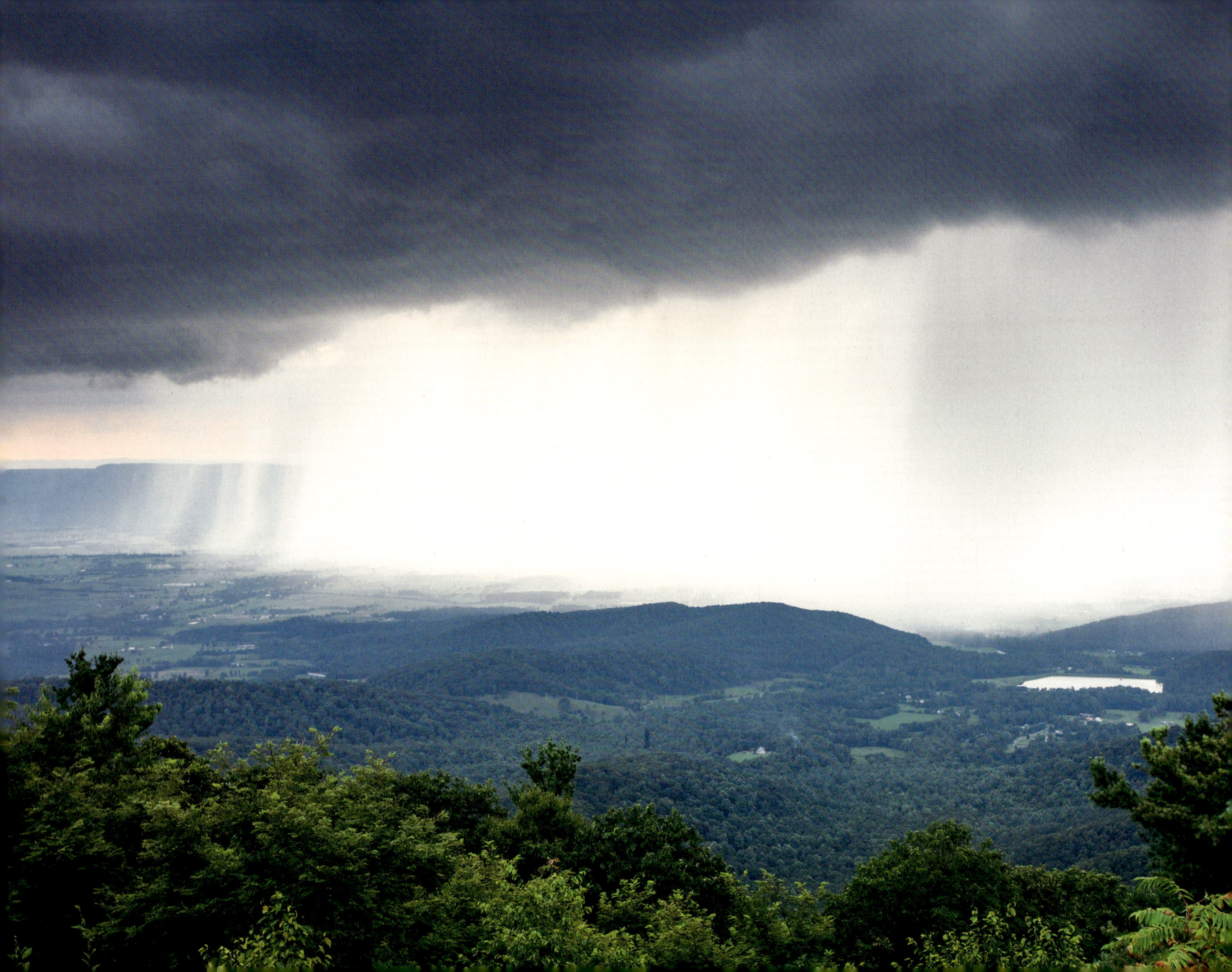

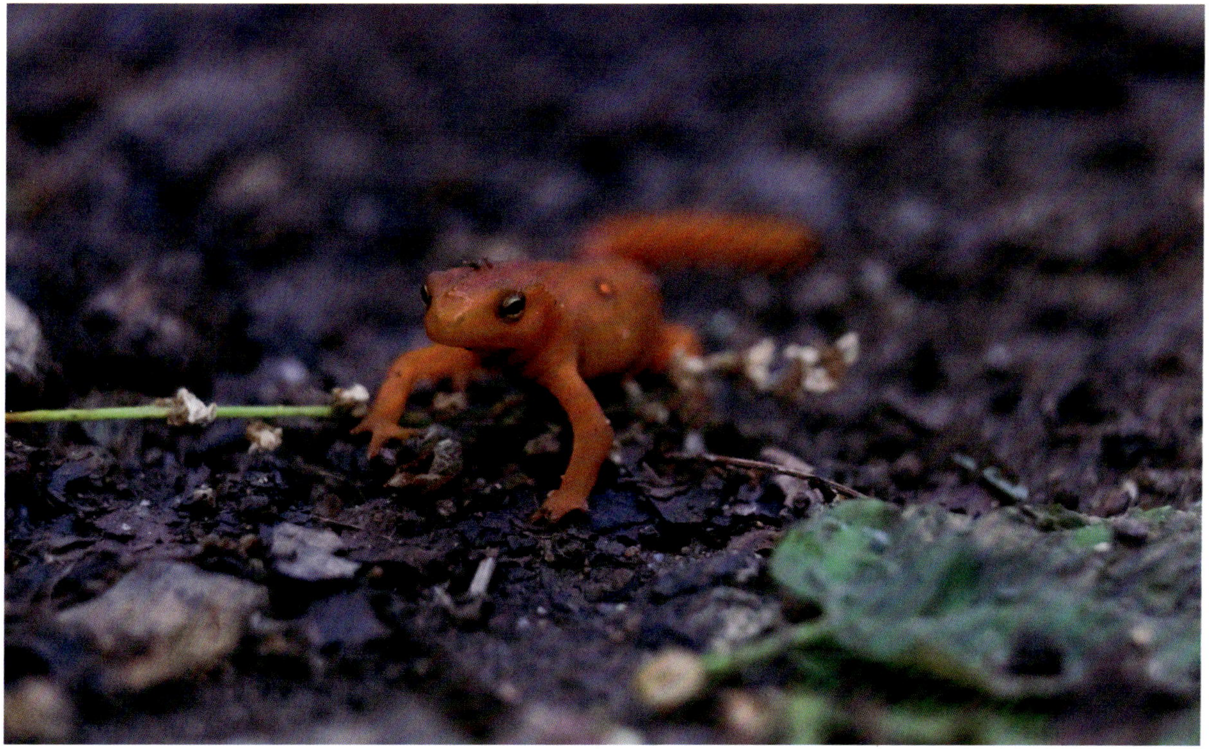

Afternoon storms are common by July and cool the landscape burdened with heat and humidity. The forest floor, now cool and soft, is all the red eft needs to wander. The eft is a juvenile eastern newt, and its red color is an example of aposematic coloration and serves as a warning for would-be predators to stay away. Shenandoah National Park, Virginia, and Seneca Rocks–Spruce Knob Recreation Area, West Virginia.

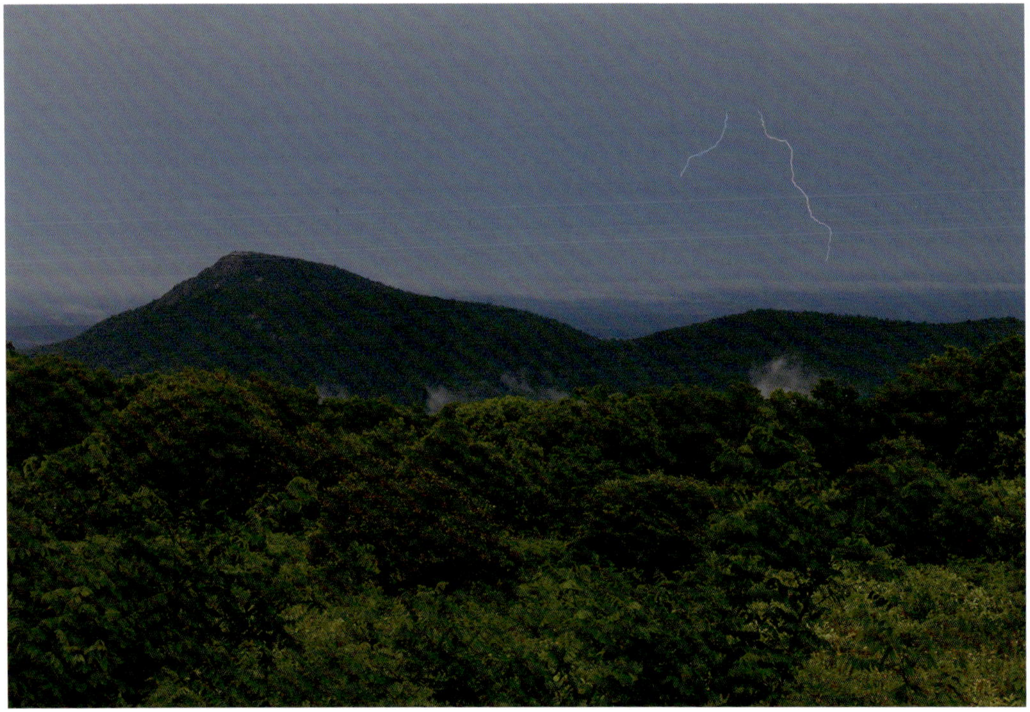

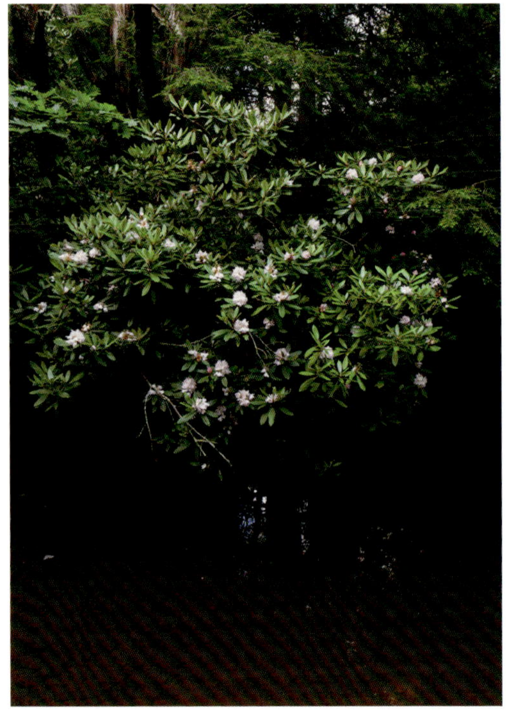

Lightning strikes near the summit of Old Rag Mountain in Shenandoah National Park.

Great rhododendron blooms above the Savage River in Garrett County, Maryland.

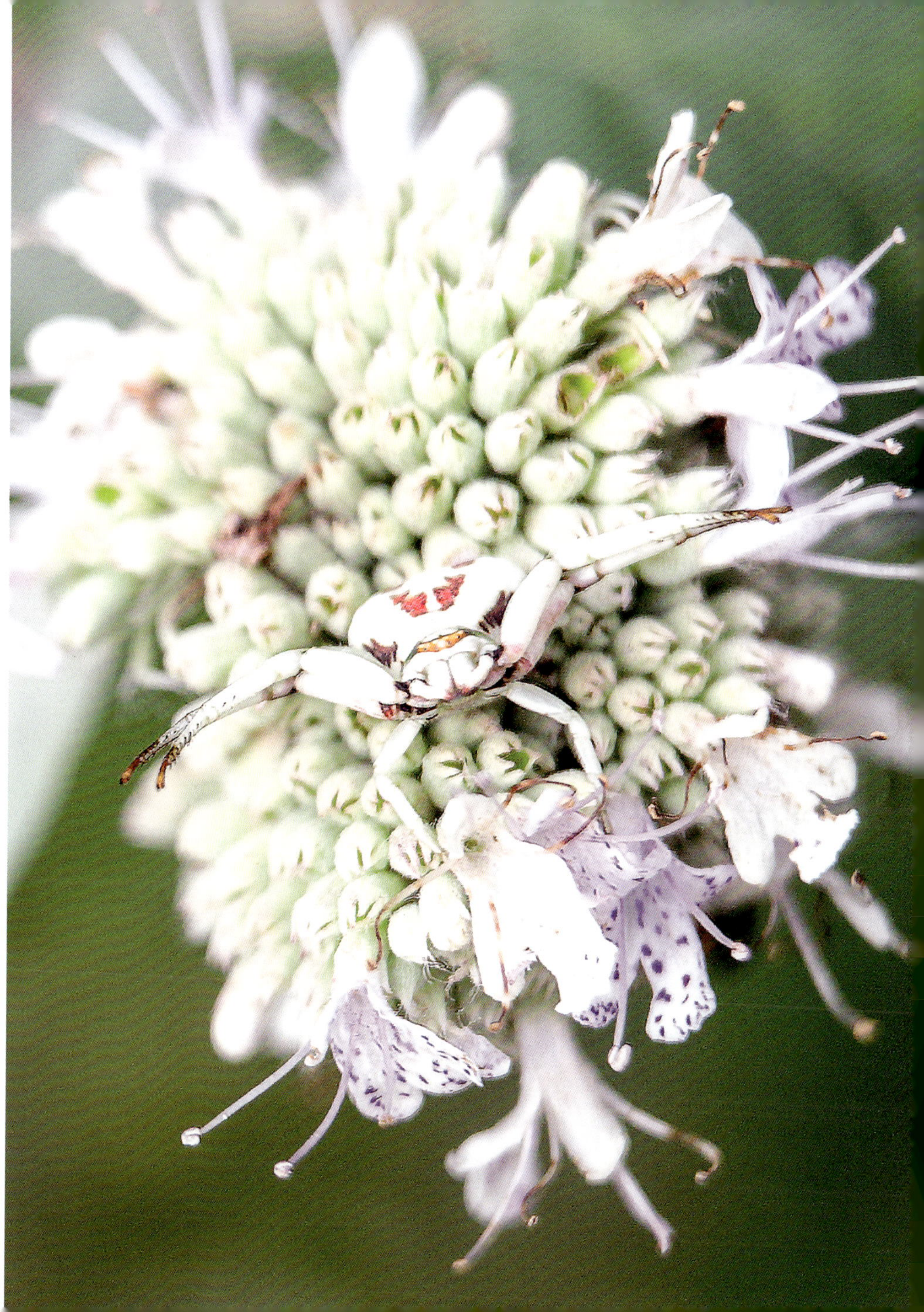

The stealthy crab spider is both beautiful and deadly. Well camouflaged among the flowers it occupies, it will attack any unsuspecting pollinator. Sideling Hill Creek Preserve, Maryland.

Underwater in the Mountains

On a late July morning, I meandered down a stretch of the Savage River in Garrett County, Maryland. Even for midsummer, the water was remarkably cool due to the mature eastern hemlocks that shade the river on both sides. These high-elevation, cool mountain waters are home to brook trout, the only native trout to the eastern United States, and an Appalachian specialty. "Brookies" are quite attractive fish. Their green marbled pattern is interspersed with red dots that are surrounded by blue halos. A reddish-orange belly completes the color spectrum, and during the autumn spawning season it becomes even redder.

Among the rocks and substrate, I can clearly observe sculpin, blue gills, crayfish, and, after some time, a beautiful brook trout. A mink patrols the edge of the river, swimming from rock to rock in search of a meal. Eventually it encounters what must be the very odd sight of a half-submerged human coated in a neoprene wetsuit. But the mink isn't impressed and continues its hunt.

Under the surface of the river is a world rich in biodiversity. By surveying the rivers, creeks, and streams, one can get a glimpse into the health of these Appalachian waters. And, pun intended, it's so cool to explore the freshwater world like one would diving along a coral reef.

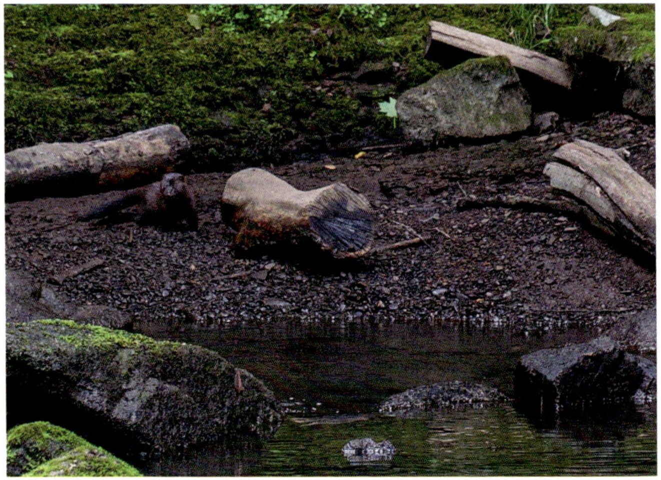

A mink along the river shore. Savage River State Forest, Maryland.

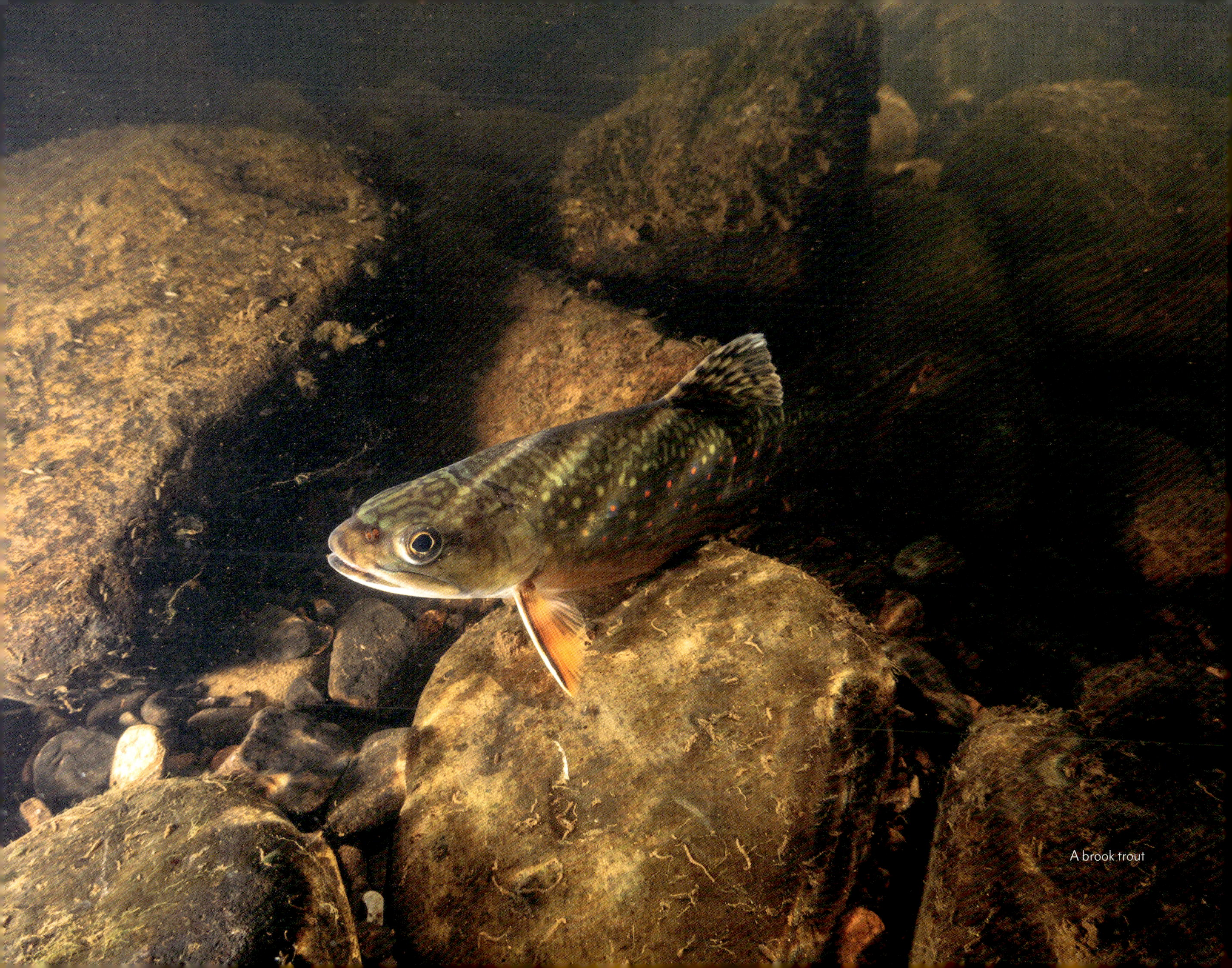

A brook trout

Some unique animals live in these waters, but perhaps none more than the eastern hellbender, North America's largest salamander. Living an entirely aquatic existence, this "canary in the coal mine" is extremely sensitive to disturbances in water quality, from excess sediment due to road construction, to fertilizer runoff, acid mine drainage, and other pollutants. Excess sediment fills in the substrate of rocks under which the hellbenders seek shelter. Consequently, it is reliably found only in a few areas within the Chesapeake Bay watershed and within all the Appalachians. In most of its range, the population is dangling by a precarious thread. Nesting boxes have now been deployed to allow the hellbenders to breed and lay eggs where the males will watch over them into the early autumn.

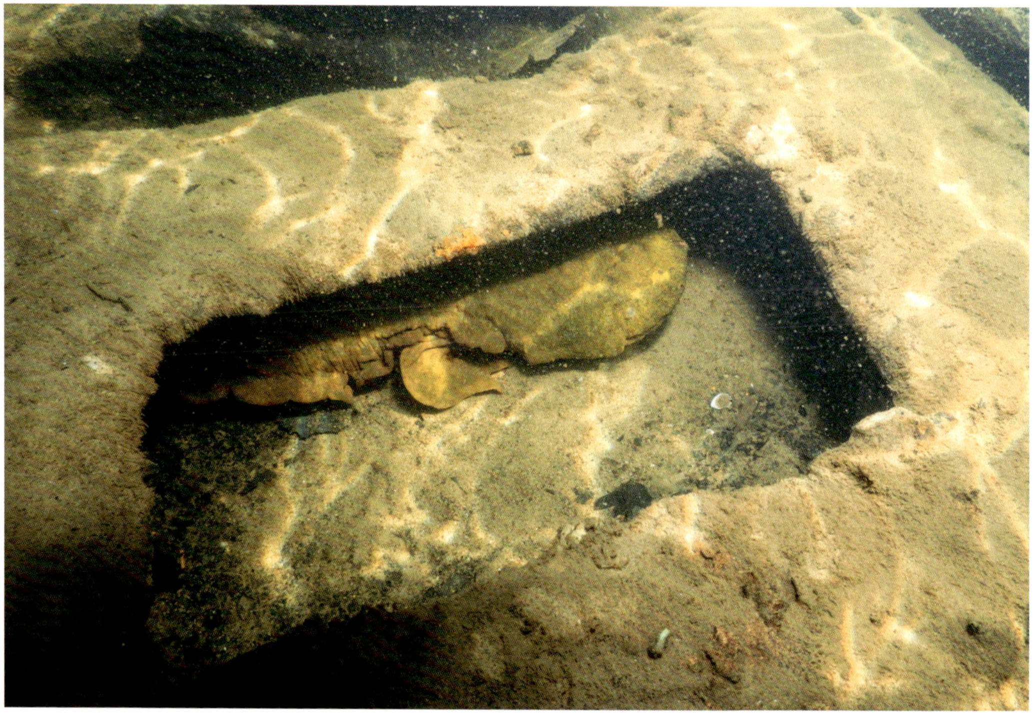

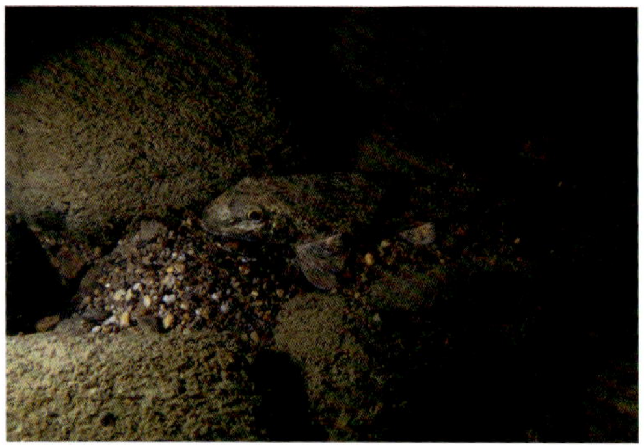

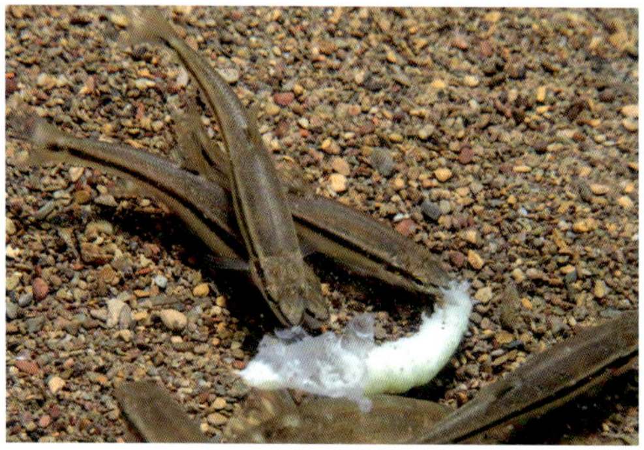

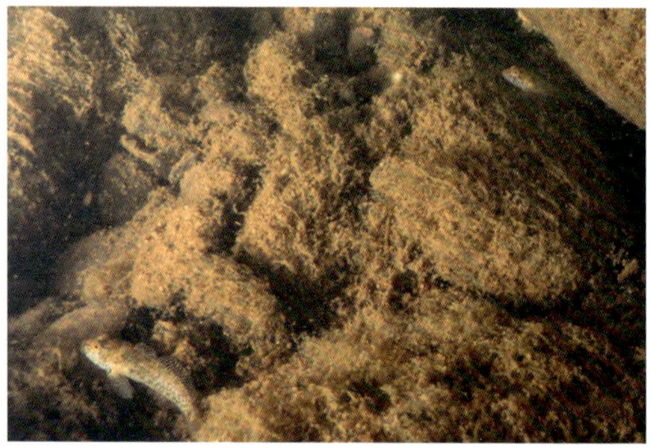

An eastern hellbender in a man-made nesting box

A northern hogsucker

Minnows devour a fallen caterpillar.

Sculpins

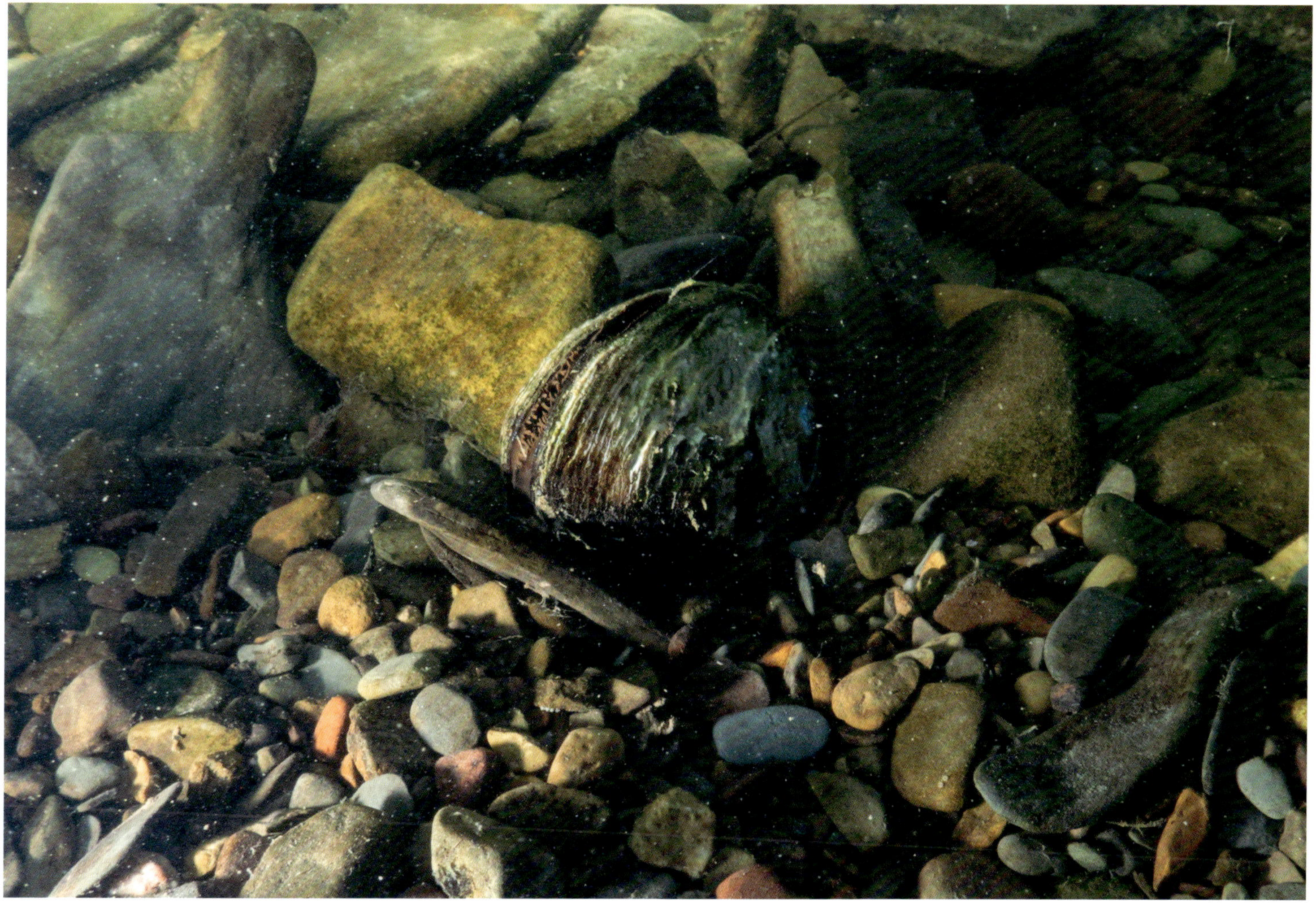

Freshwater mussel species have experienced severe decline throughout their range, and many species are endangered. They occupy an important ecological niche: improving water quality by filtering out silt, heavy metals, and bacteria.

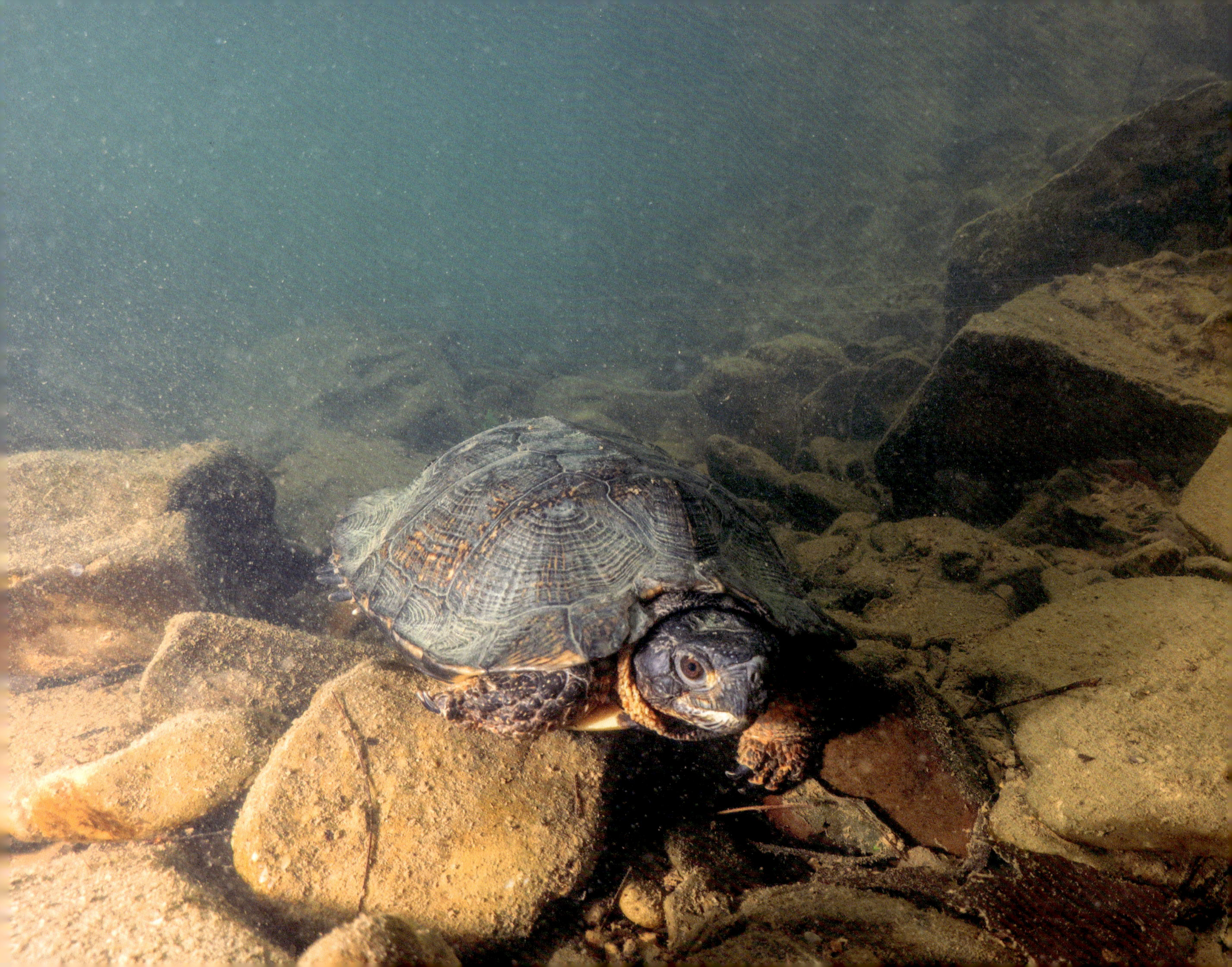

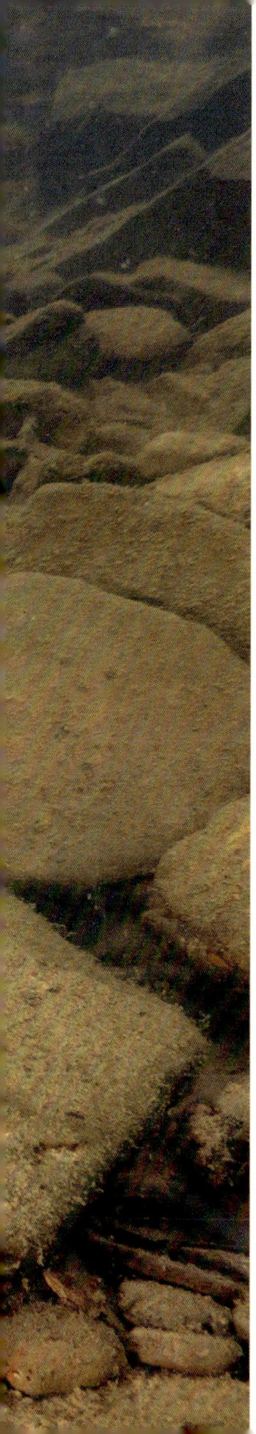

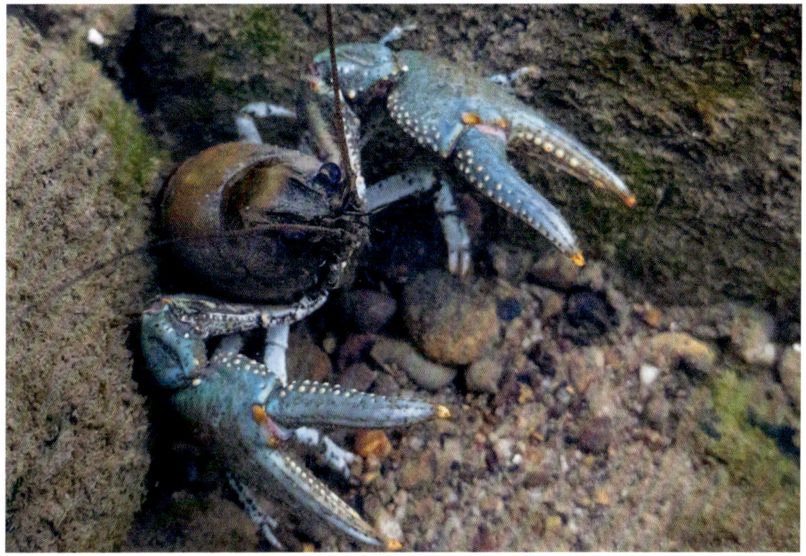

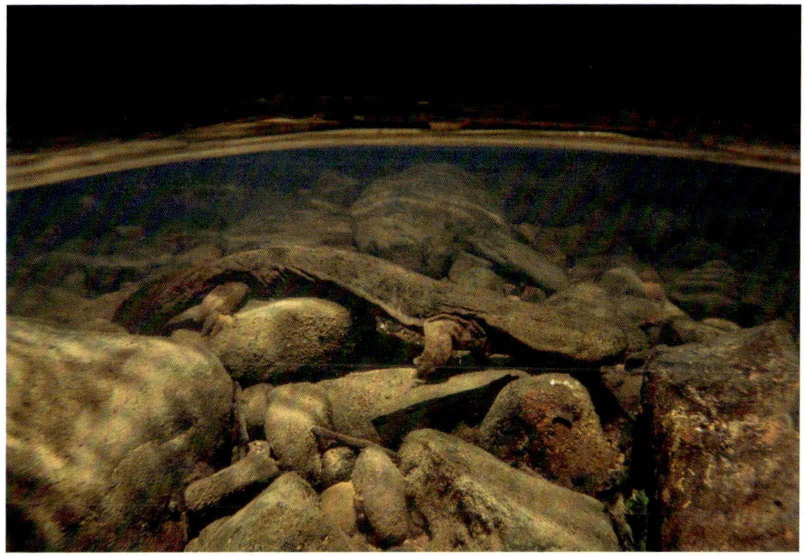

A wood turtle swims through a clear mountain stream.

Crayfish

A rare eastern hellbender

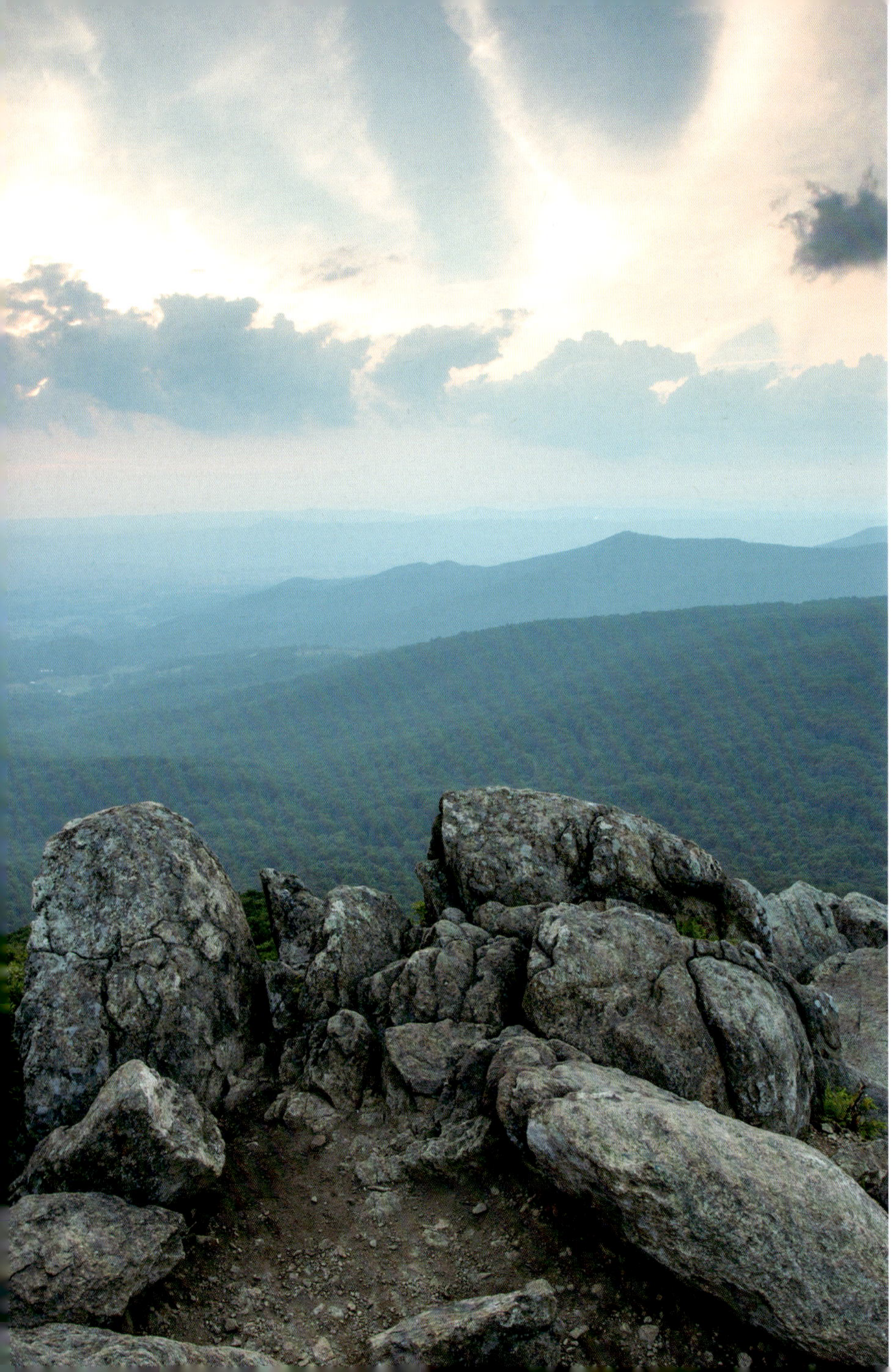

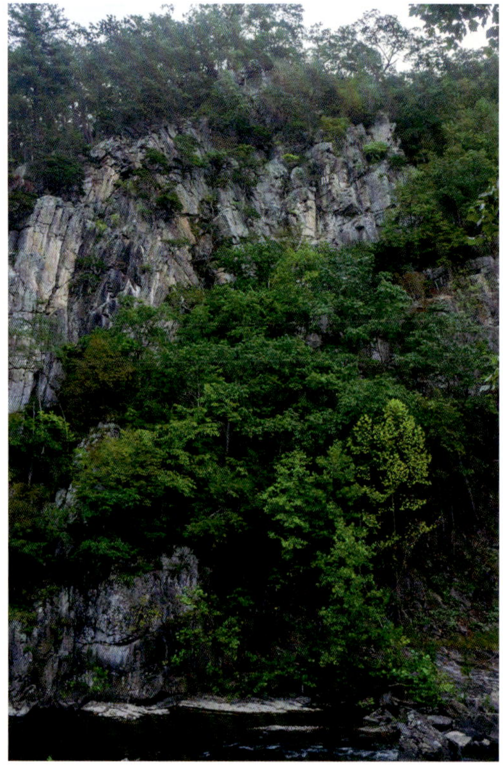

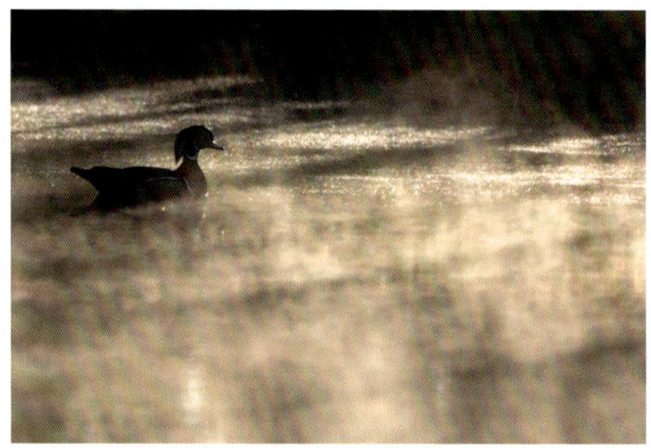

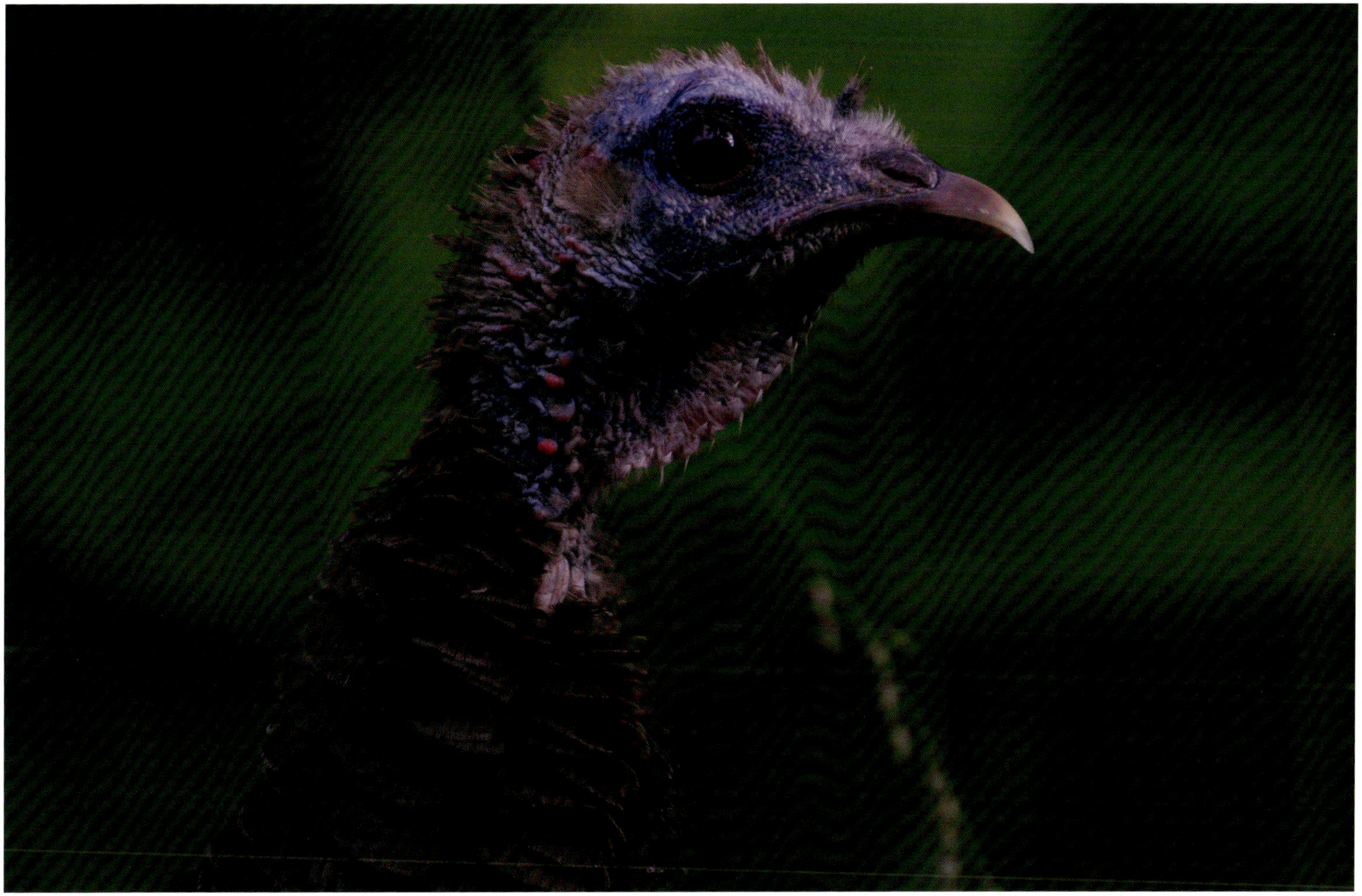

The summit of Mary's Rock overlook showcasing billion-year-old basement rock. A testament to the mountains' primordial history. Shenandoah National Park, Virginia.

One of the most remote areas in the region is the Smoke Hole Canyon. The canyon is a 20-mile gorge at the South Branch of the Potomac River in the Allegheny Mountains of West Virginia. Morning mist is very common in the summer; hence the "Smoke Hole" name.

A wood duck drake

A wild turkey near the canyon

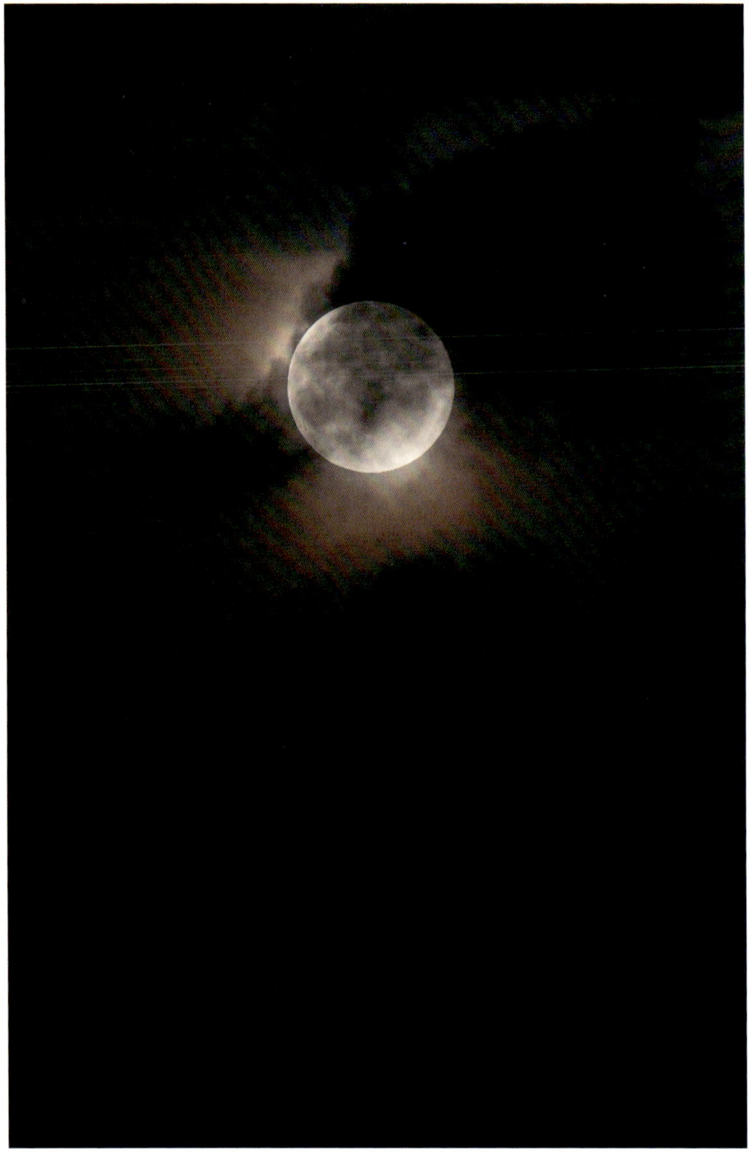

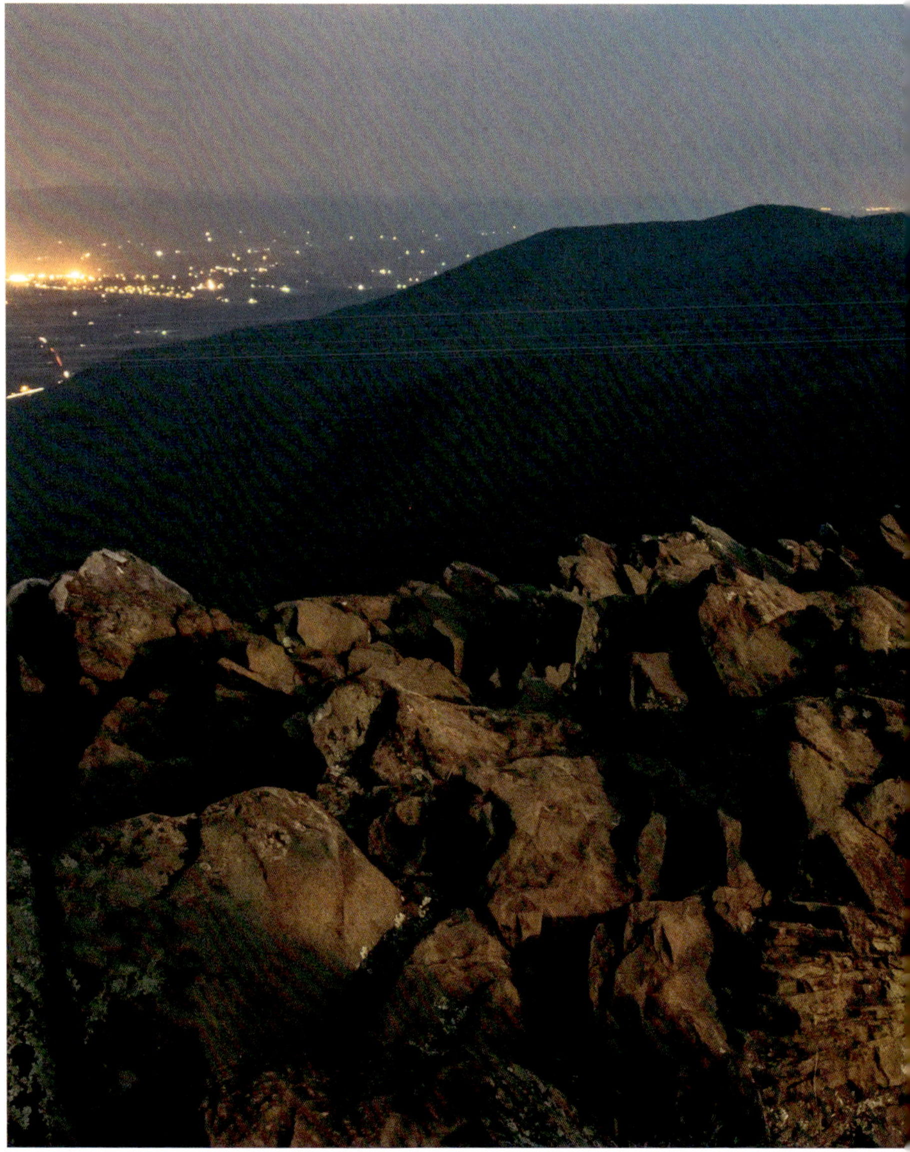

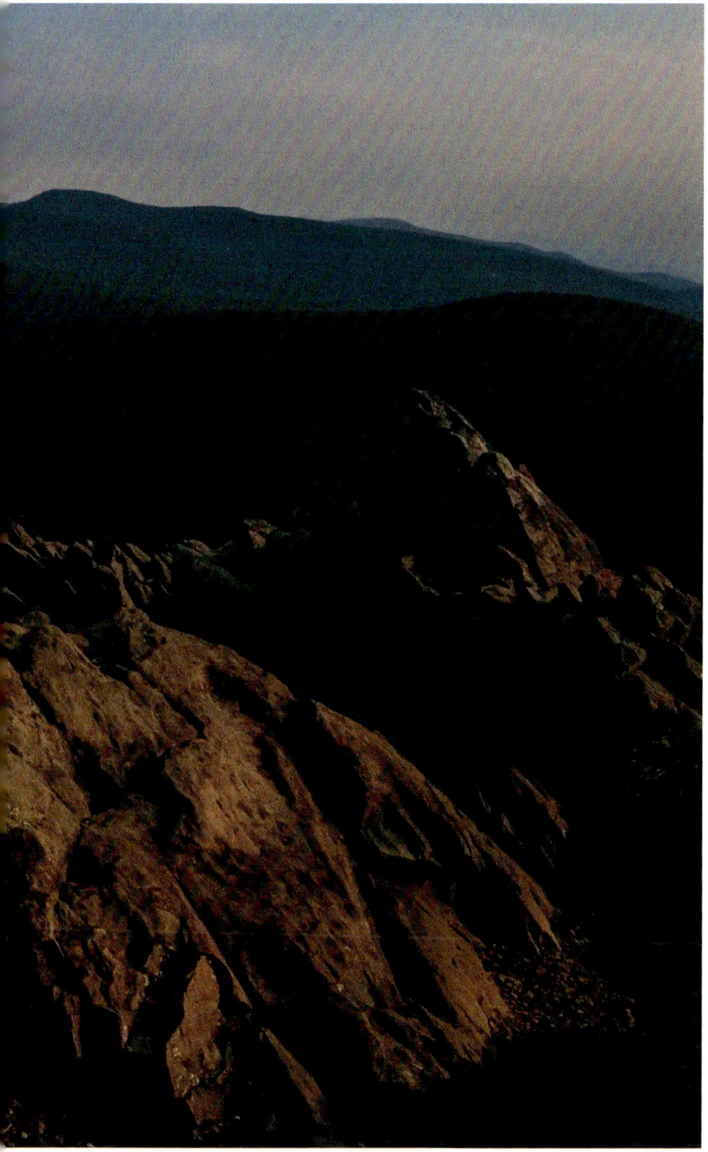

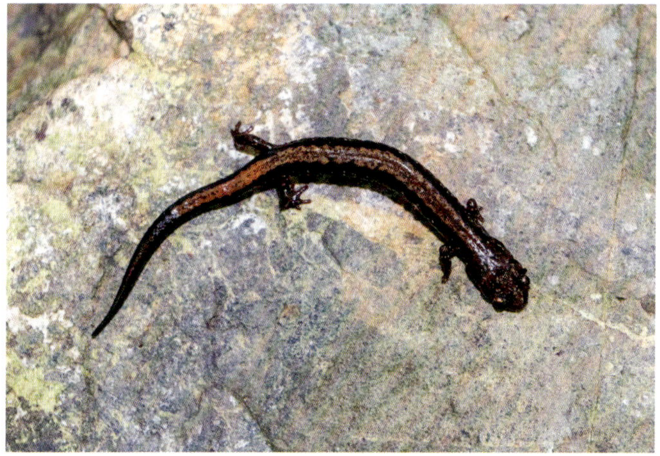

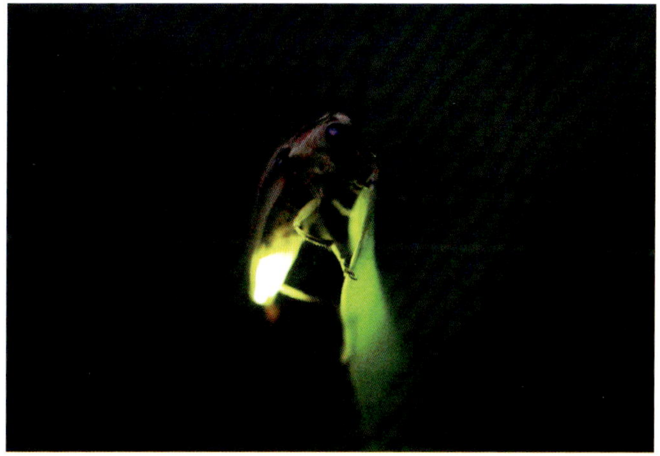

The summit of Hawksbill Mountain, Shenandoah National Park's highest peak, is illuminated by a full moon (*left*). The nocturnal Shenandoah salamander moves along a talus slope (*center and right*).

No summer is complete without the dazzling display of fireflies. Their golden bioluminescence is used to communicate and find mates.

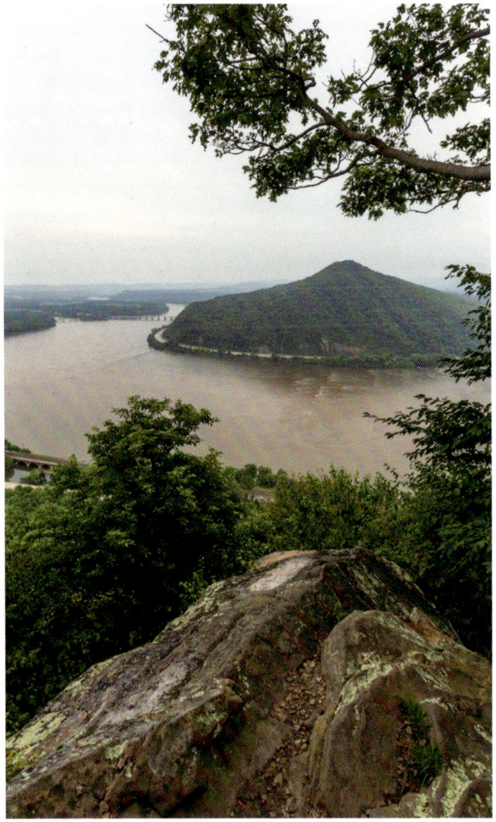

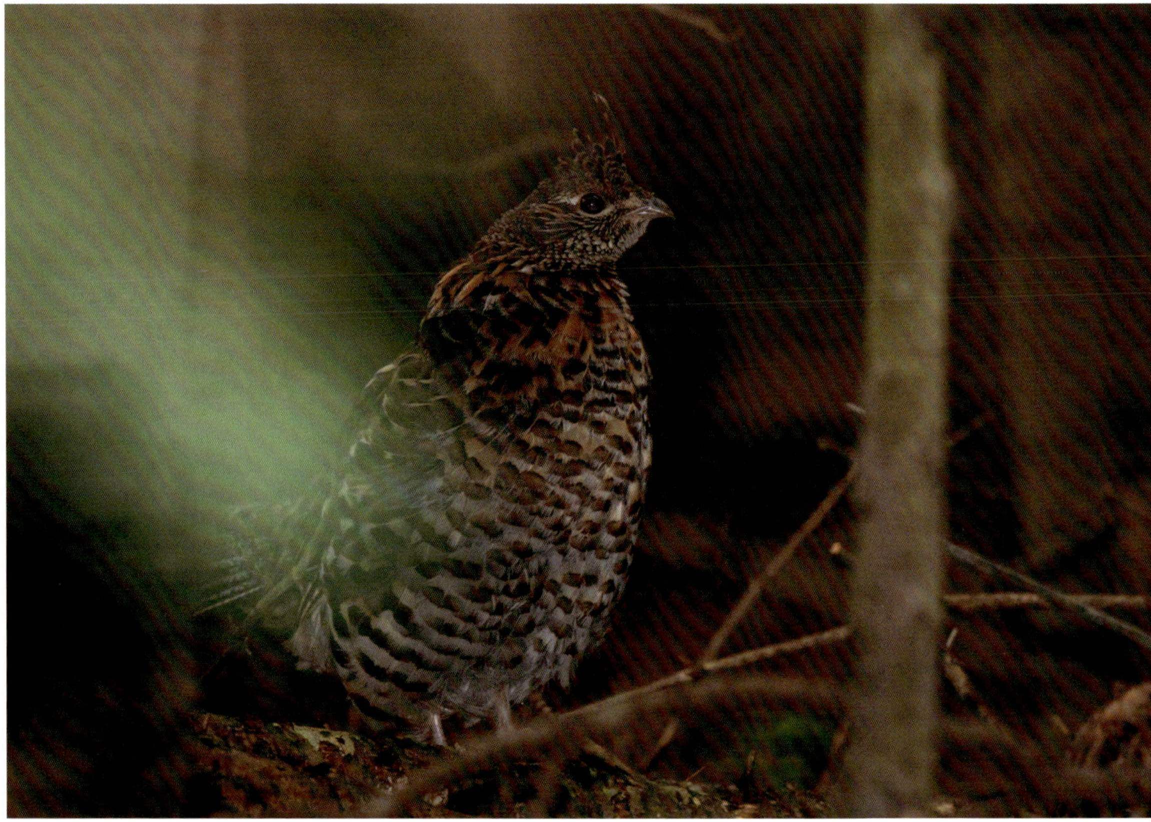

The Susquehanna River curves around Cove Mountain on its way to the Chesapeake Bay. Perry County, Pennsylvania.

The secretive ruffed grouse stays still below a stand of conifers. The species' population has dropped significantly in many areas in the Central Appalachians due to the loss of young forest and West Nile virus. Elk State Forest, Pennsylvania.

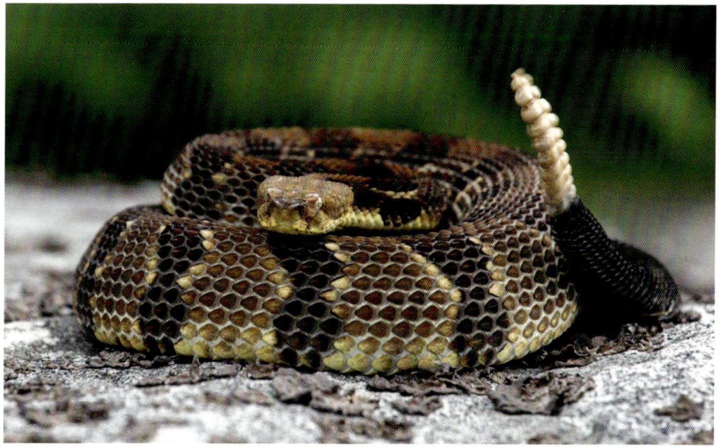

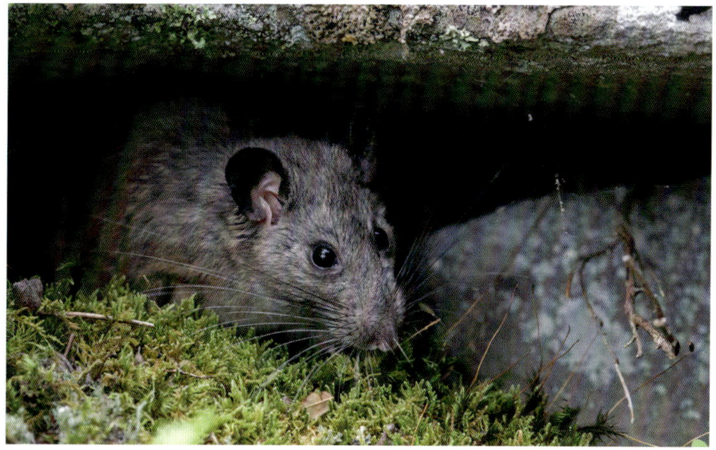

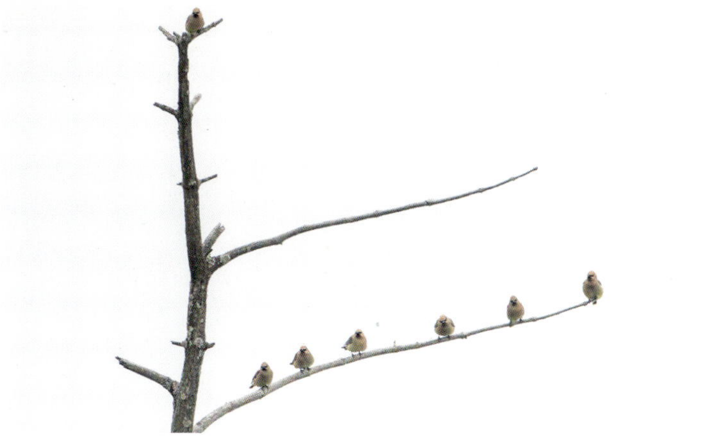

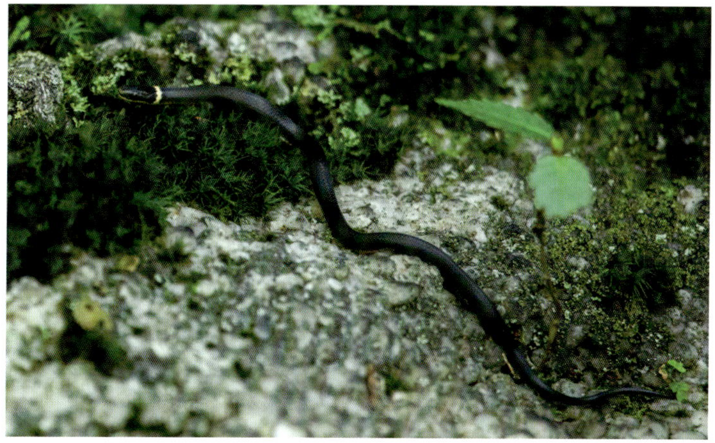

In July, male timber rattlesnakes are ready to breed and are hyperaggressive, warning any threat with a harsh rattle.

A flock of cedar waxwings perched on a snag in the Finzel Swamp Preserve. Garrett County, Maryland.

The Allegheny woodrat is an imperiled species of the central Appalachians. The loss of the American chestnut as well as the parasitic raccoon roundworm are the main culprits of its decline. They keep rather clean middens (homes) and have a curious nature that has endeared them to researchers.

A ring-necked snake suns itself on exposed rock outcrop. Savage River State Forest, Garrett County, Maryland.

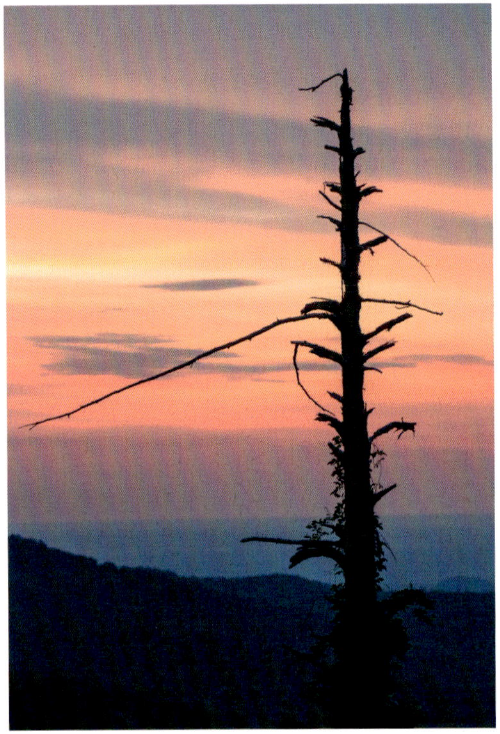

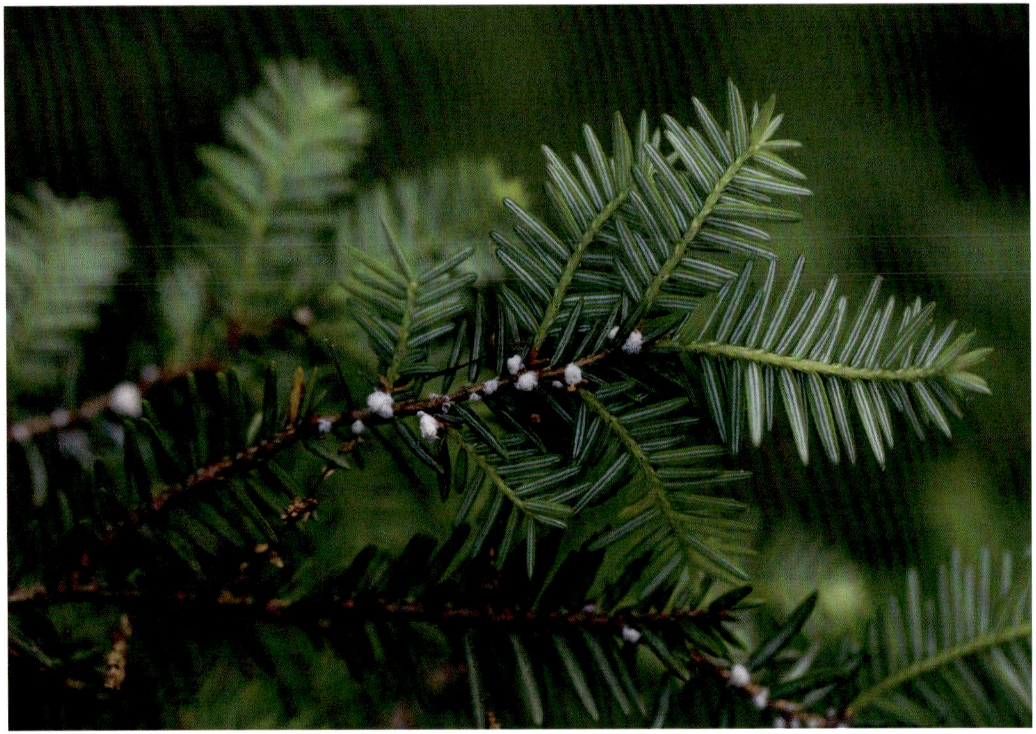

A dead hemlock in Shenandoah National Park. This park, as well as other areas within the Central Appalachians, has been infested with the invasive hemlock woolly adelgid. This insect feeds on the trees' starch reserves. If left untreated, the tree will eventually succumb. An important tree, hemlocks are essential to keeping brook trout streams cool and preventing erosion in stream banks.

A red spruce tree in the Dolly Sods Wilderness. Even in the heat of the summer, the high-elevation plateau that makes up the wilderness provides a much-cooler habitat than the valley below.

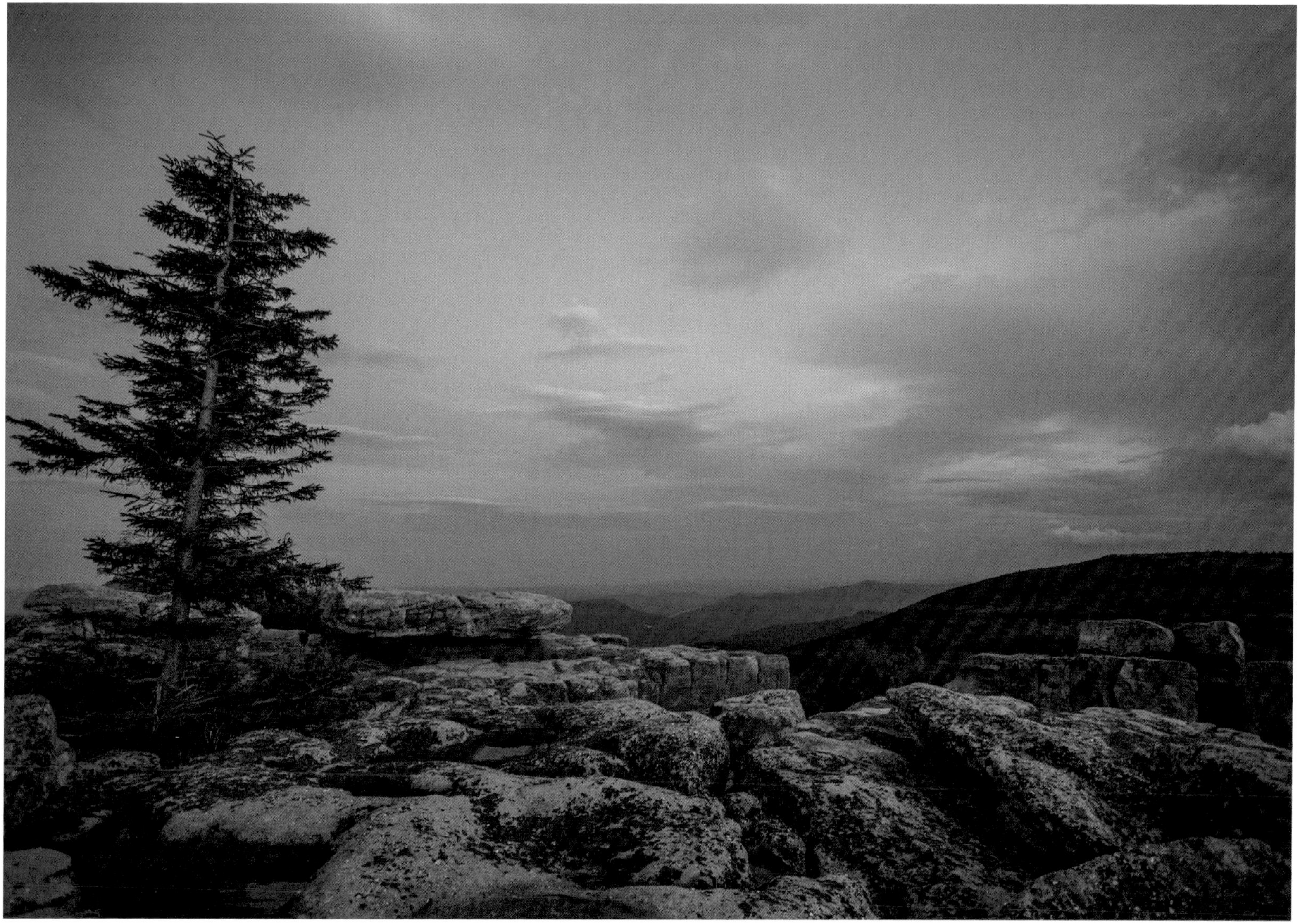

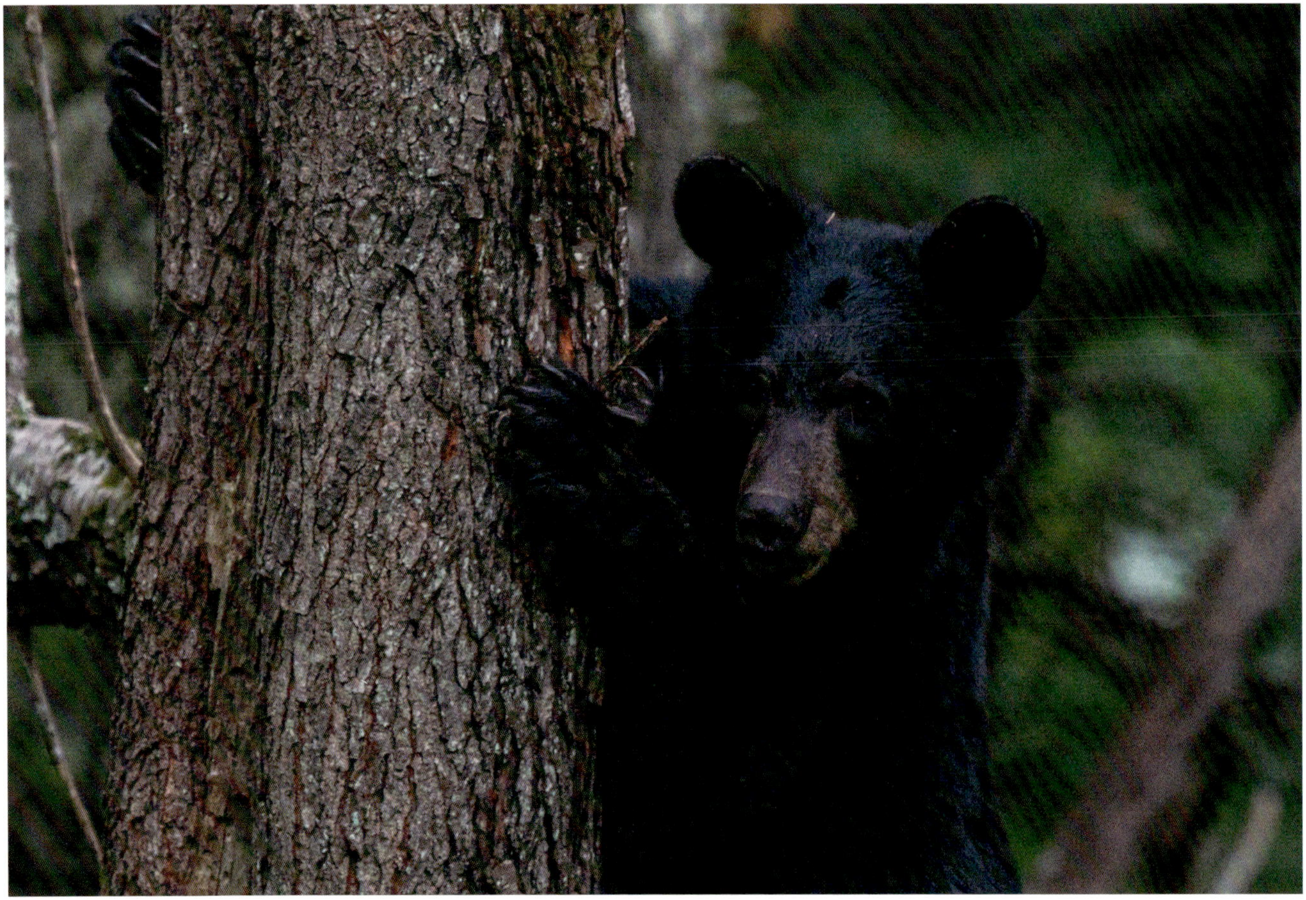

A black bear marks its territory as night falls on the mountains.

A starry night over Seneca Rocks, one of West Virginia's most spectacular landmarks. The sculpted rock rises 900 feet over the North Fork River below.

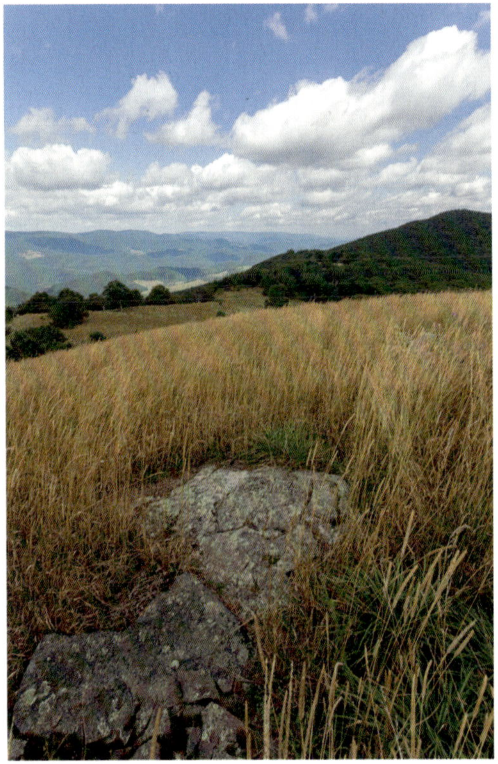

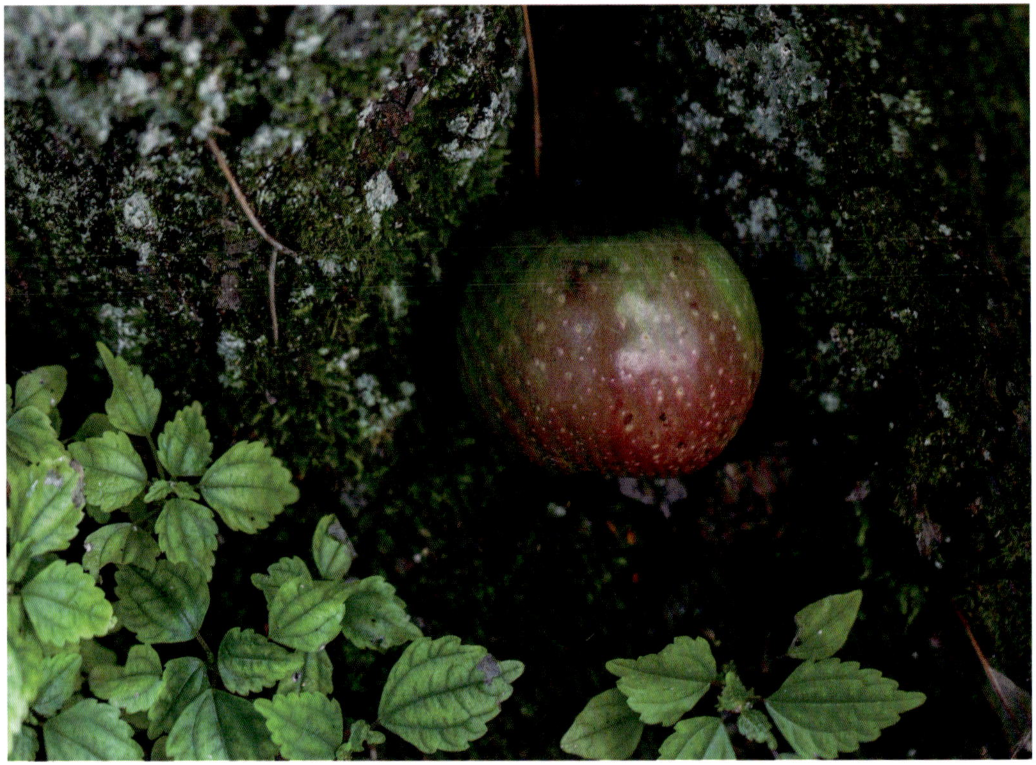

By August, grasses on West Virginia's Nelson Sods have transformed to an earthy brown and the first apples begin to fall, a gentle reminder that autumn will return soon. Pike Knob Preserve, West Virginia.

Azalea caterpillars in defense posture. Shenandoah National Park, Virginia.

A large black swallowtail caterpillar in George Washington National Forest, Virginia

A mosquito meets its demise in a mountain meadow.

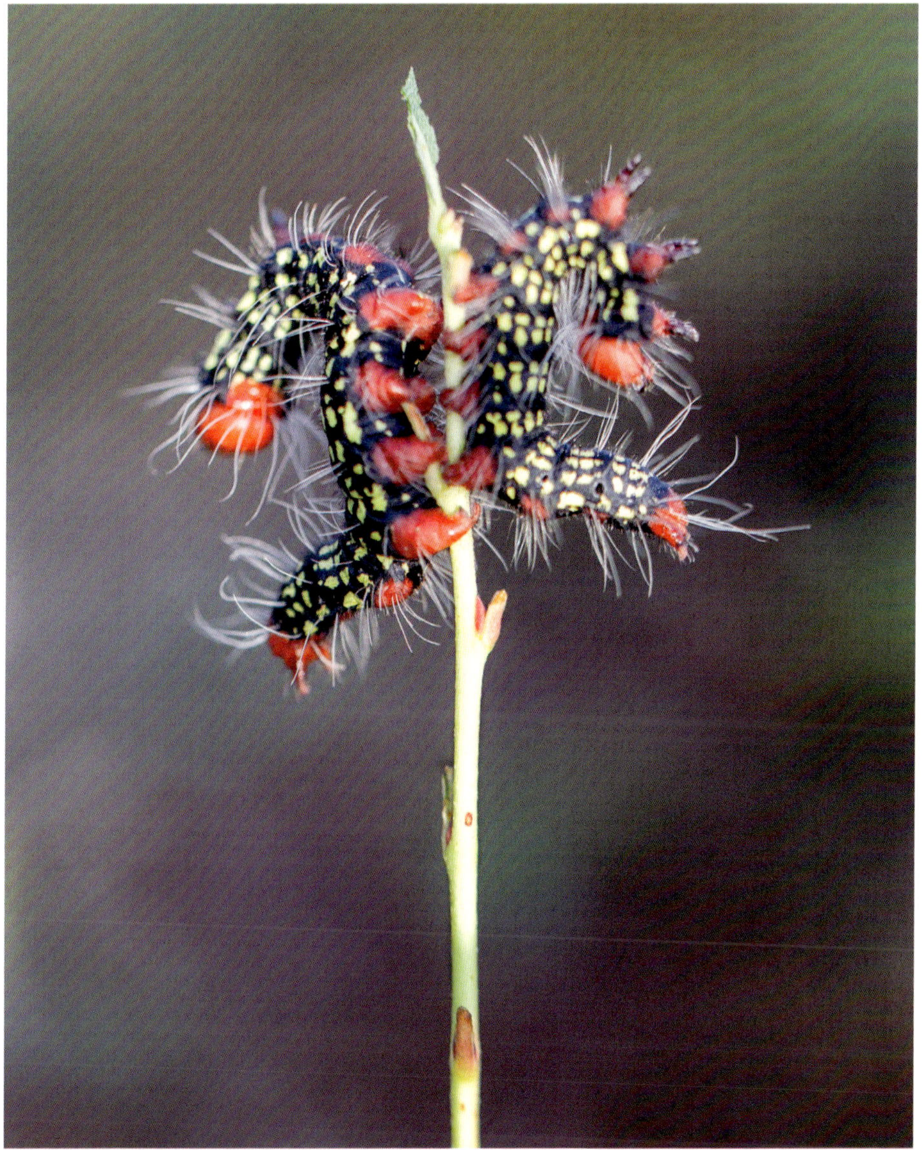

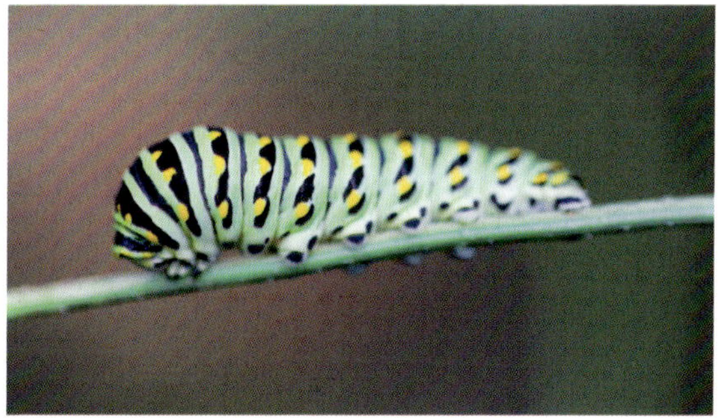

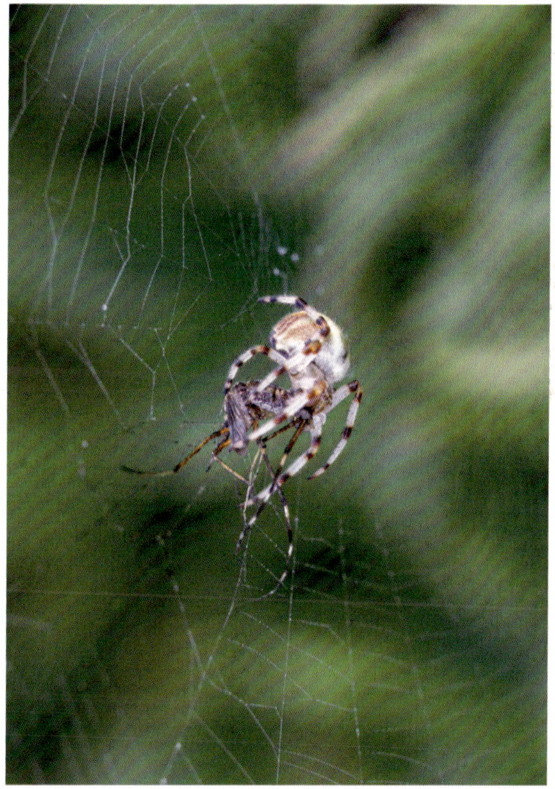

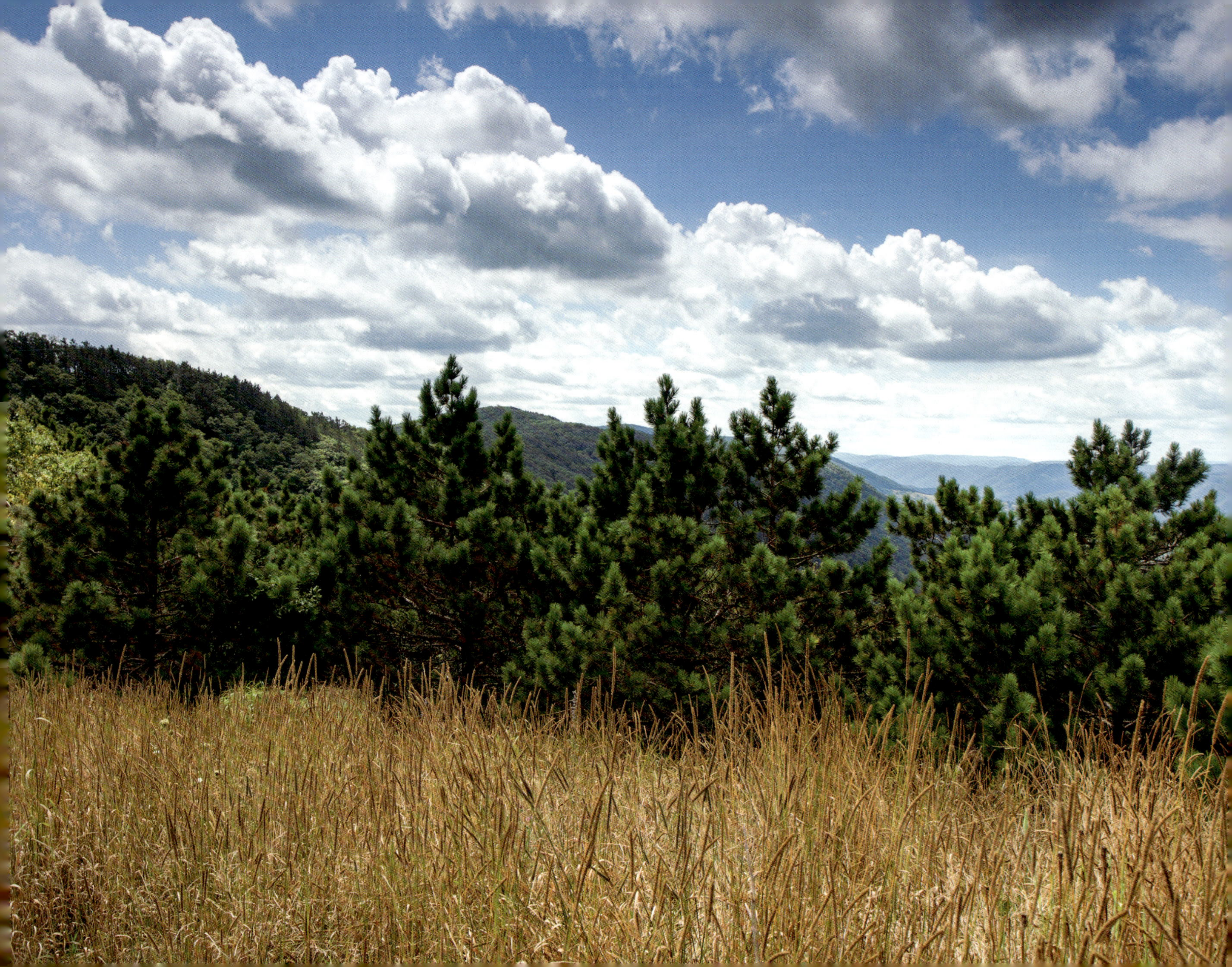

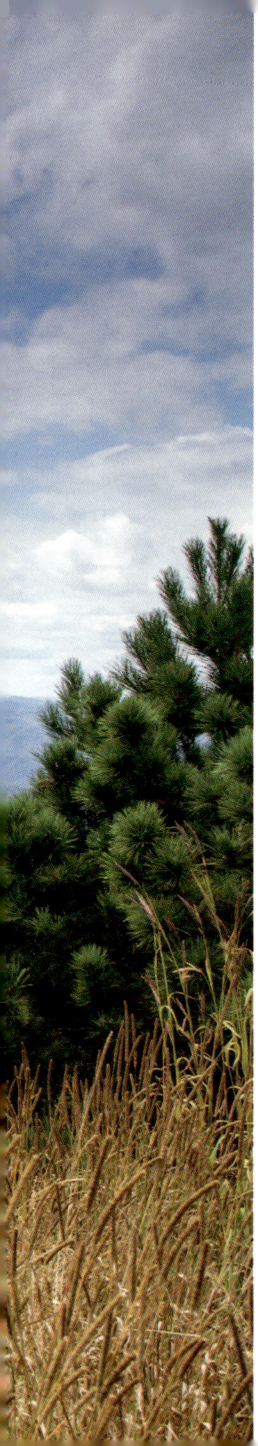

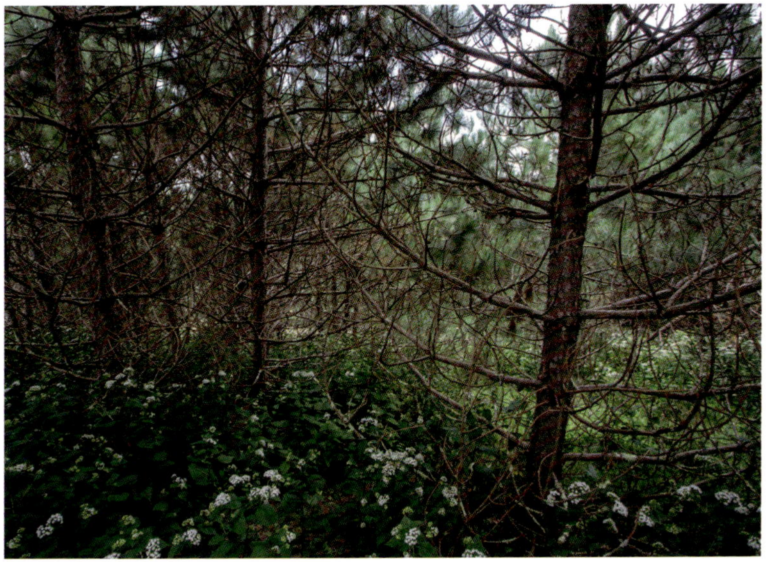

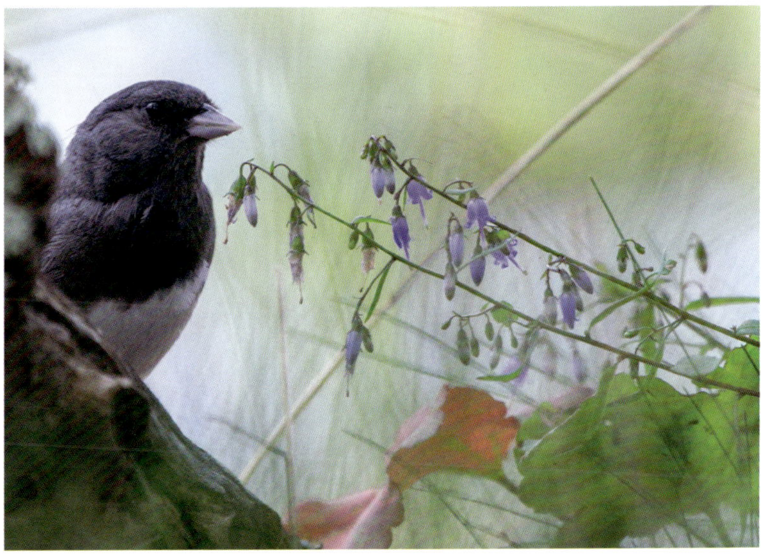

The southernmost population of red pine is found on the summit and surrounding mountainsides of Pike Knob on West Virginia's North Fork Mountain. Pike Knob Preserve, West Virginia.

A junco among late-summer wildflowers. Catoctin Mountain Park, Maryland.

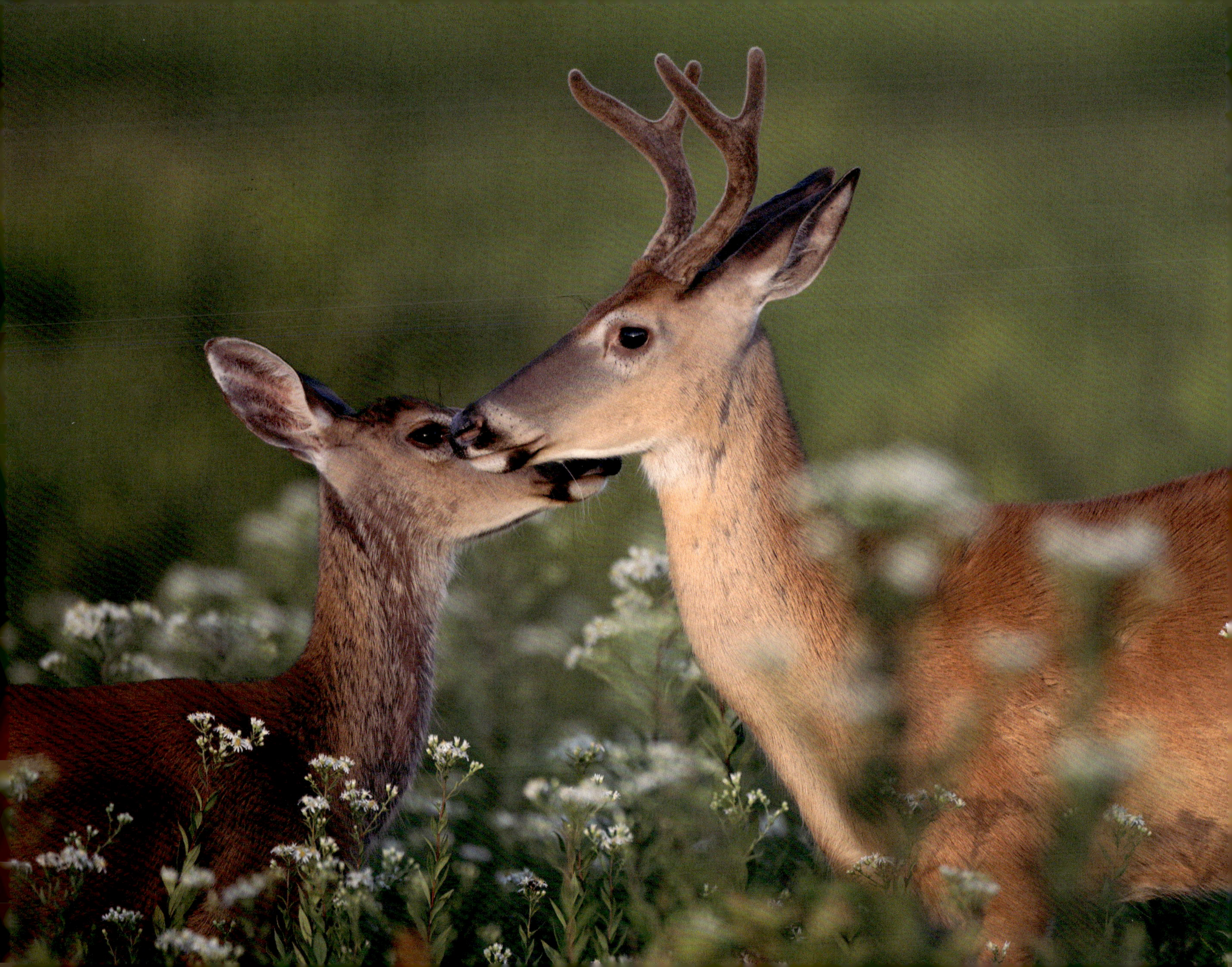

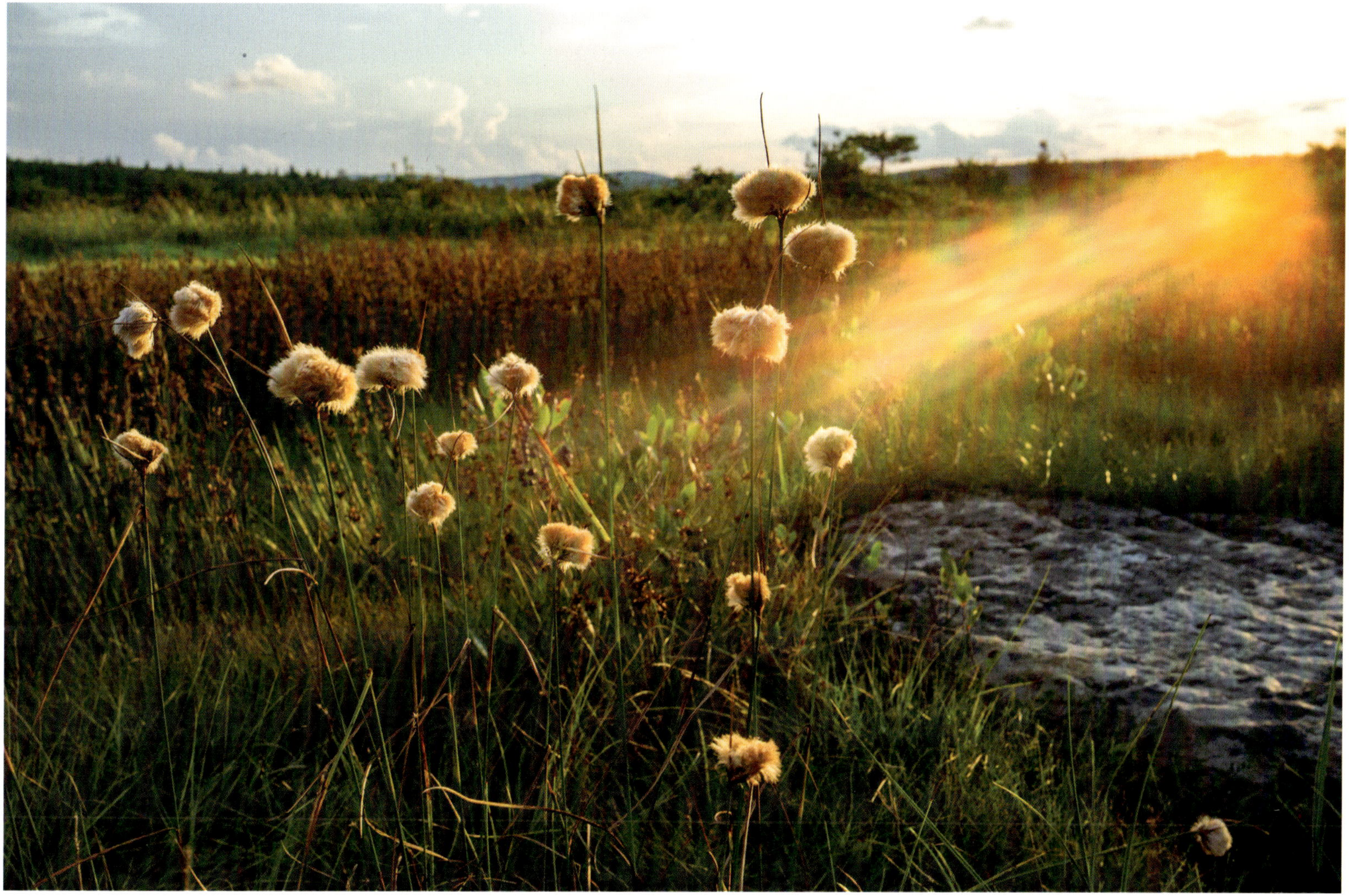

A white-tailed deer doe nuzzles a young buck, his velvet still adorning his rack. Shenandoah National Park, Virginia.

Typically associated with the Arctic, cottongrass also makes its home in the Allegheny Highlands of West Virginia.

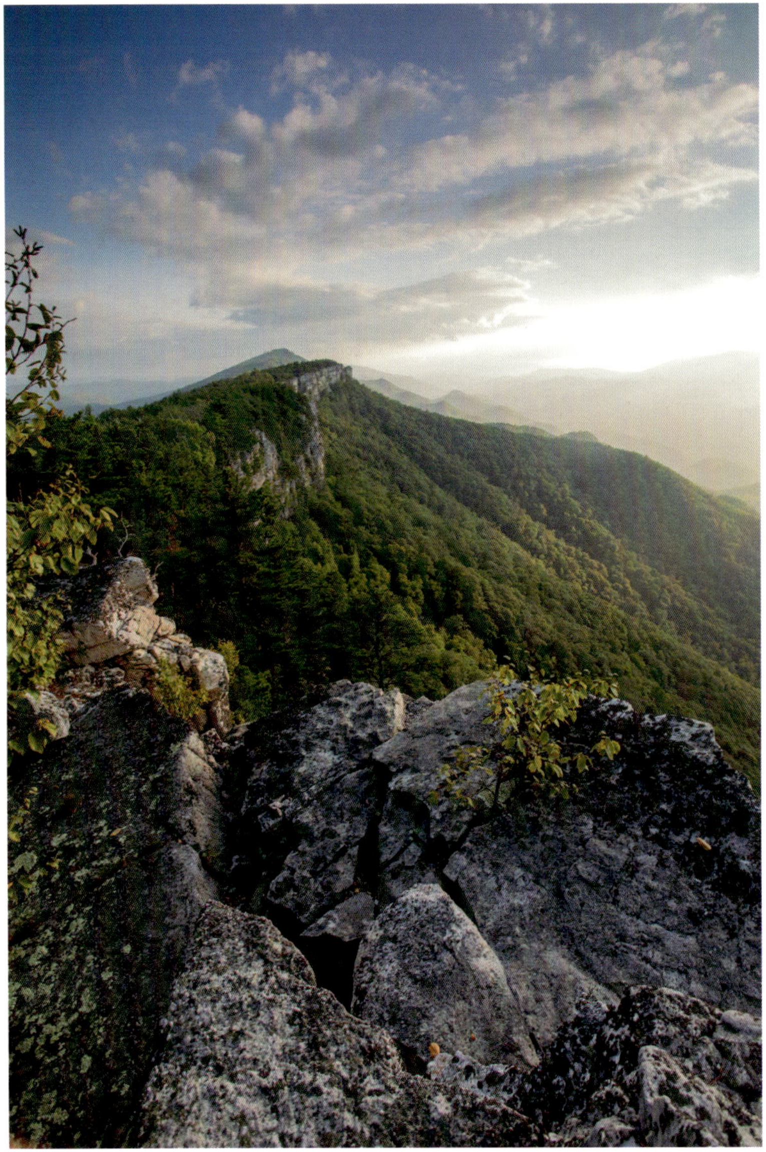

West Virginia's North Fork Mountain is one of the driest mountains in the Central Appalachians, since higher peaks to the west, such as Spruce Mountain, cause storms to drop most of their precipitation at those higher peaks. Here, the quartzite ridge can be easily viewed from the summit. Monongahela National Forest, West Virginia.

Late-summer goldenrod carpets the landscape with yellow and is an important food source for migrating monarch butterflies. For the birch (*following spread*), it is the first dash of color to surround it before its leaves change. Shenandoah National Park, Virginia.

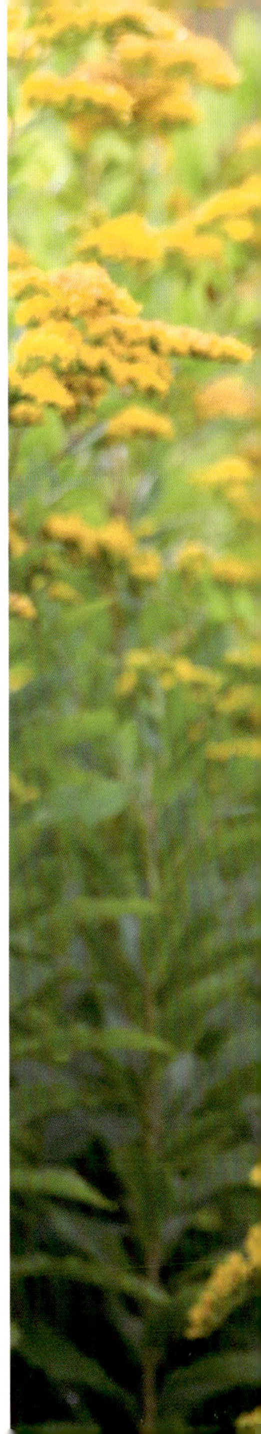

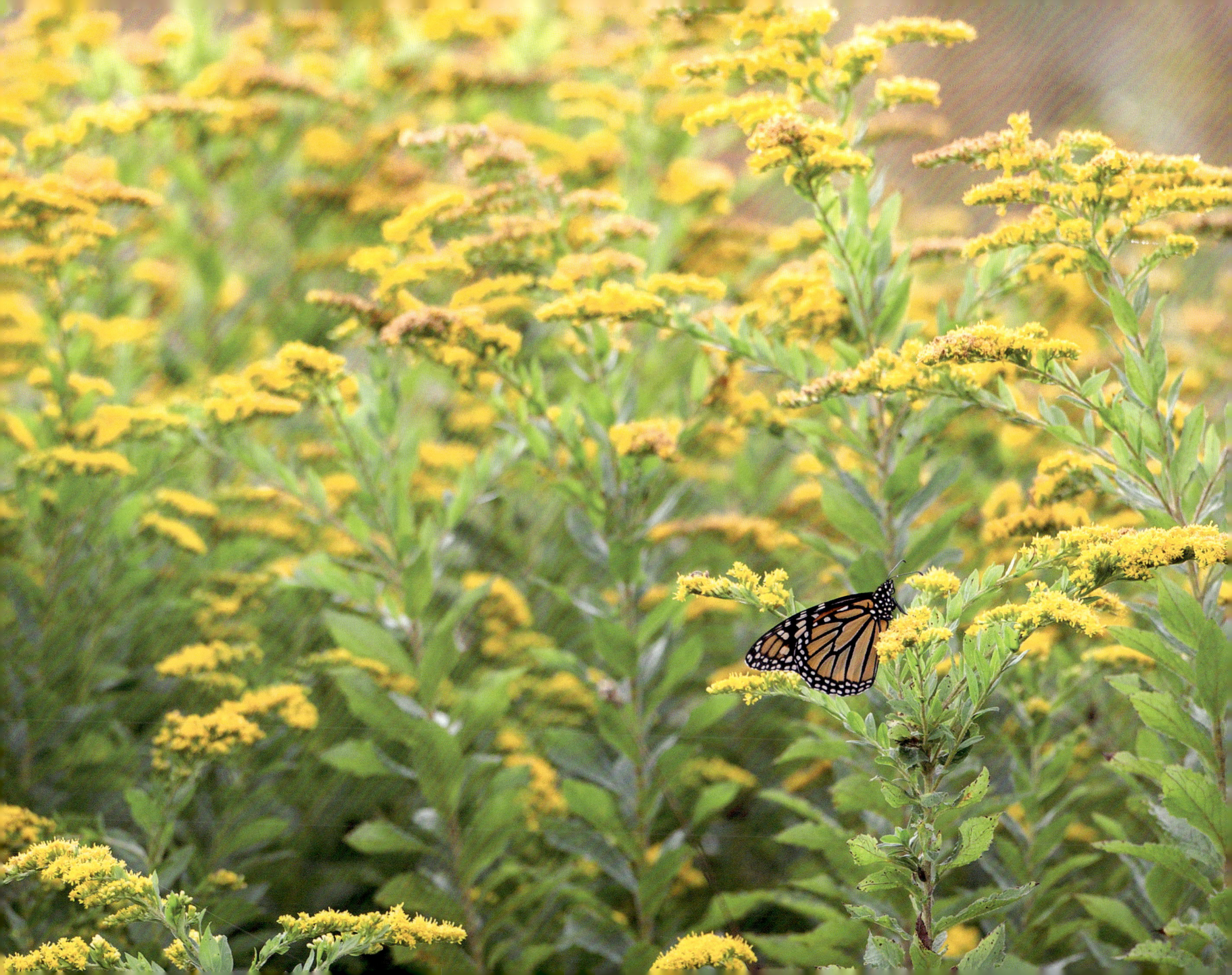

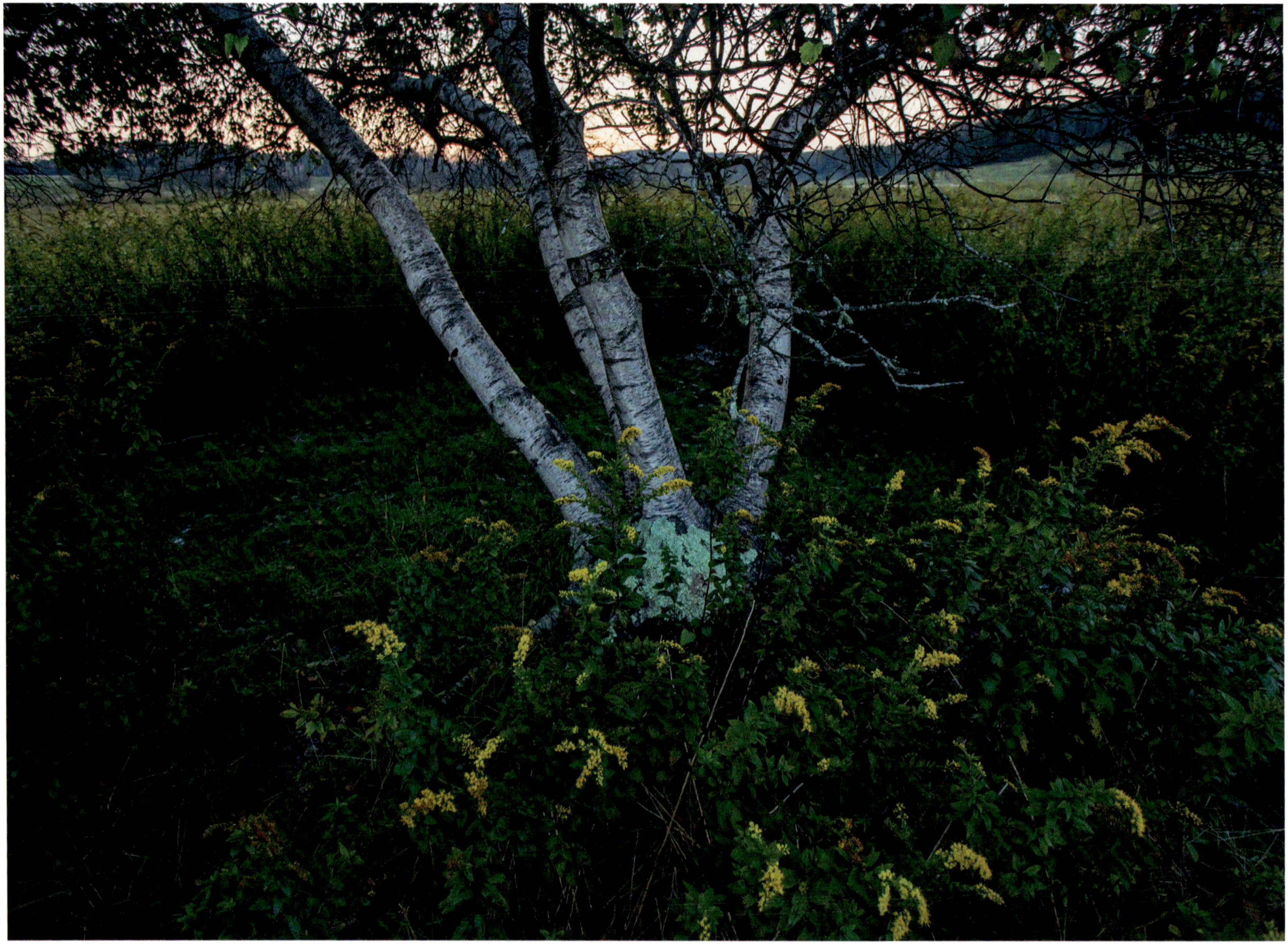

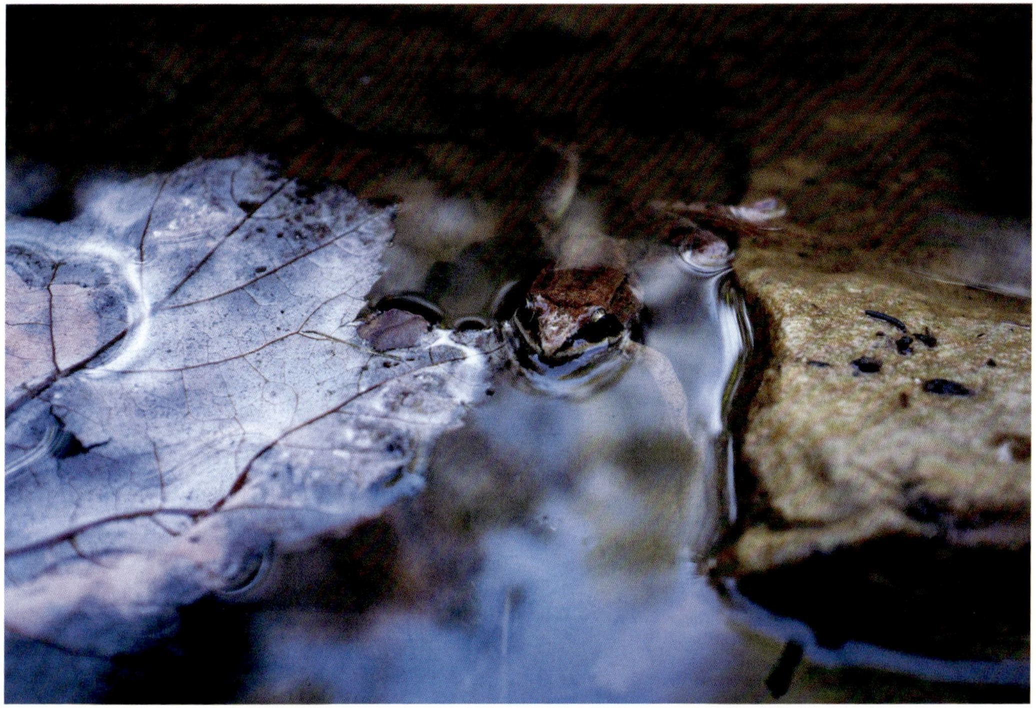

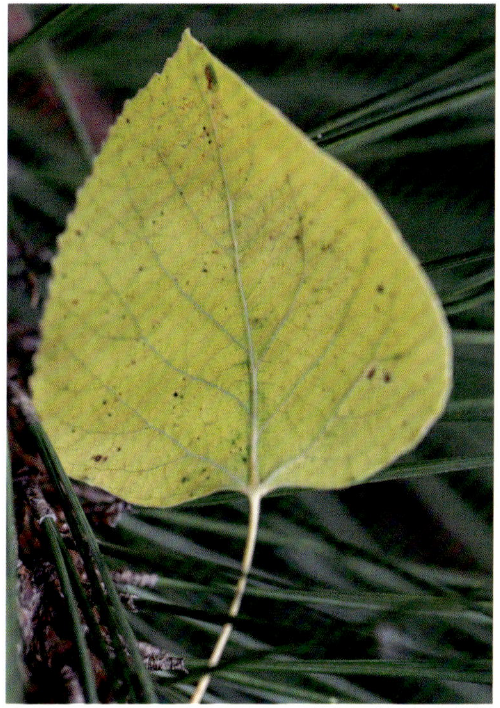

A wood frog in the Dolly Sods Wilderness, West Virginia. Wood frogs are freeze tolerant and thrive in the mountains.

A yellow quaking aspen falls along red pine needles. While still summer, in the higher elevations it is not uncommon to see shrubs and trees transform before the autumnal equinox passes. Pike Knob Preserve, West Virginia.

Chasing Leaves

Autumn is a benediction, a season that is equal parts celebration and reflection. For me, no other season conjures such nostalgia and excitement as the sweet melancholy of falling leaves. In between hayrides, apples, and pumpkin-flavored everything, there is chasing the color transformation of autumn leaves. When I say I'm chasing this color change "in between," I really mean it's an all-encompassing, deep-rooted obsession . . .

I am an unapologetic "leaf peeper." Come August, a longing for autumn leaves takes a strong hold on my psyche, and by September I am in the throes of a full-blown obsession that I cannot alleviate without—you guessed it—leaves. I annually implement a self-imposed assignment to capture as many images as possible of the fleeting leaves each day, the paradox being that there's never enough time to capture all the images I want. It is October's agony: it is beautiful, it does not last long, and it always ends too soon.

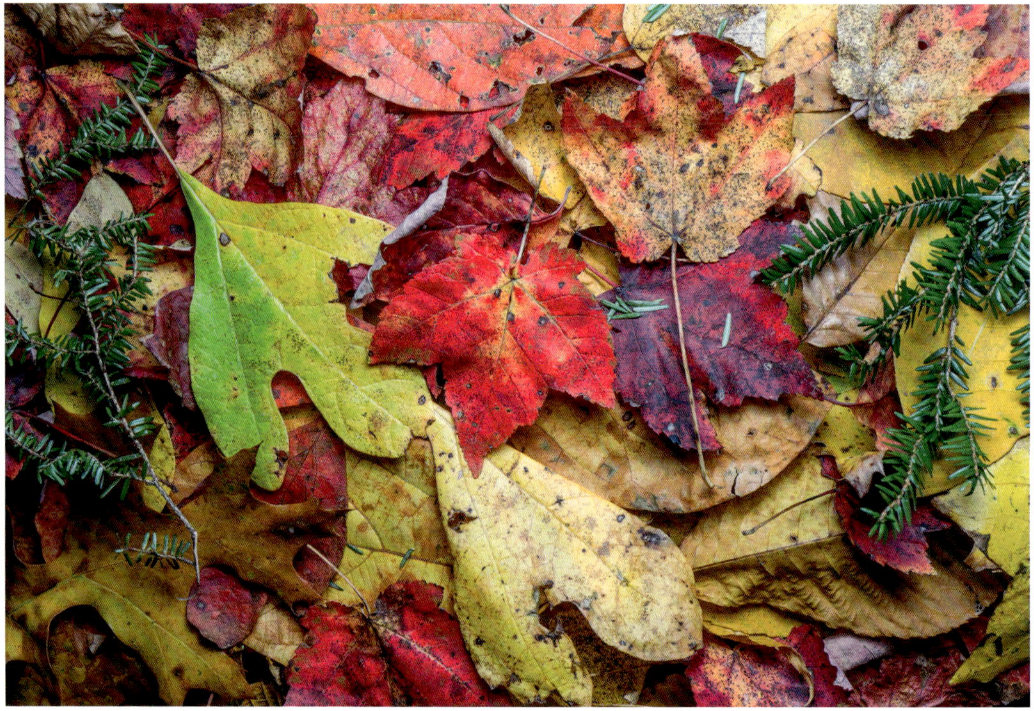

Autumn leaves and hemlock boughs. Rothrock State Forest, Pennsylvania.

Aerial view of early-autumn forest at Saint John's Rock. Garrett County, Maryland.

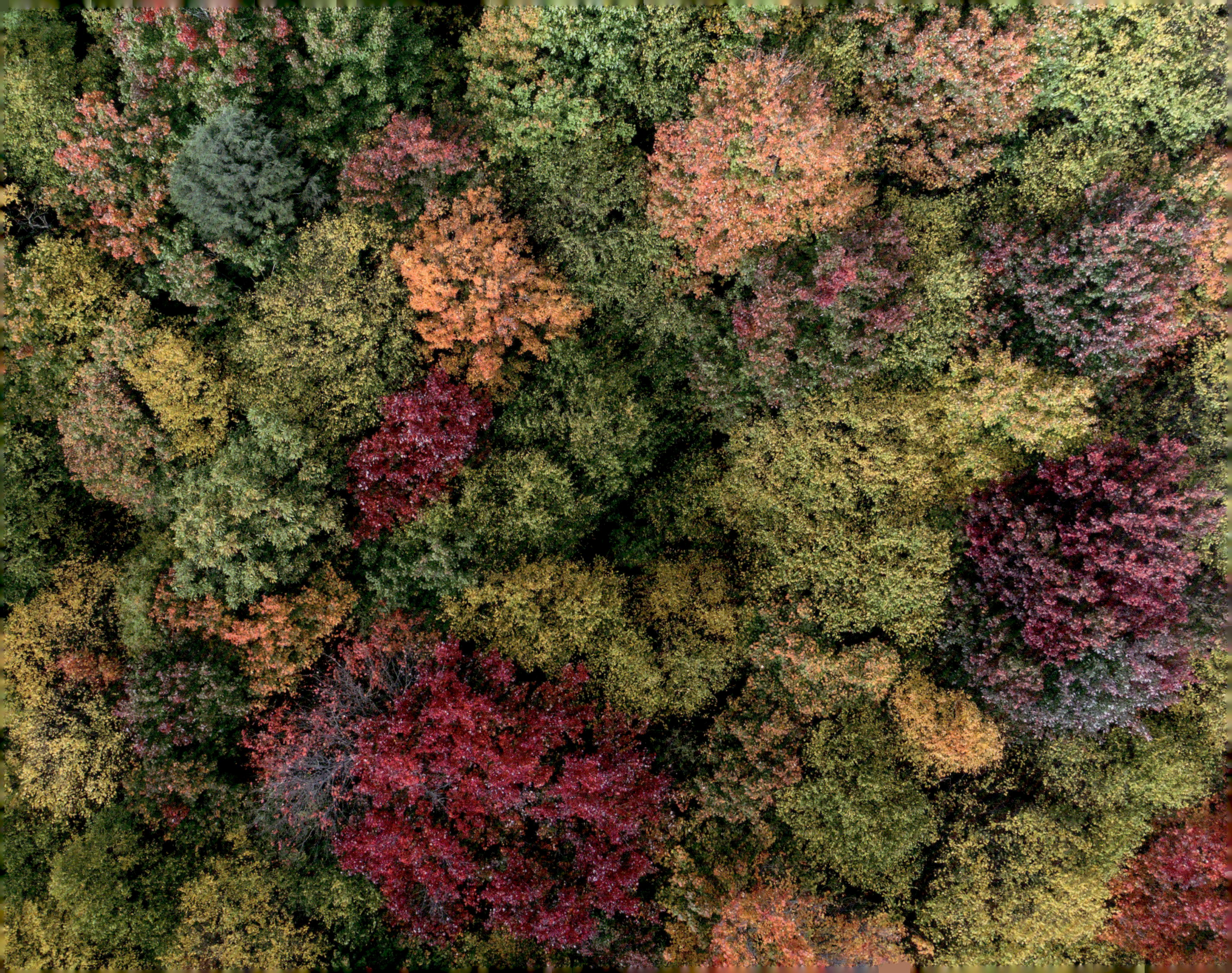

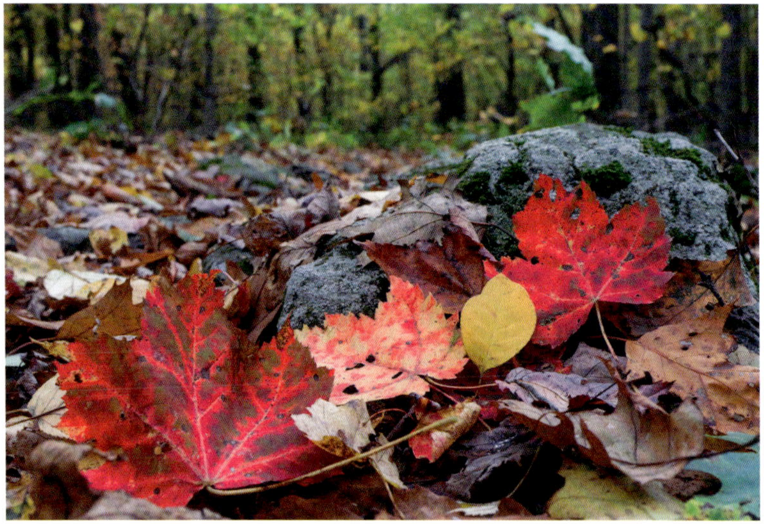

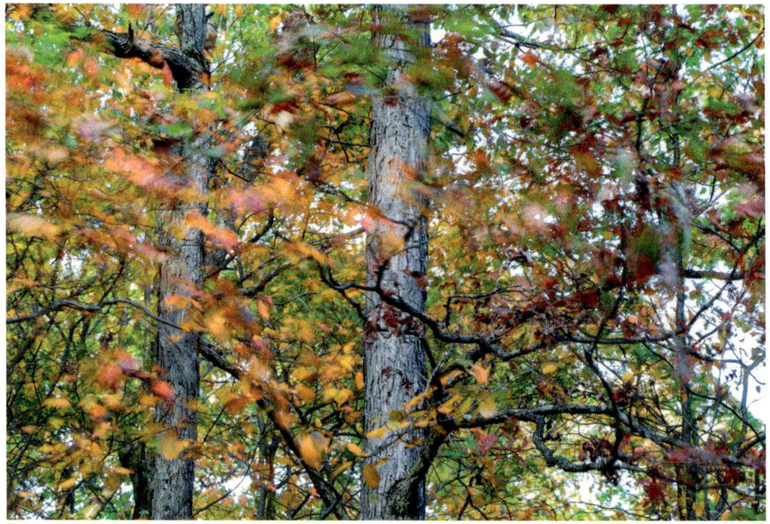

A forest hike during an Appalachian autumn is the ultimate meditation for the senses. The crisp, cool morning air condenses my breath into a cloud. High above the canopy, I hear the wilderness call of a raven; from below, the crunch of dry leaves under my boots. The intoxicating aroma of hay-scented fern permeates the forest floor. And, of course, my eyes are treated to a visual symphony of color: a kaleidoscope of hues of red, orange, purple, and yellow.

The Central Appalachian forests host my favorite "leafing" locations. Between the four states that are profiled in this book, I reliably find color from late September into early November. Predicting autumn foliage is not an exact science, since summer temperature, rainfall, and warm days and cool nights in September are variables. Not every year is vibrant, but a less vibrant autumn is better than no autumn.

While oaks are the most represented tree, a diversity of hardwoods including maples, poplars, dogwoods, and birches all contribute to the autumn canvas. Color and timing changes with elevation. But these moments are short lived. What begins as a sprinkle will soon become a splash as the mountains reach "peak" color, the time when most of the trees have changed.

As when there is no more firewood to keep a blaze alive, the season concludes

Savage River State Forest, Maryland

Wind impression of autumn leaves. St. John's Rock, Maryland.

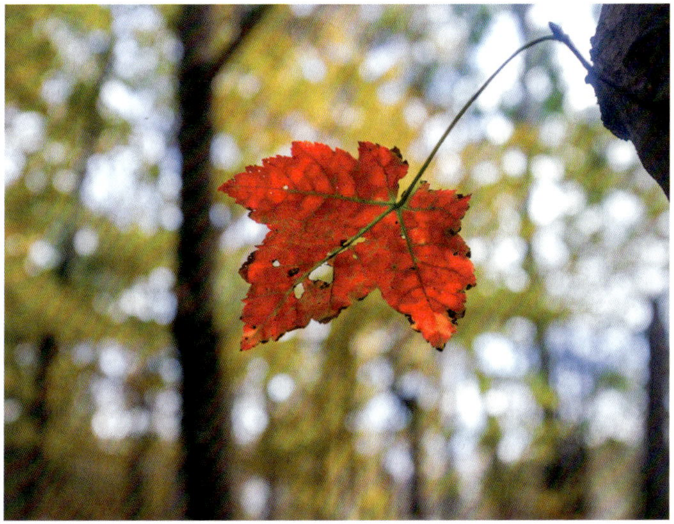

much too soon. Sure, patches of color will linger like a single hot, orange coal, but soon the last leaf will detach from its branch, floating to the forest floor. Returning to the ground from which the tree was born, the sojourn is finished. I hate to see it depart, but maybe that is why I love the experience so. In the brevity lies a deep appreciation, because it is never long enough.

A red maple leaf hangs in a yellowish oak-hickory forest of Shenandoah National Park.

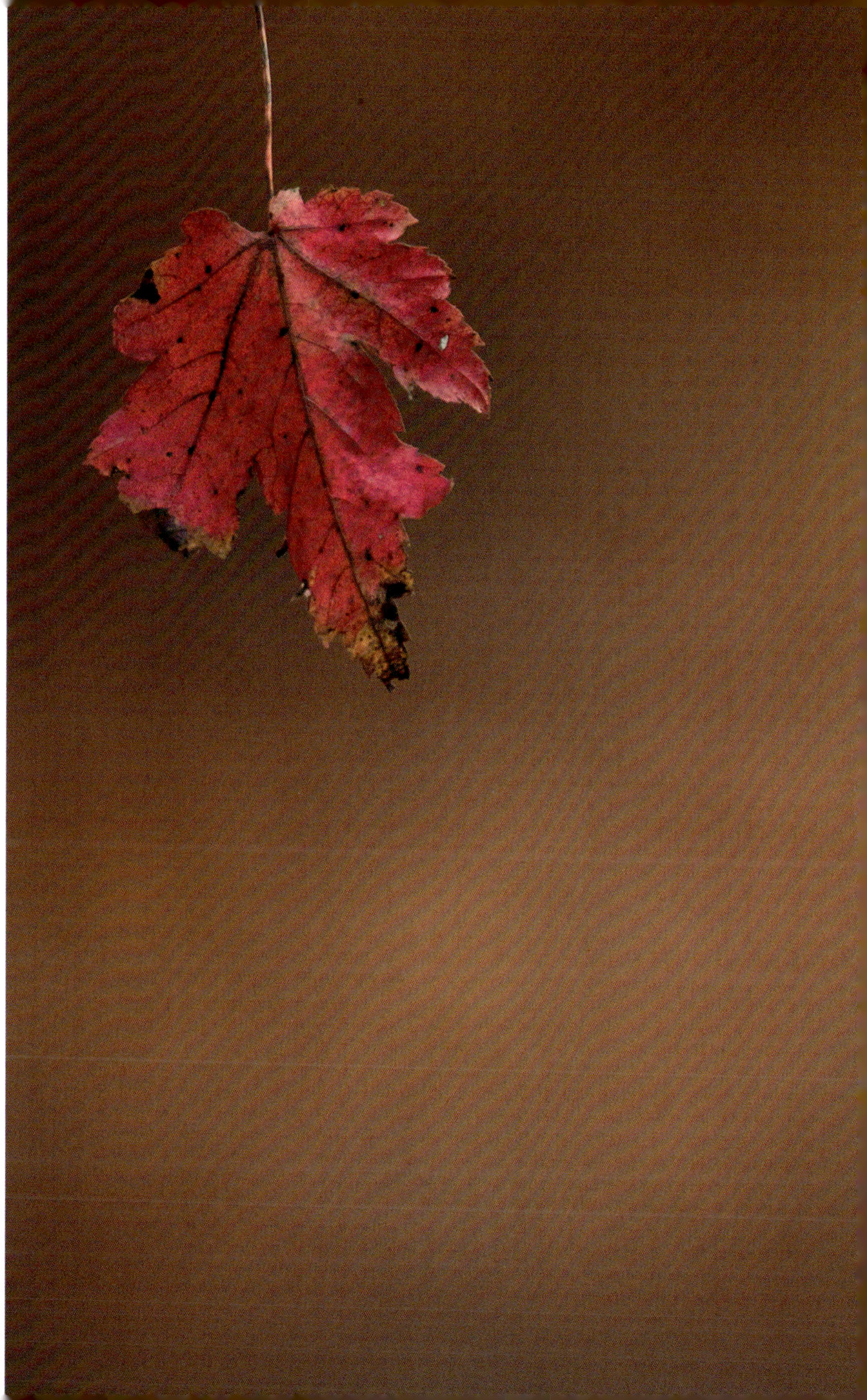

AUTUMN

Before the sun reaches its zenith, signaling the autumnal equinox, hints of the season's arrival began advancing weeks prior.

The shorter days and cooler nights of September have alerted the black gum and tulip poplar trees: breaking down the chlorophyll held within their leaves, they advance to vibrant hues of red and yellow. Postequinox wildlife must prepare for winter in the never-ending struggle to survive in the Appalachian wilds. From atop Spruce Knob, the highest peak in West Virginia and the Chesapeake Bay watershed at 4,863 feet, fallen leaves are covered in morning frost. Up in the high-elevation forest, a red squirrel caches spruce and pine cones, moving in an erratic fashion across the forest floor.

Deer and elk have traded their summer coat for a thicker pelt in preparation for the cooler weather. Antler-adorned males will engage in often-violent encounters during the rut. Black bears forage as much as twenty hours a day on acorns and soft mast crops, in an effort to pack on as much fat as possible before hibernation. Migratory songbirds leave the mountains for much-warmer wintering grounds.

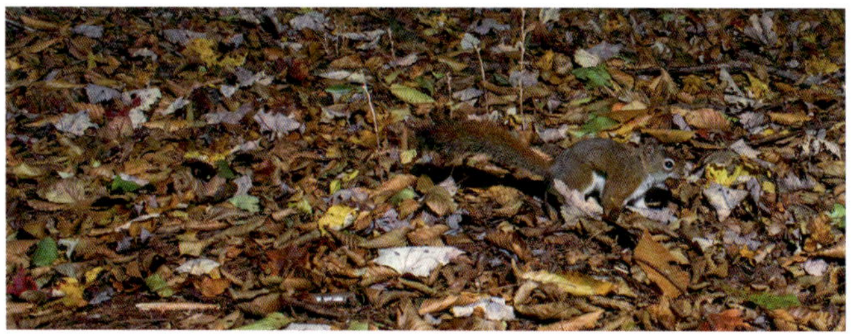

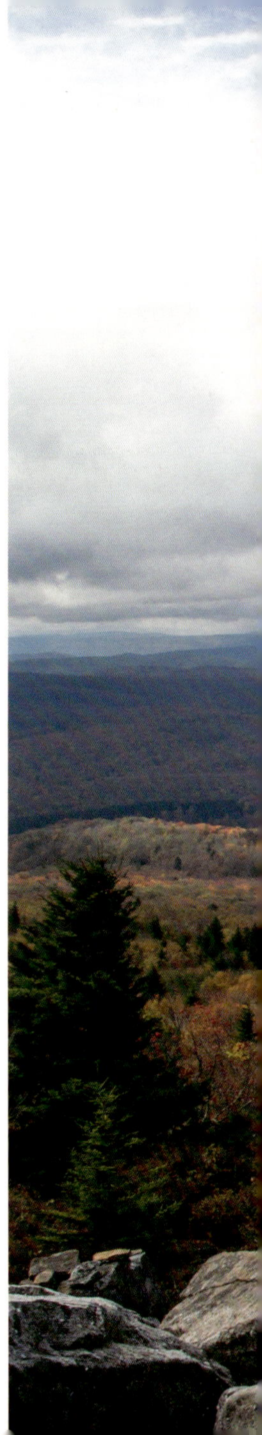

The summit of Spruce Knob, the highest peak in West Virginia and the Chesapeake Bay watershed

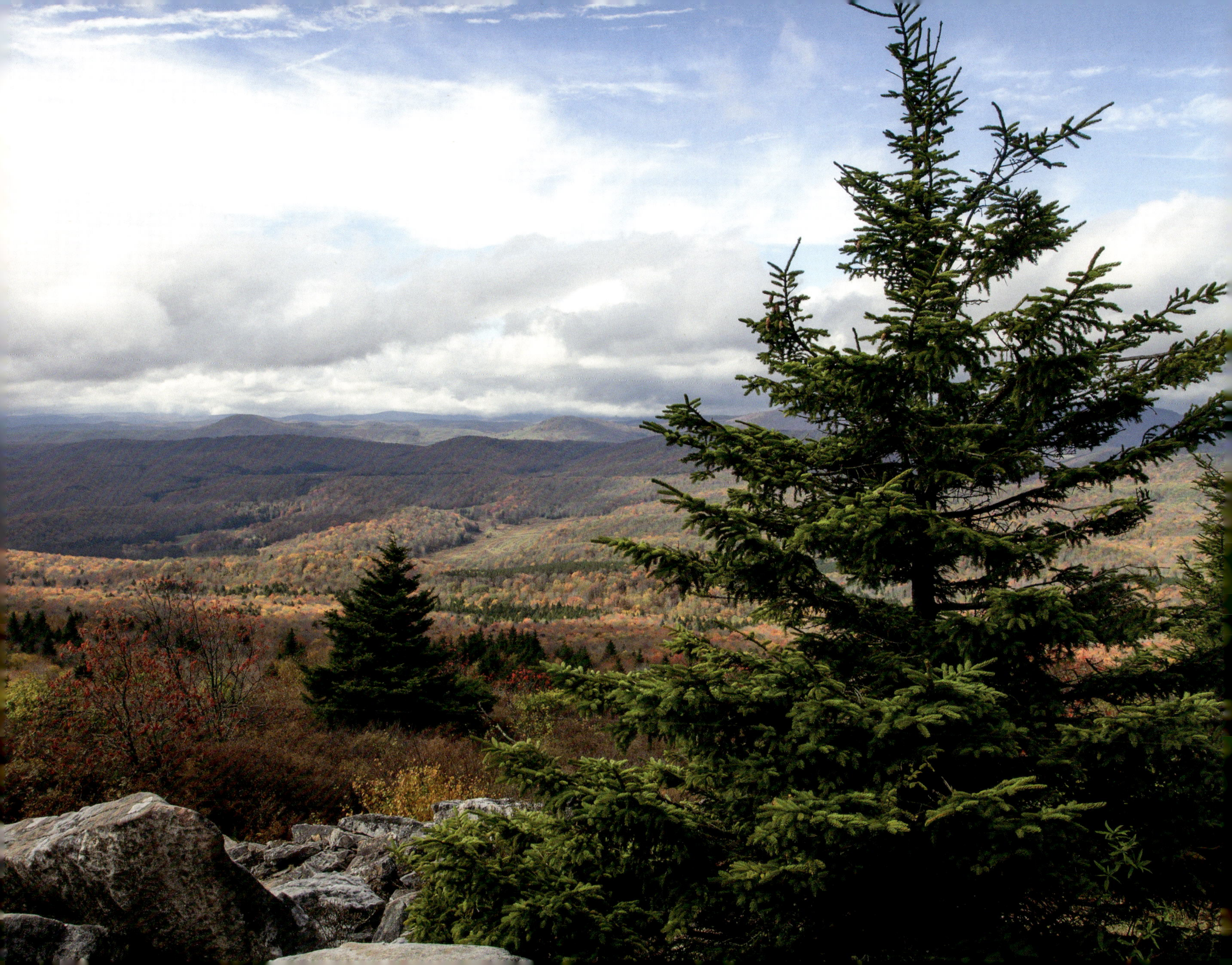

Parks and public lands receive their highest visitation as "leaf peepers" flock to the Central Appalachians to view the stunning showcase of autumn color that decorates the mountains.

Nature's color palette is now on full display. The green vigor of the summer ferns transitions into earthier tones of gold and brown. The vast acres of blueberry and huckleberry shrubs of the heath grounds of the Dolly Sods Wilderness, a high-elevation plateau along the Allegheny Front in West Virginia, are brushed in crimson red beginning in late September. By mid-October the northern hardwood forests of beeches, sugar maples, and birches illuminate the higher elevations of the mountains. The leaf change continues into late October through November, as the dominant oak-hickory forests of the mountain slopes will be the site of autumn's last stand before the arrival of winter.

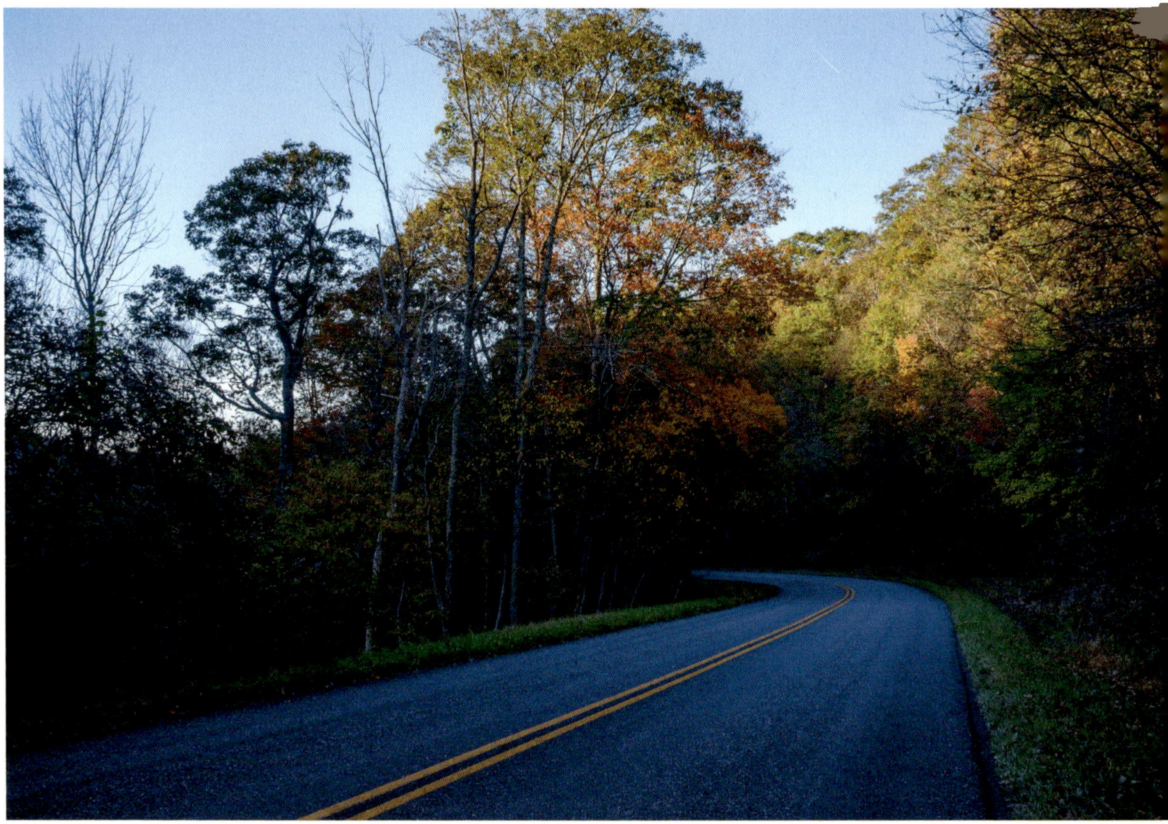

The Dolly Sods Wilderness comes aflame in late September. A combination of elevation, topography, and weather patterns have created a microclimate more akin to Canada than elsewhere in the Central Appalachians, featuring sphagnum bogs, miles of heath shrubs, and red spruce forest. Bear Rocks Preserve, West Virginia.

The Blue Ridge Parkway in autumn

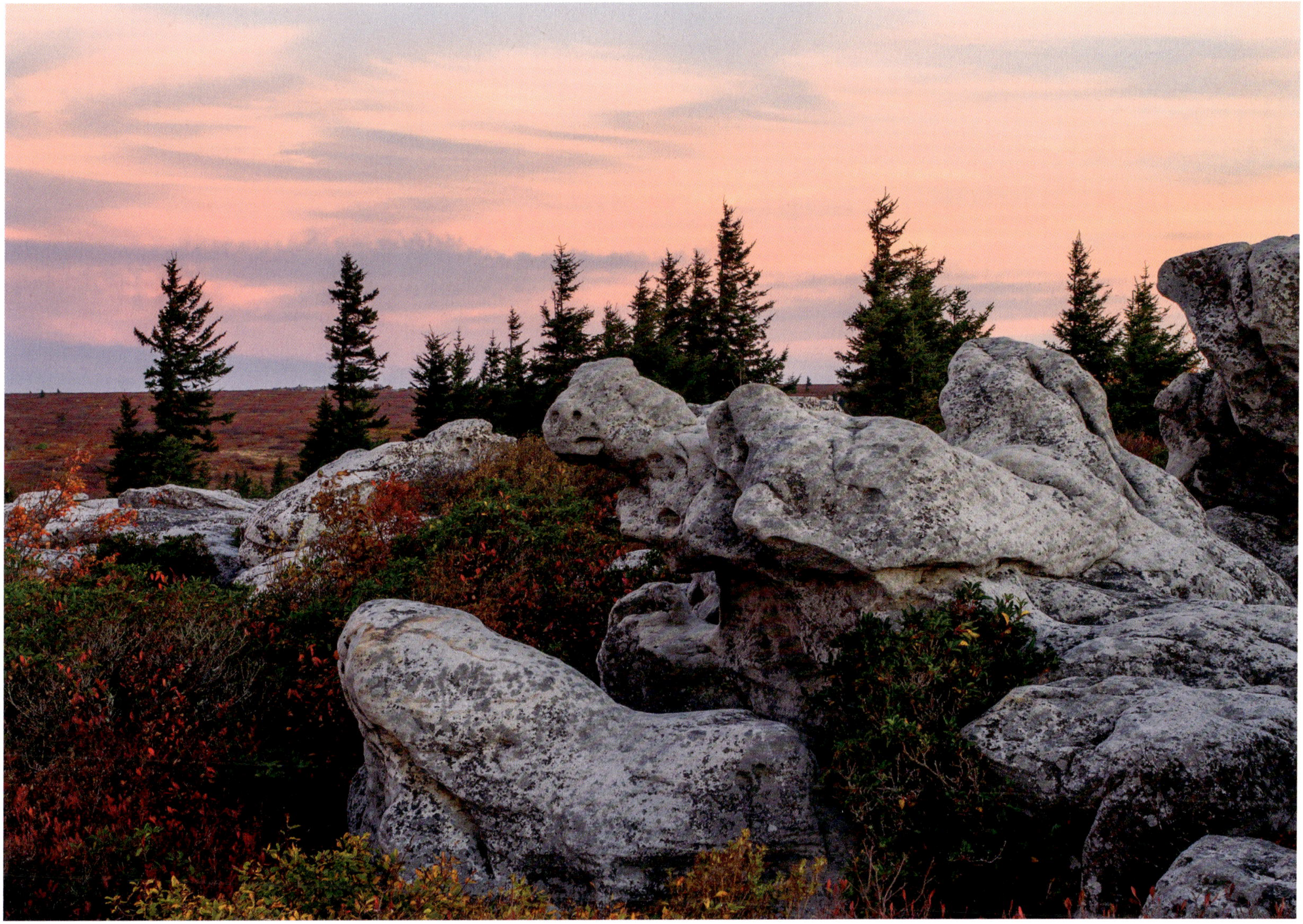

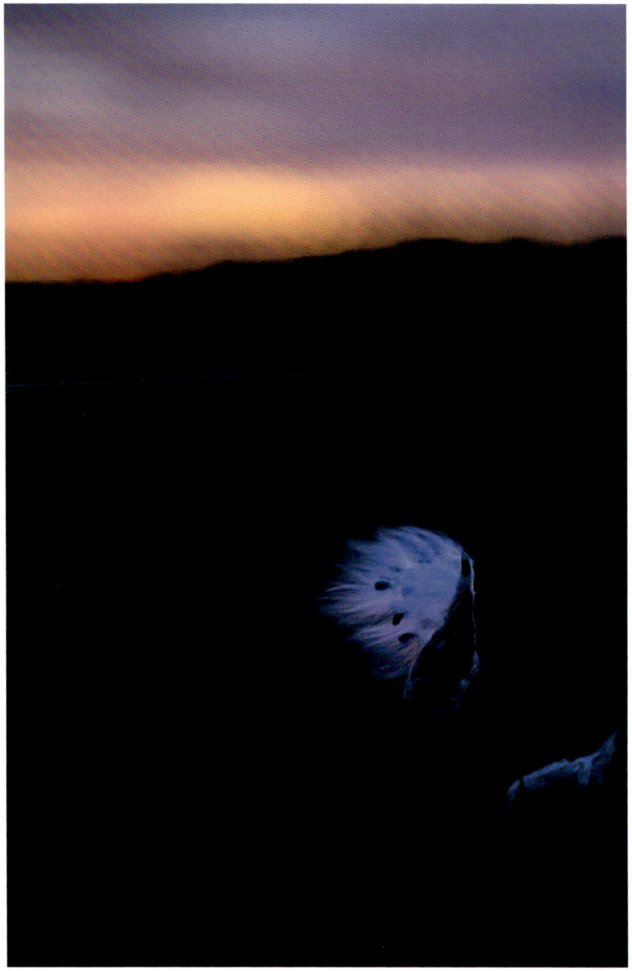

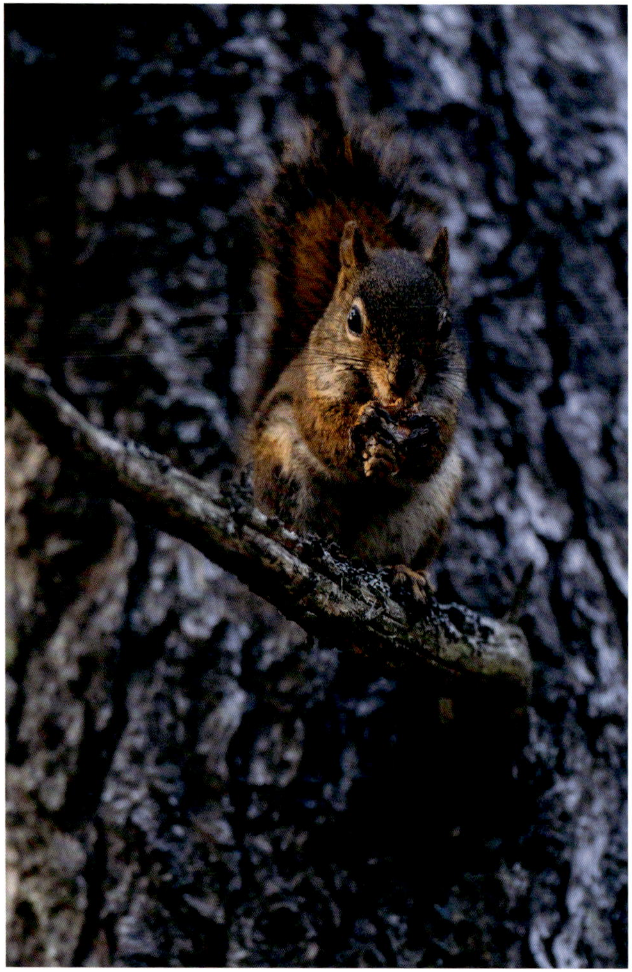

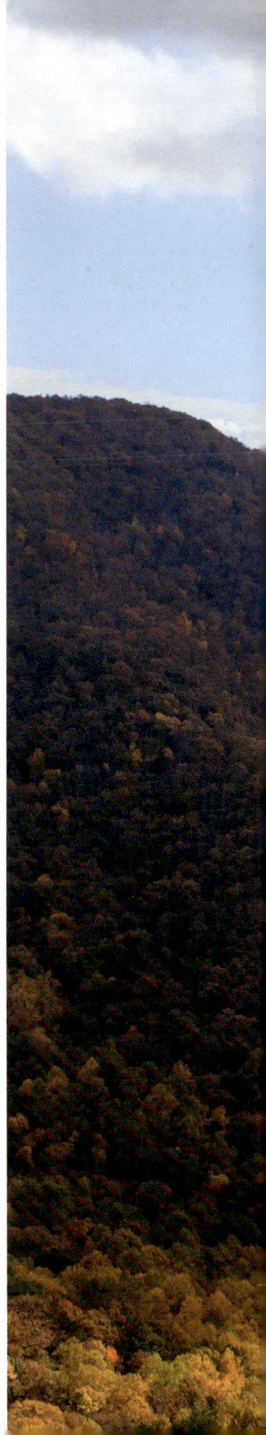

Milkweed pods open in autumn. Their seeds carry in the fall breeze. Shenandoah National Park, Virginia.

A red squirrel with its prize, a red spruce cone. Garrett County, Maryland.

The oak-hickory forests of the Blue Ridge Mountains in Shenandoah National Park are painted in hues of orange and yellow.

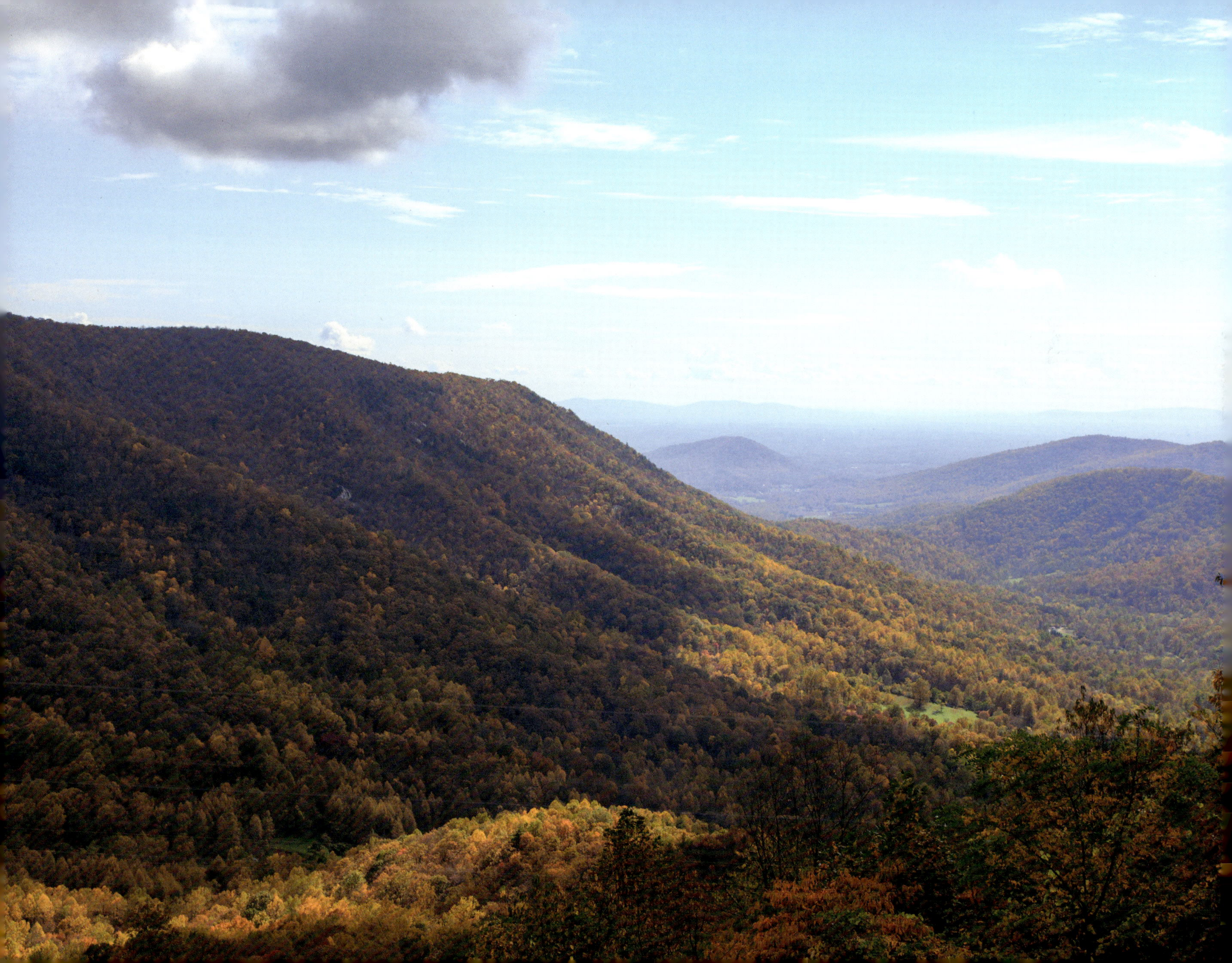

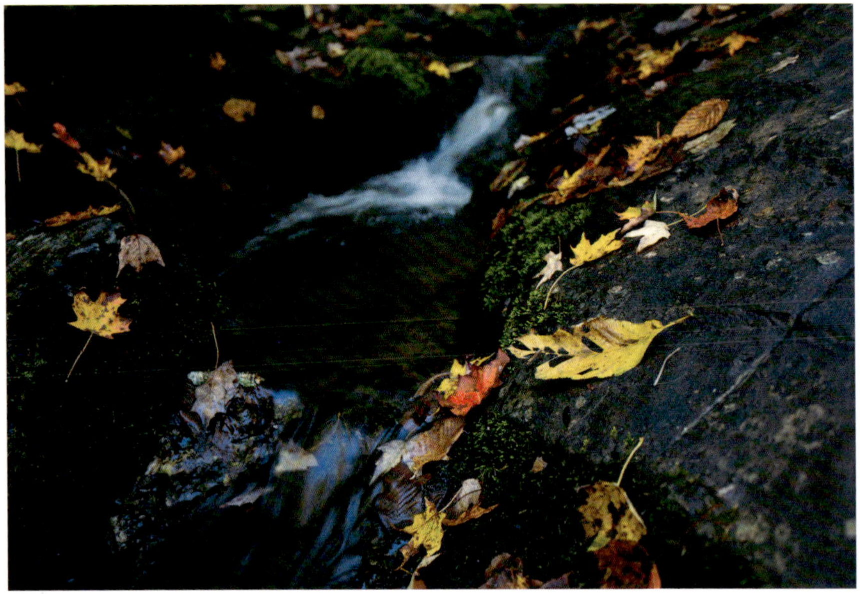

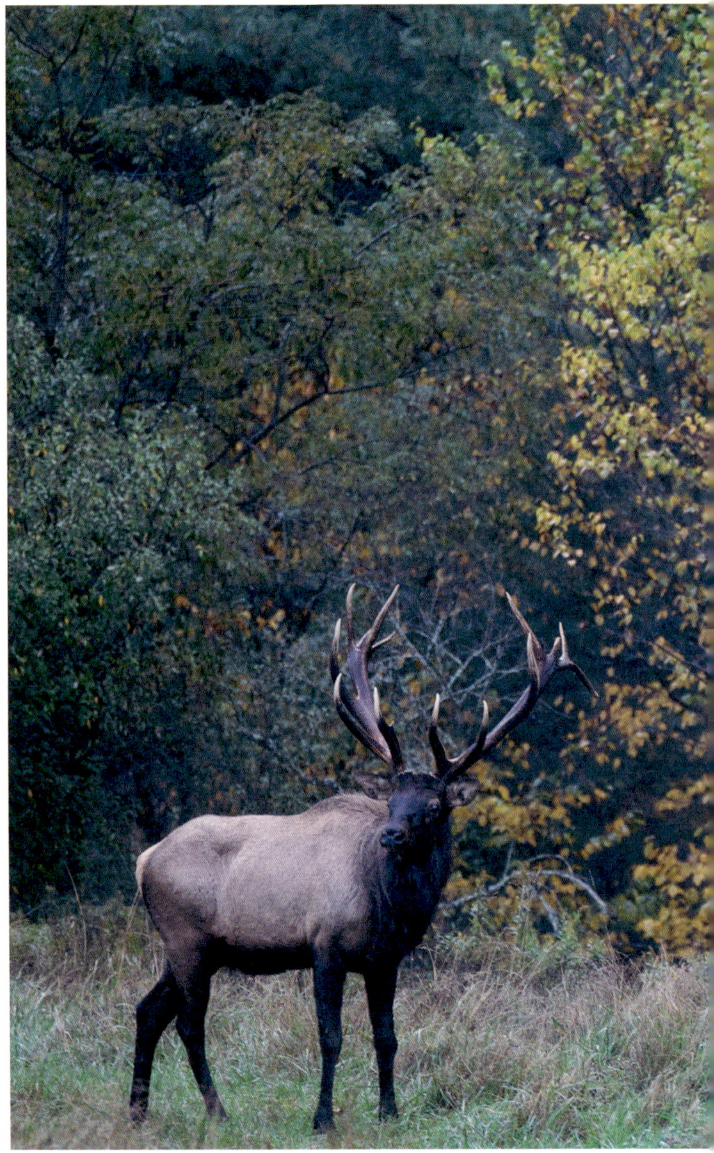

Fallen leaves on Dark Hollow Falls. Shenandoah National Park, Virginia.

A bull elk watches over his harem as autumn signals the start of the rut. Benezette, Pennsylvania.

Golden ferns carpet the forest floor. George Washington National Forest, West Virginia.

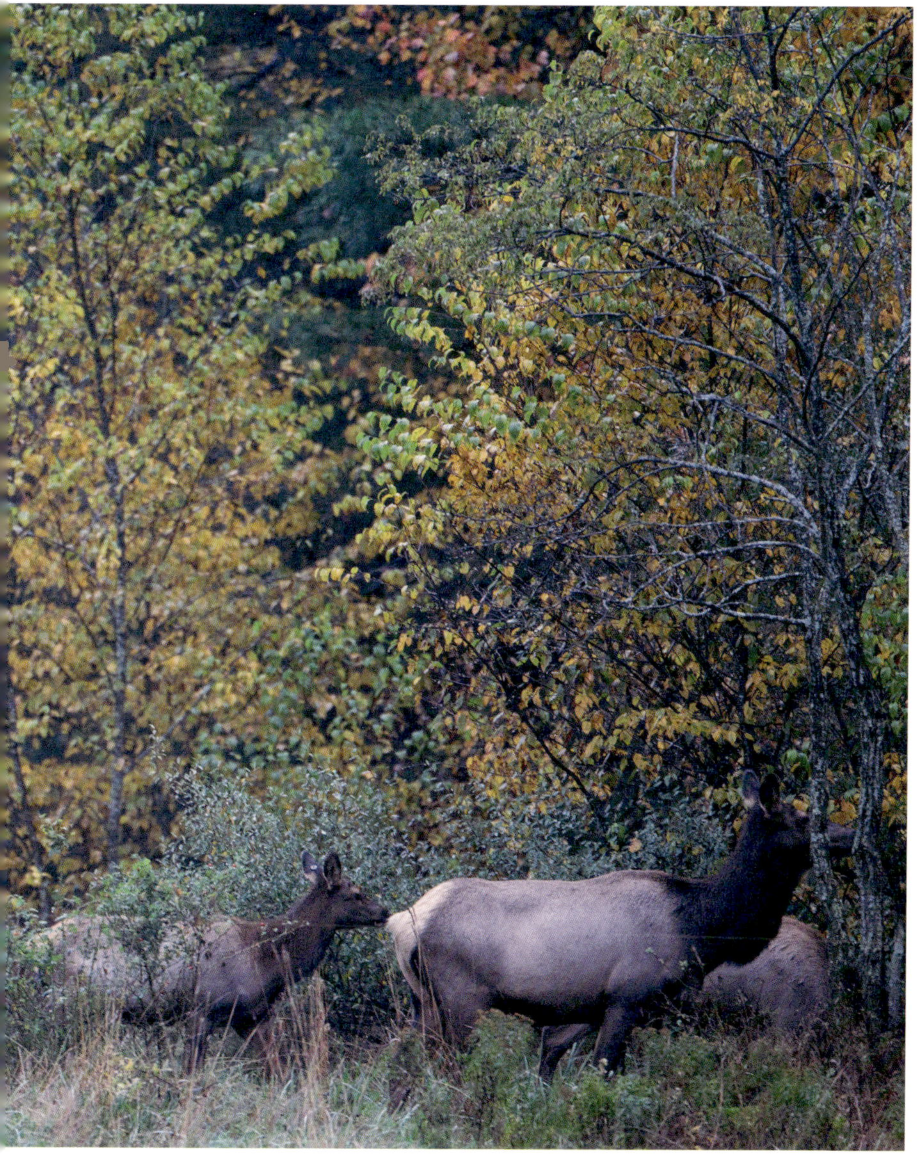

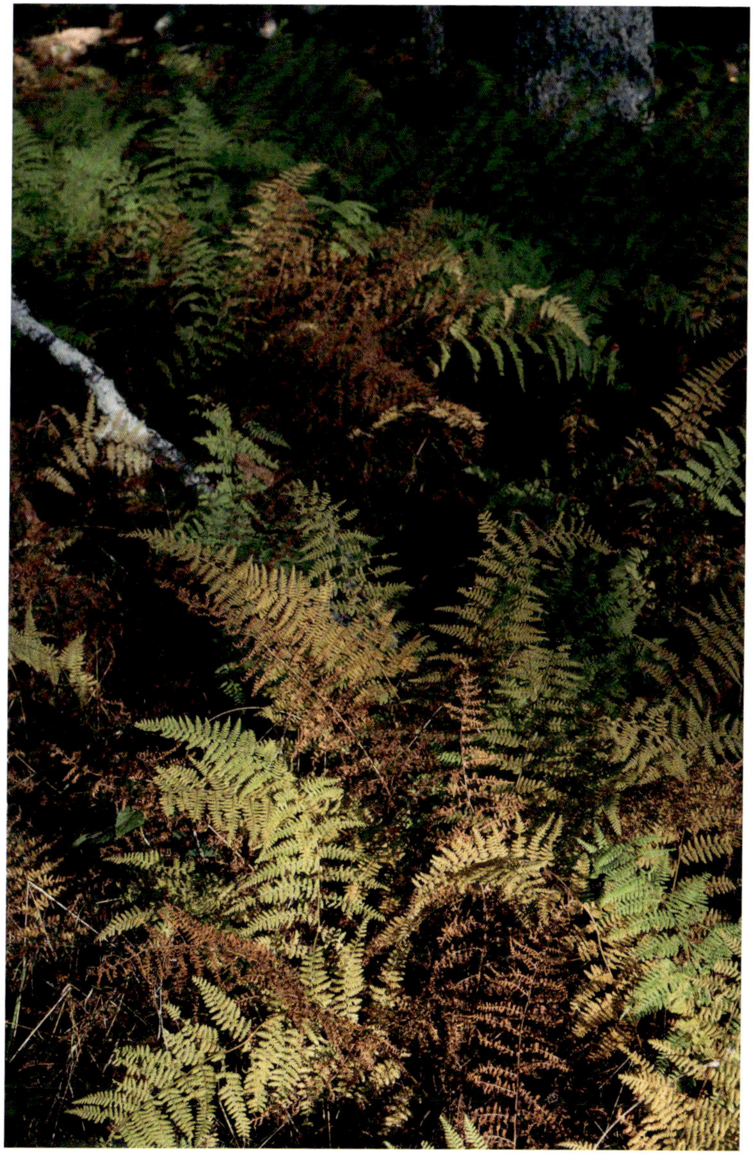

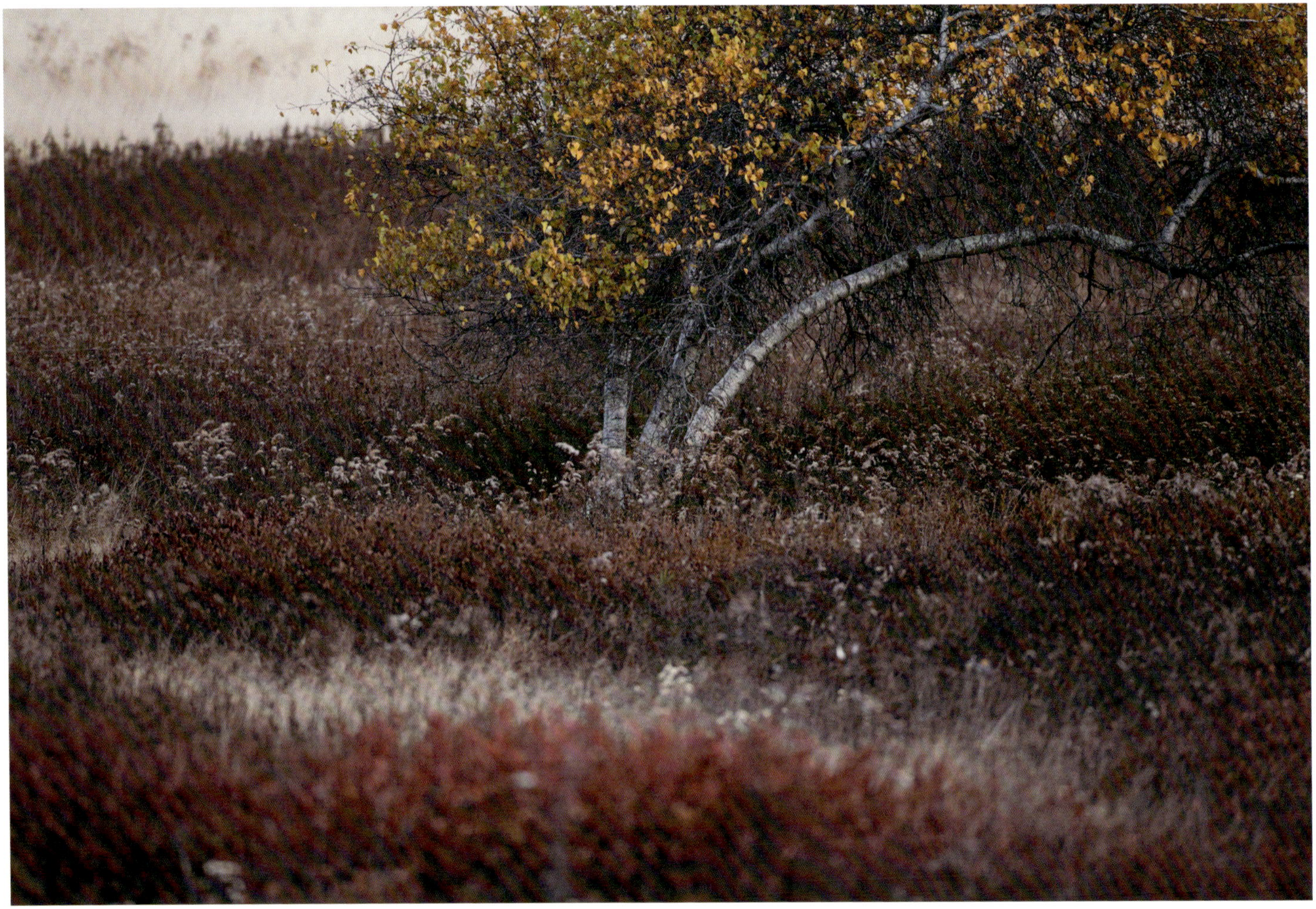

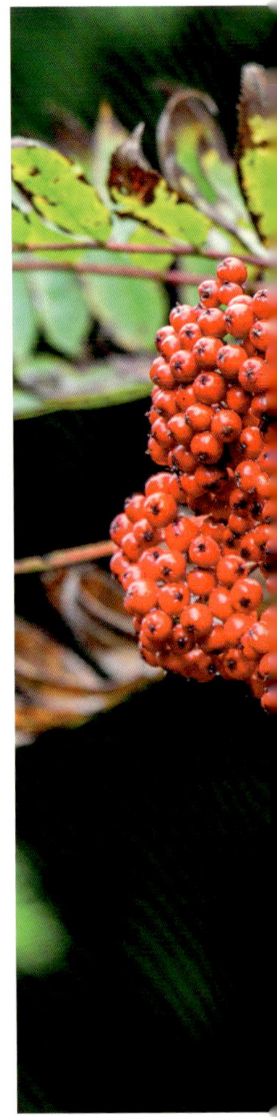

Autumn has painted the birch and the surrounding blueberry shrubs in a kaleidoscope of color, but autumnal red is also found in the berries of a mountain ash and in a mushroom on the forest floor. Shenandoah National Park, Spruce Knob, West Virginia, and Frostburg, Maryland.

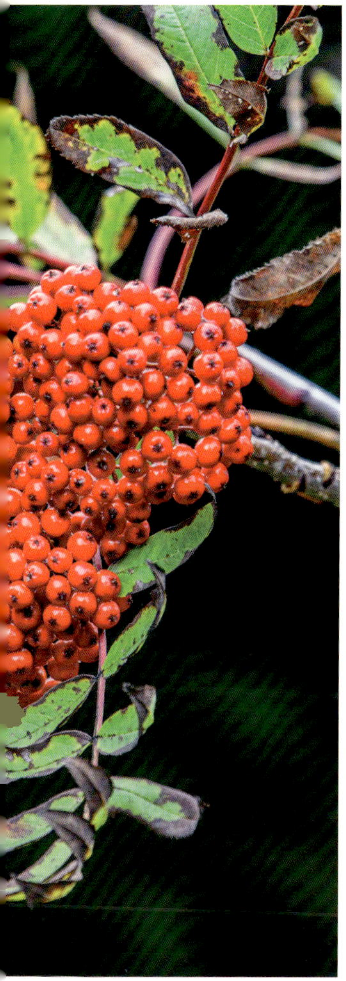

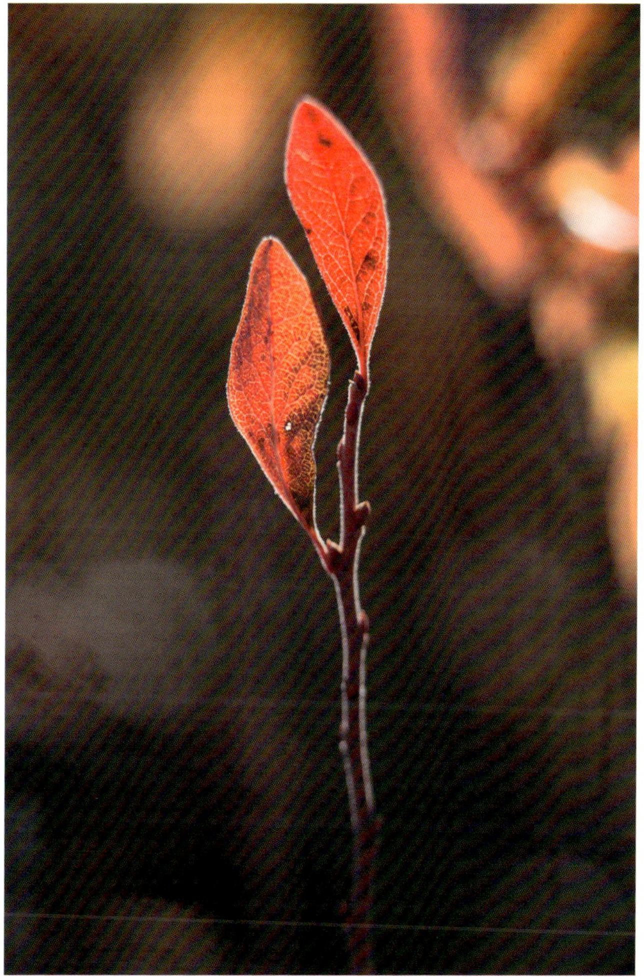

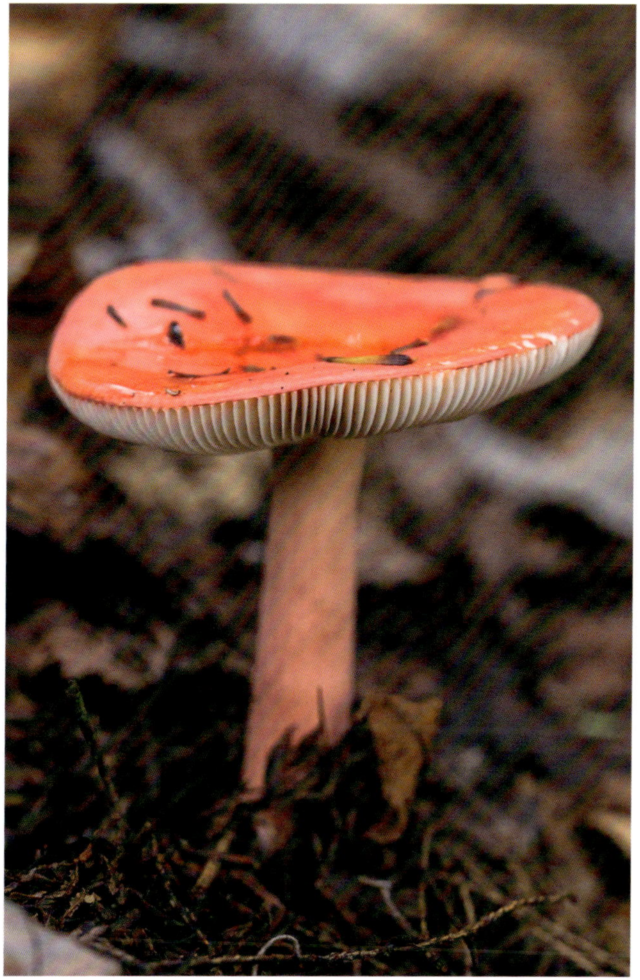

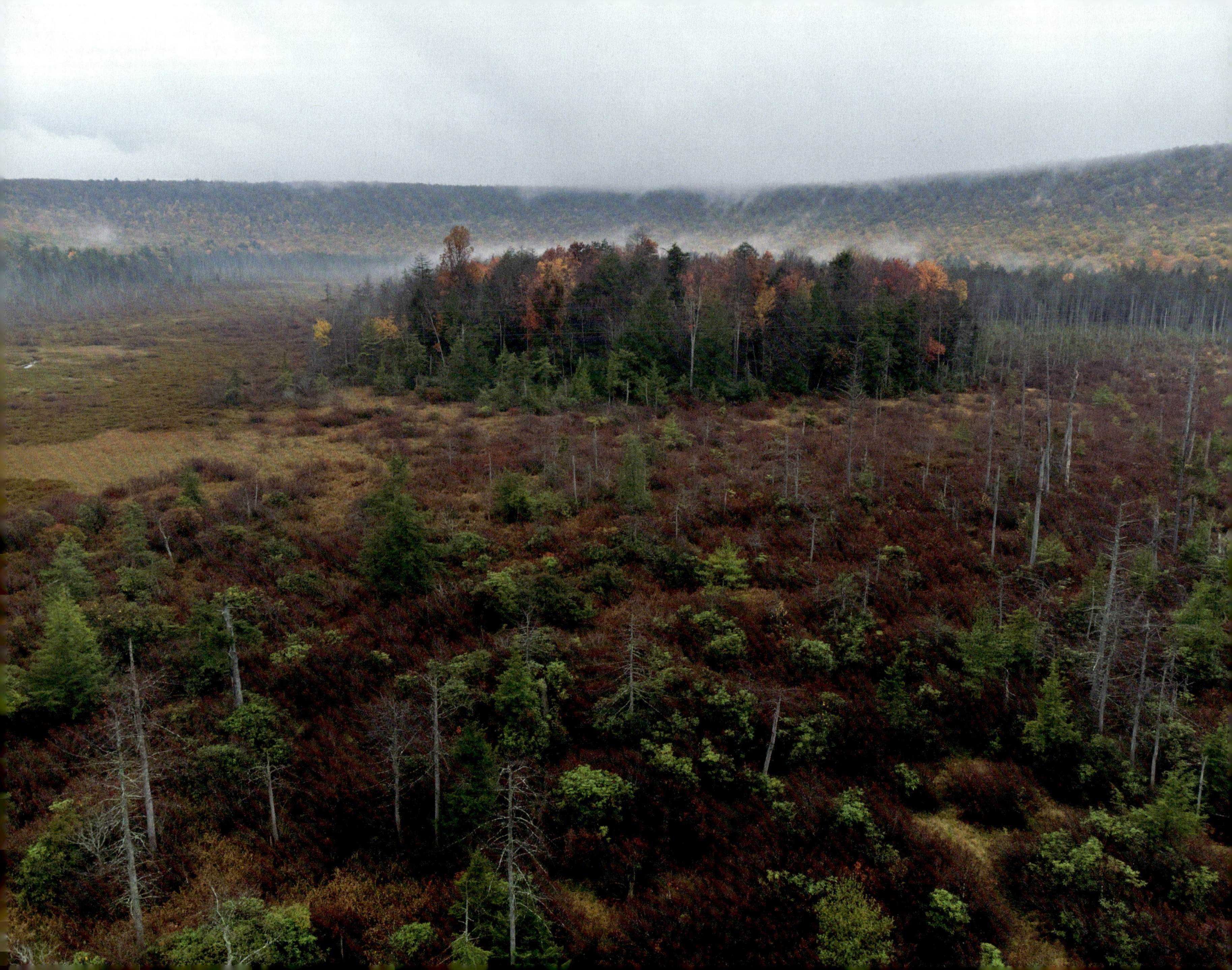

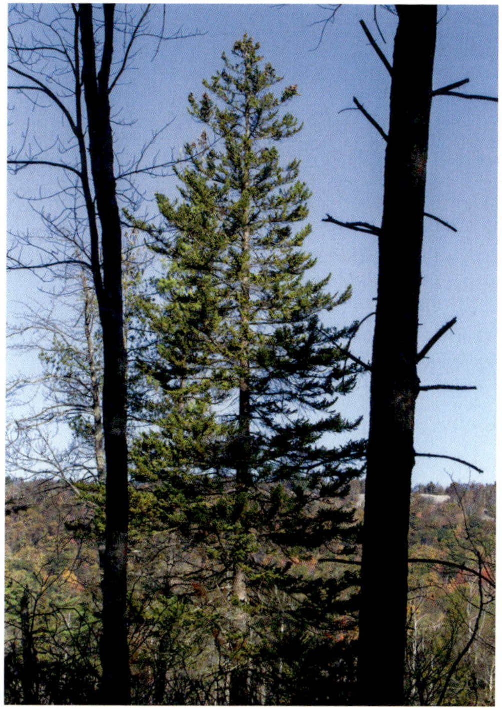

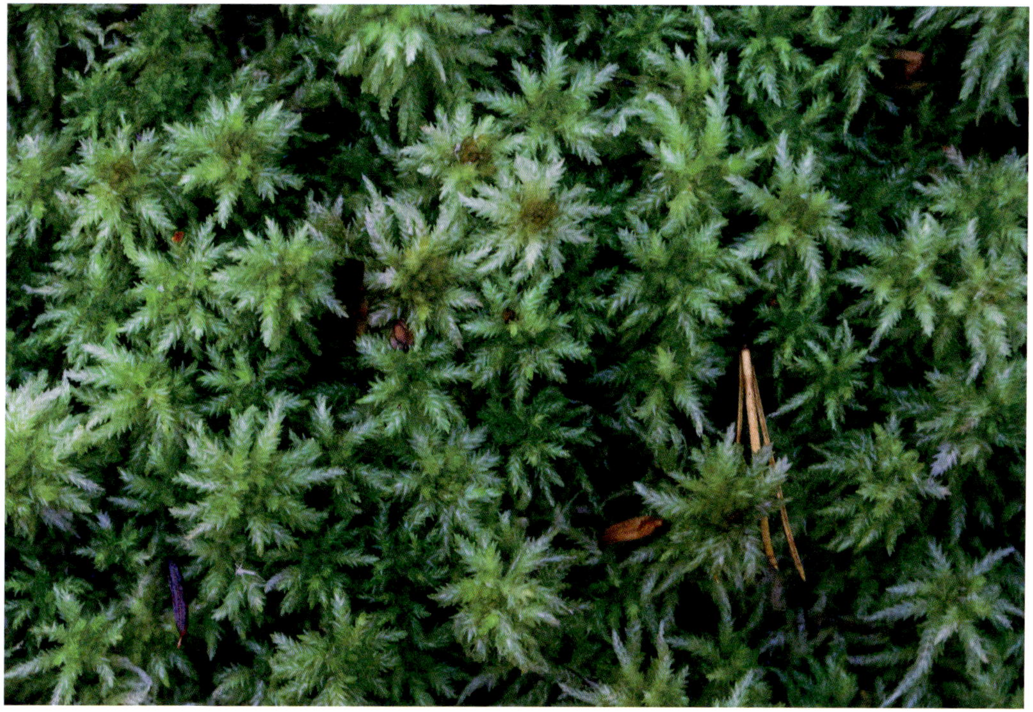

The Bear Meadows Natural Area is a relic boreal bog that formed at the end of the last glaciation period. Northern species, such as black spruce, balsam fir, and sphagnum mosses, are still found here. Centre County, Pennsylvania.

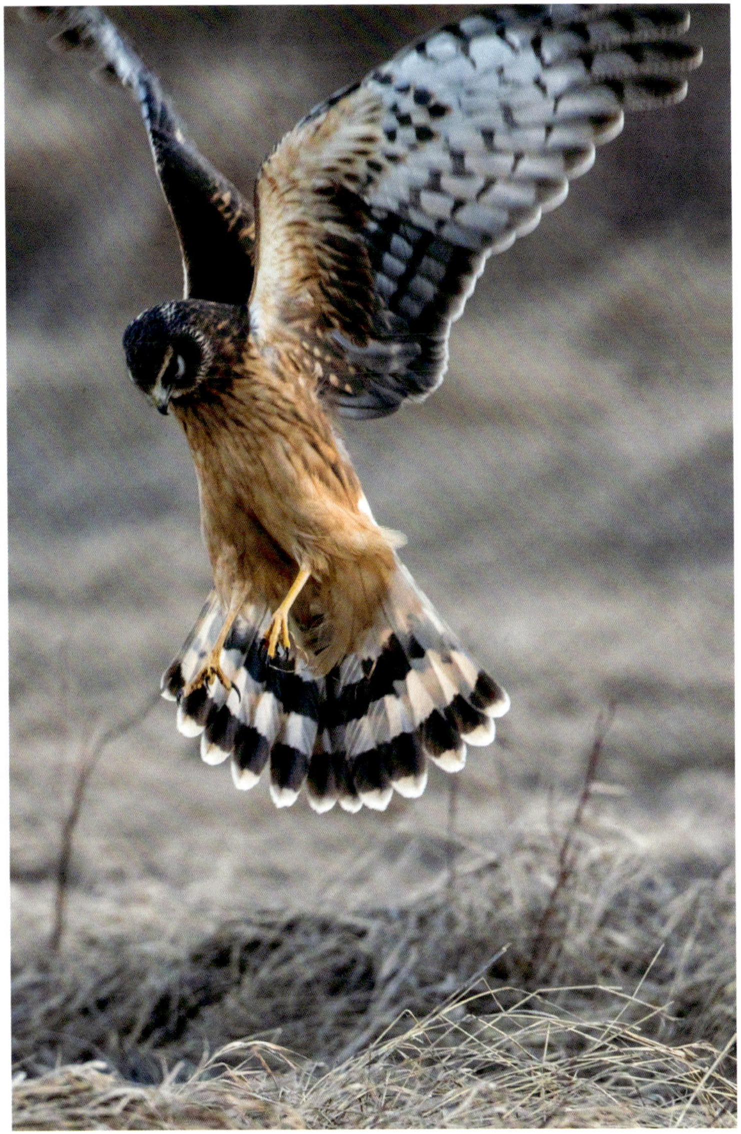

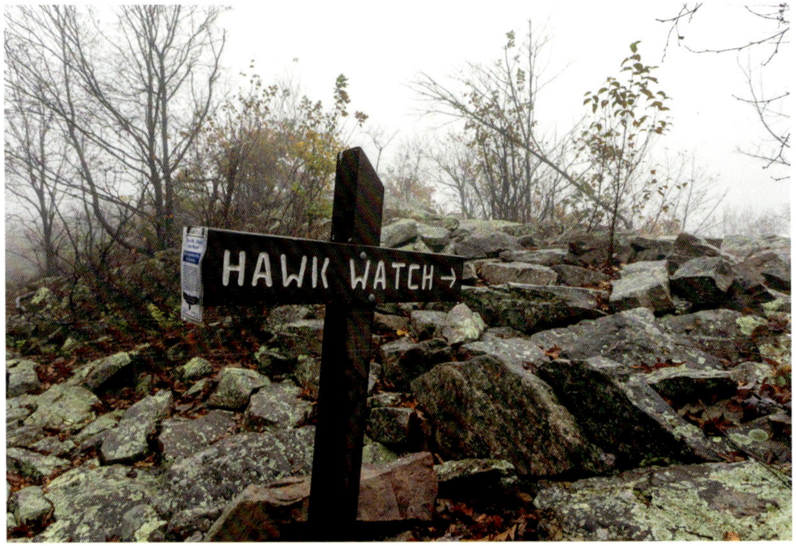
HAWK WATCH →

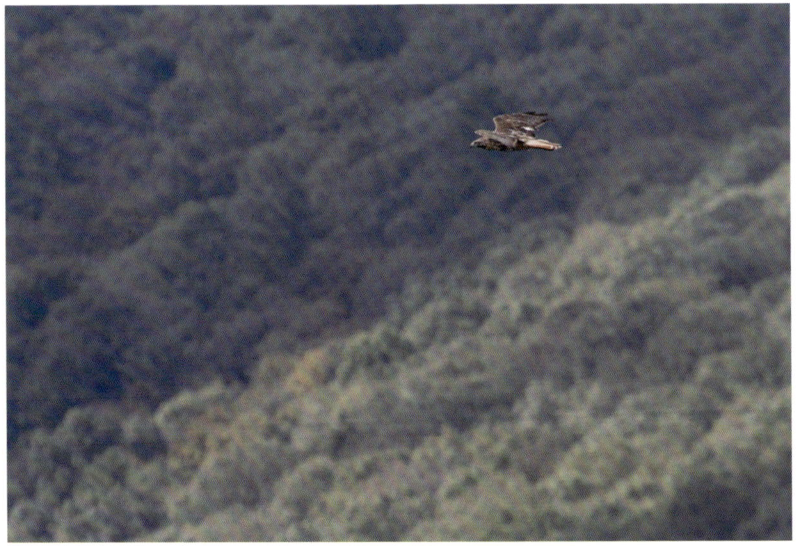

A northern harrier hunts in Big Meadows, Shenandoah National Park.

The Central Appalachians include important flyways for bird migration, especially raptor migration, in autumn. Stone Mountain, which overlooks the Juanita Valley of Pennsylvania, is a popular and important area. Mifflin County, Pennsylvania.

A red-tailed hawk seen from Stone Mountain

Maple and birch leaves on a fallen gray birch. Shenandoah National Park, Virginia.

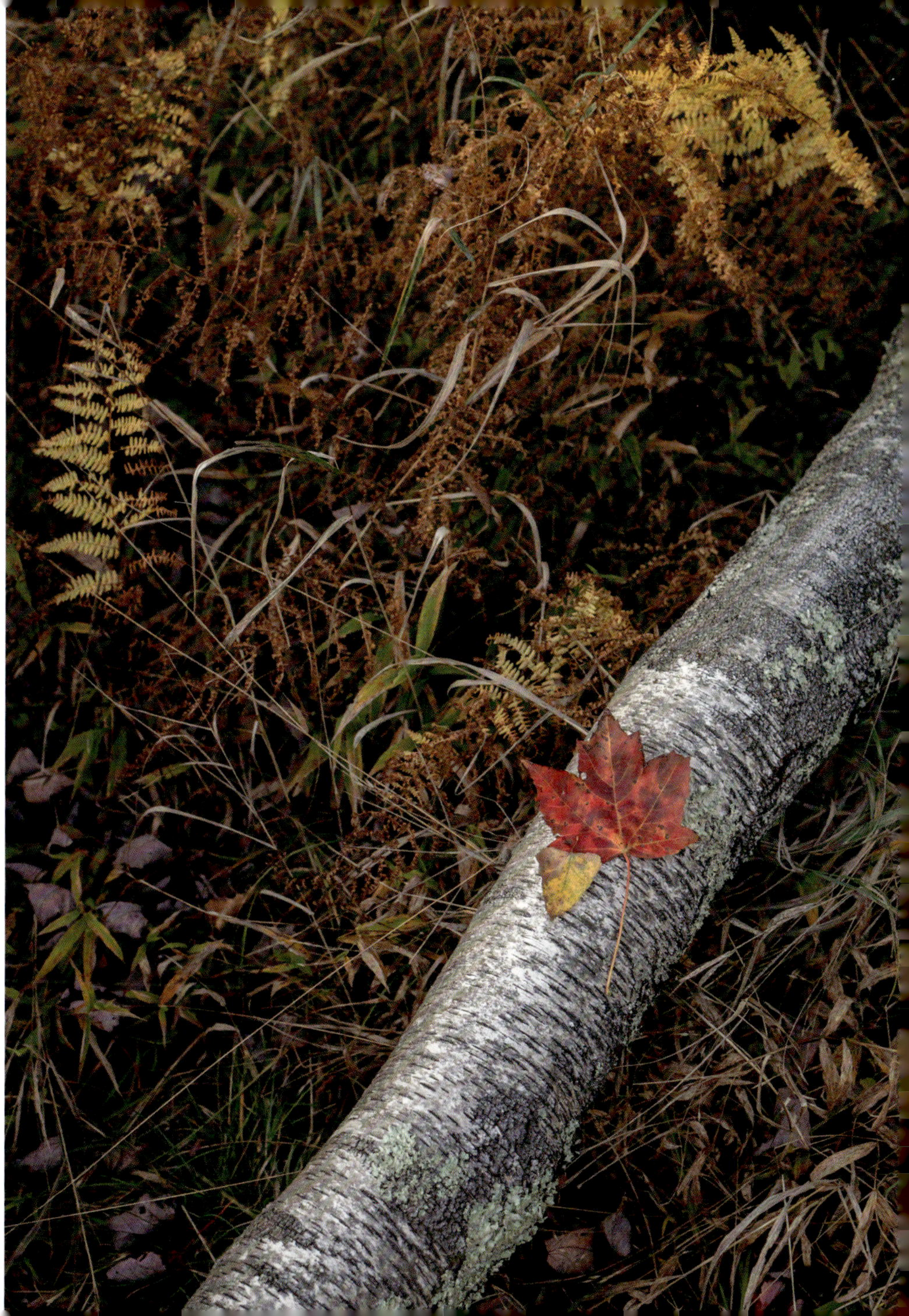

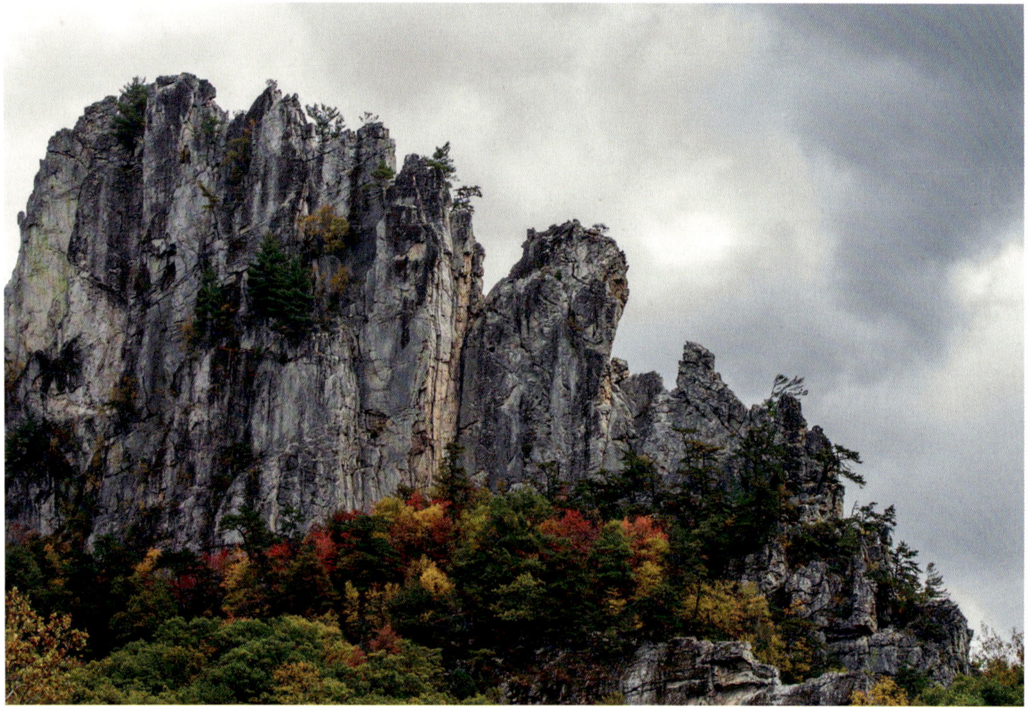

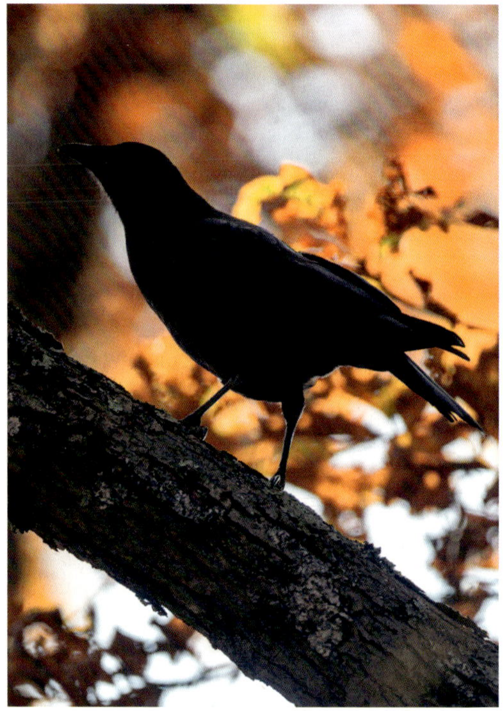

Seneca Rocks in autumn

A crow perched in front of illuminated oak leaves. George Washington National Forest, Virginia.

Great blue herons fly over the James River. Monroe, Virginia.

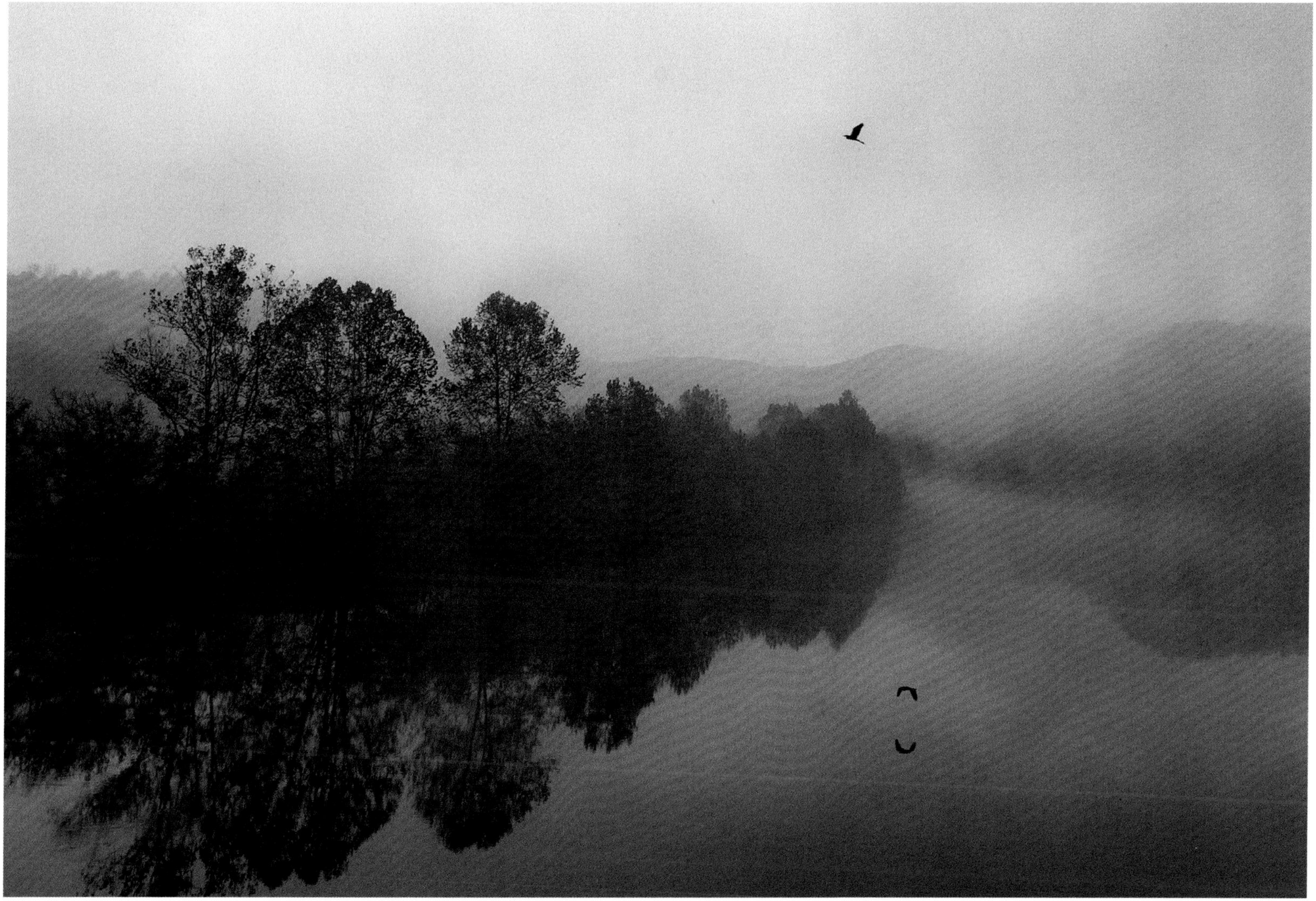

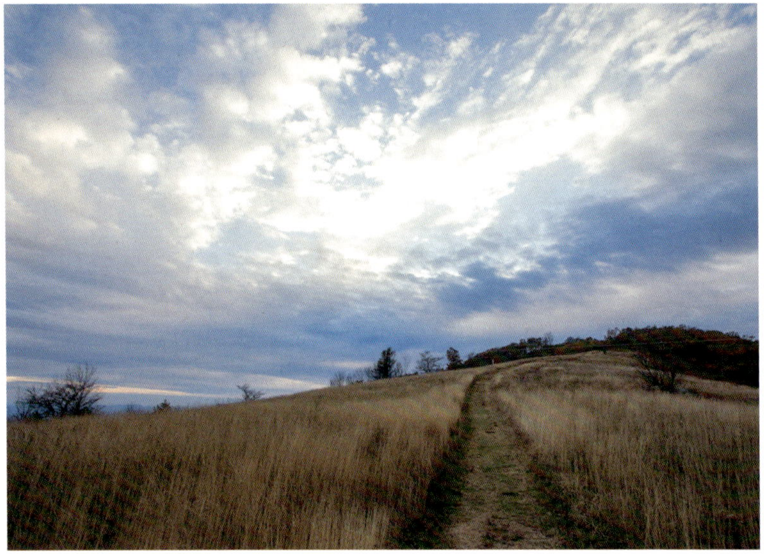

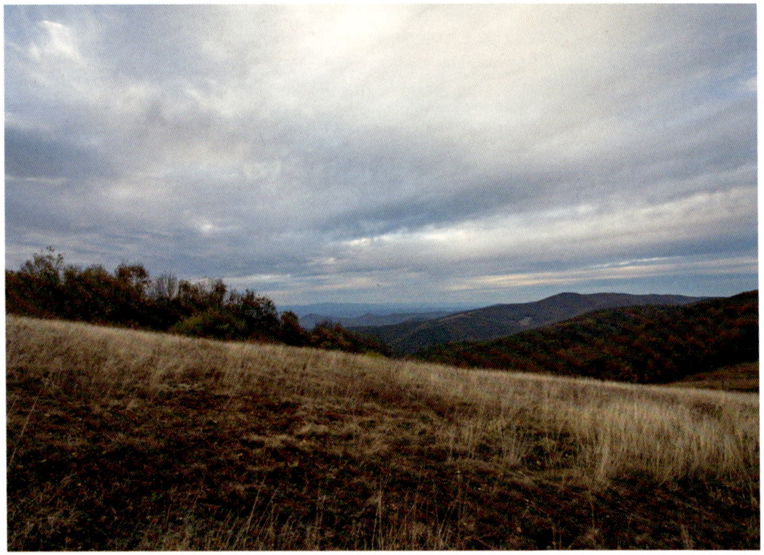

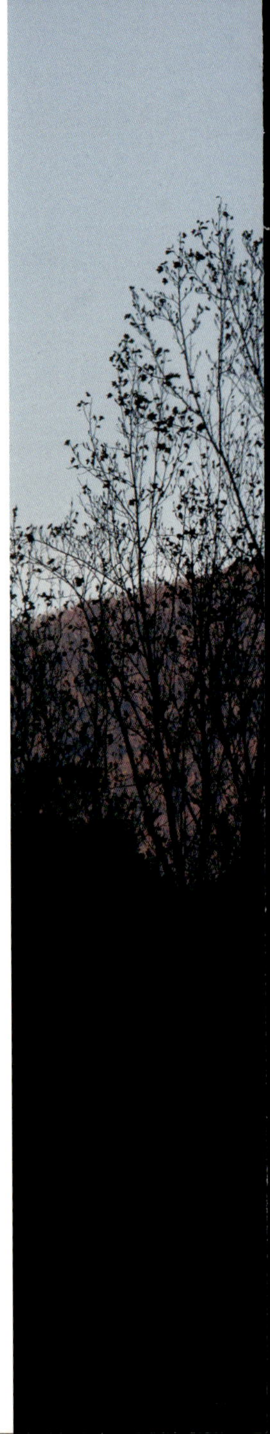

More common in the Southern Appalachians, the open summit of Cole Mountain in Virginia features a grassy bald that the Appalachian Trail goes through. The mountain is also known as "Cold" Mountain.

Terrapin Mountain, seen from the Blue Ridge Parkway in autumn, Virginia.

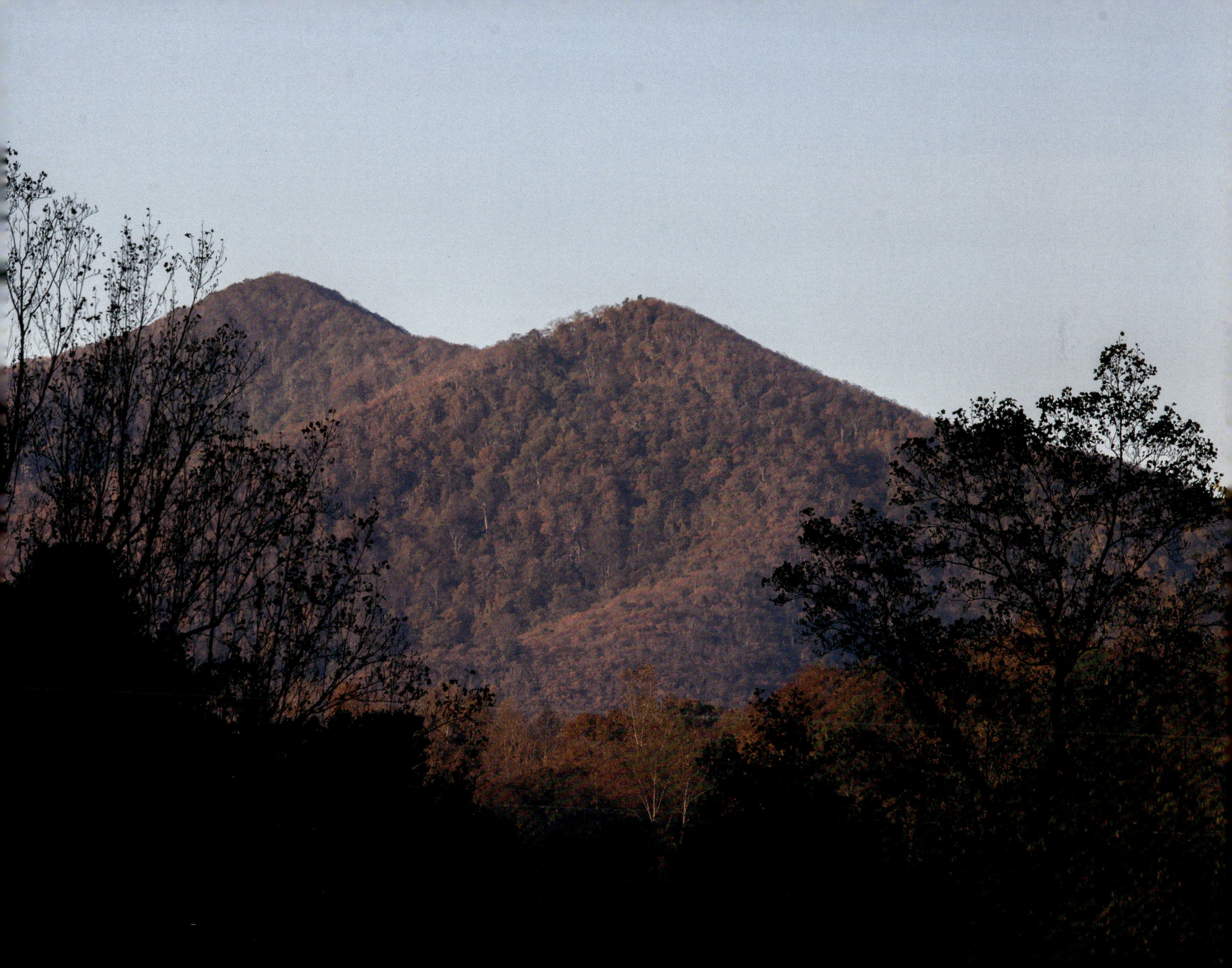

Where Elk Still Roam

On a late September morning, in the chill of the predawn, I stand on the edge where a hemlock forest meets a meadow in north-central Pennsylvania.

The landscape has only just begun to reflect the azure tint of twilight but had already been pulsing with a different type of natural symphony. A haunting and ancient song encompassed the hills and valleys around me, a song at once mournful and chilling: the bugle of the elk. A large bull, a 7×7, emerges like an apparition from the fog. As if shrouded in mysticism, it opens its mouth and emits a resonant, multitoned, high-pitched squeal. The call signals the autumnal rut, the annual ritual when bulls challenge each other for mating rights. It is exciting and primal.

The call does not go unnoticed, as another bugle of equal intensity answers it almost immediately. Another large bull, one with a harem of over thirty cows, comes to meet the challenger. After eyeing each other in preparation for conflict, the two meet head on in what was the beginning of an epic clash. Violent exchanges between

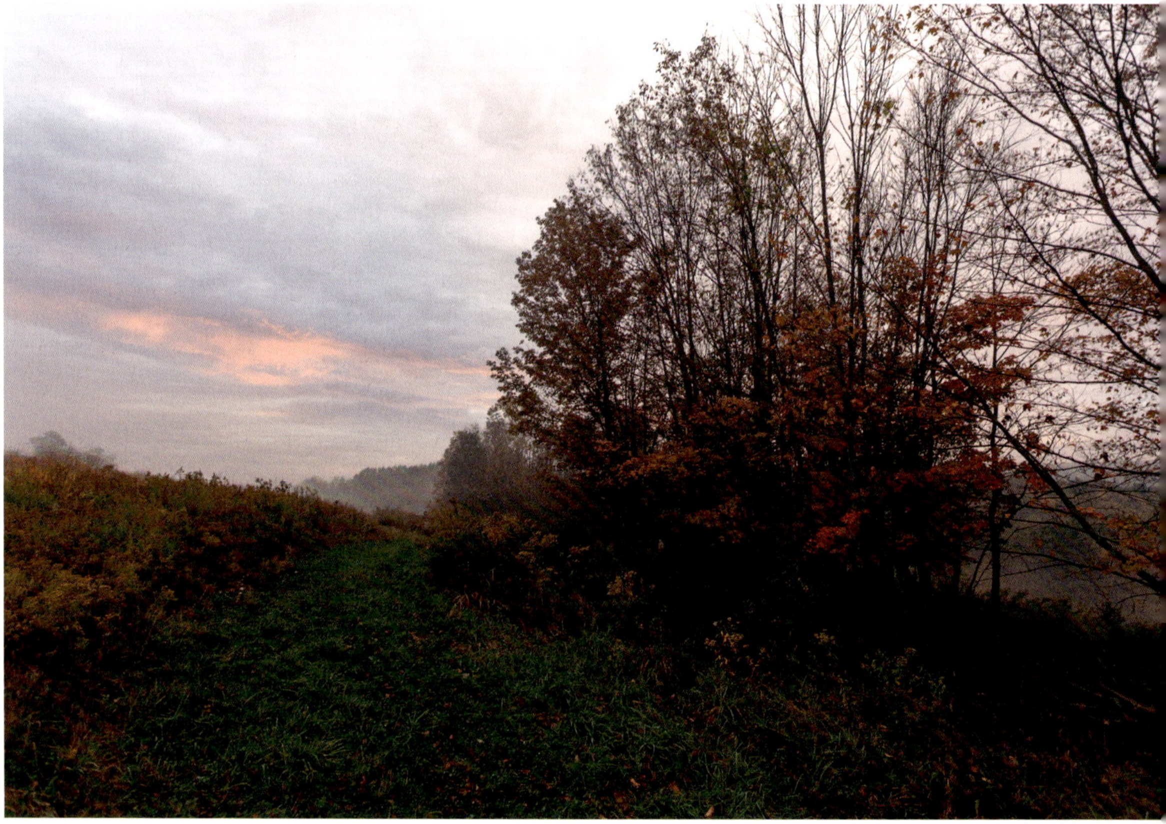

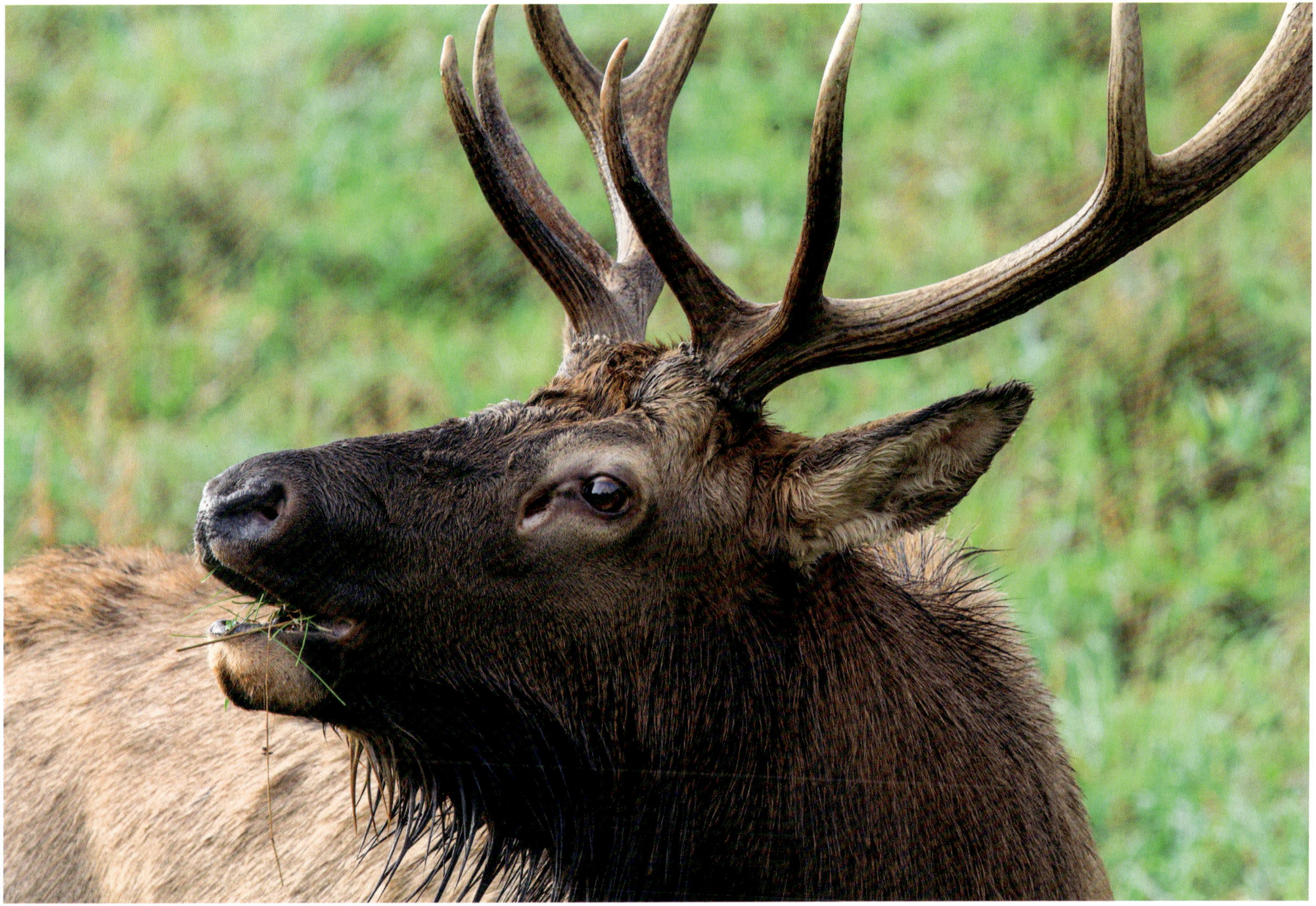

titans such as these are often swift, but on this morning the battle lasts for what seemed like many minutes. The bulls are ruthless in their aggression, and the intense battle is as exciting as any prizefight. The sound of antlers smacking and rattling echoes throughout the valley. It quickly turns into a full-contact fight as hundreds of pounds of ungulate violently thrashes with the other. The competitors push and turn with one another in almost ceremonial fashion. Finally, the bull with the harem pushes forward against the challenger. Chunks of mud fly upward from the strength of its back hooves. The challenger breaks the hold and flees in cowardice. The winner, its posture regal and majestic, now bugles louder than before: a warning to all other potential challengers in the area. The adrenaline is contagious. What an incredible moment!

The Central Appalachians were once home to charismatic North American megafauna such as the gray wolf, mountain lion, and bison. The eastern elk, once the largest subspecies of the iconic animal, was also at home here. Hunted to extinction, these elk made their final stand in the Alleghenies of north-central Pennsylvania; the last one was shot in sometime in the 1870s.

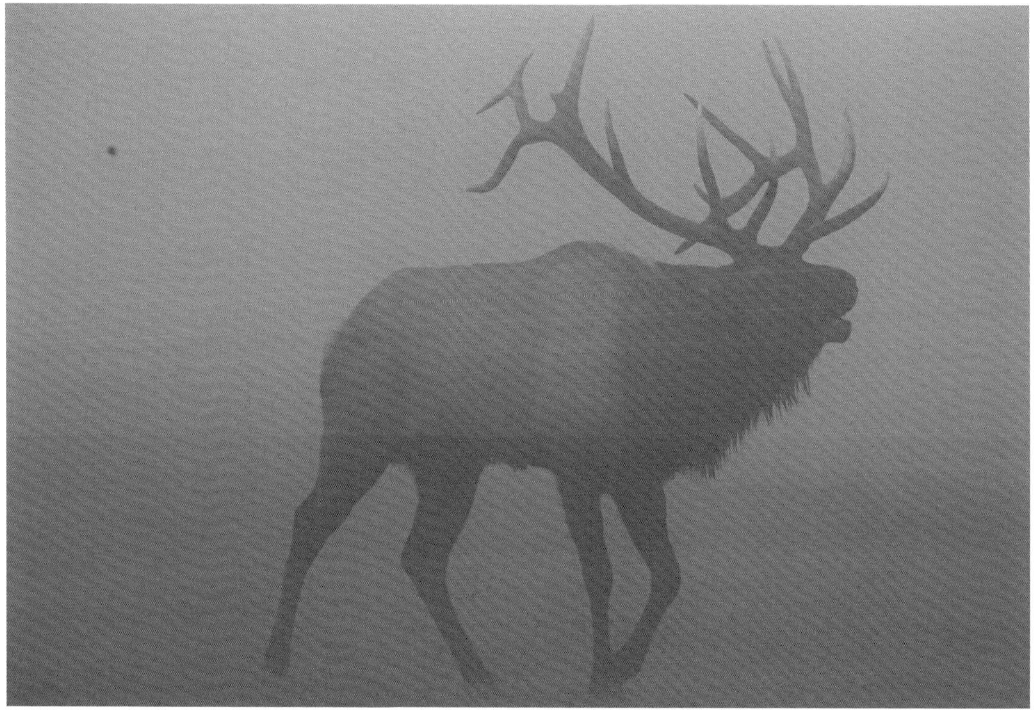

In 1913, fifty Rocky Mountain elk from Yellowstone National Park were reintroduced to north-central Pennsylvania, near where the last eastern elk was killed. More would be released later that year, and ninety-five more came in 1915. This was not like the wildlife reintroductions today; these animals were moved across the country by train and released with no acclimation period into the wilderness. The next few decades, the elk herd ebbed and flowed, and controversy followed. Conflict with agriculture, a deadly brain worm disease, and poaching all contributed to a declining herd, with extirpation to surely follow. Without intervention, these elk were on track to meet the same fate as the original species. In time, additional game lands were purchased and expanded, and habitat was managed for the elk, with support from the Rocky Mountain Elk

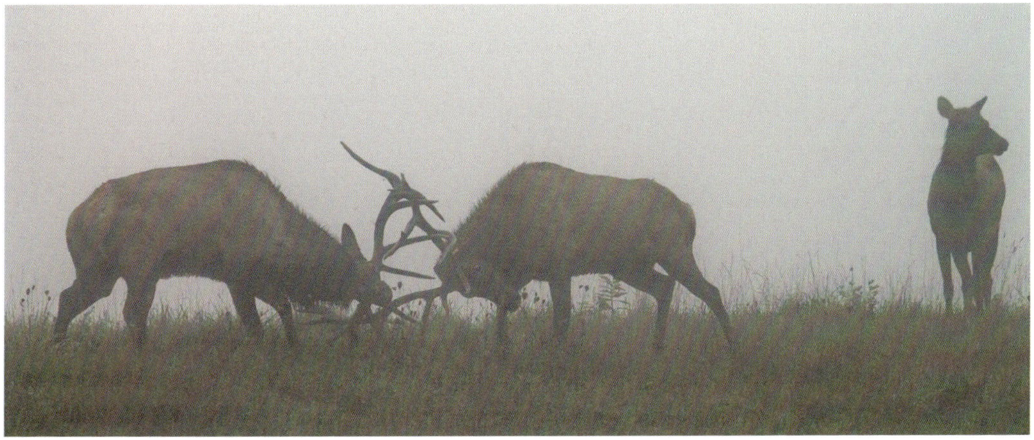

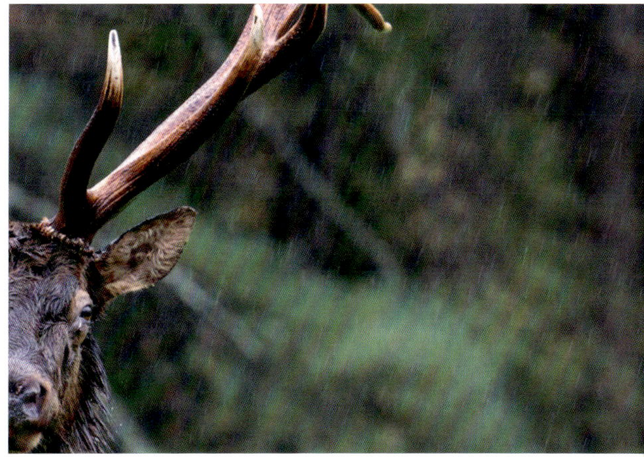

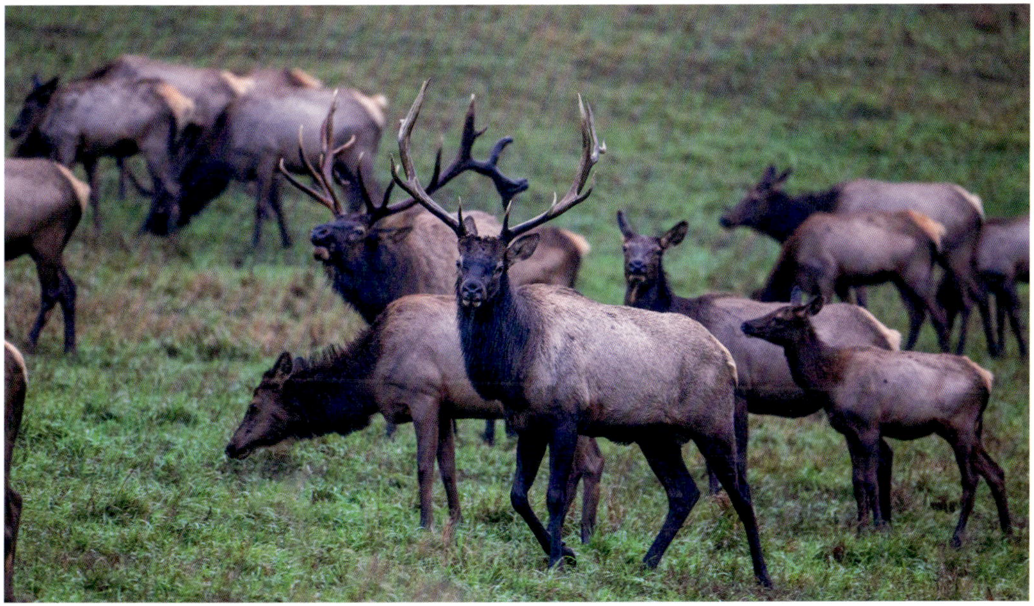

Foundation. This reduced conflicts with humans, and the habitat management benefited other species as well.

Today, the state has a thriving herd that brings thousands of tourists to these mountains, especially near the village of Benezette. For many, myself included, celebrating the rut has become an annual sojourn. For me and my friends, the rut occasions a reunion every autumn. We gather together to "camera-hunt" the elk. It never gets old. Every time I experience that haunting bugle, I am thankful that it is not emanating from the recorded history of a bygone era; rather, it is something I can hear and feel today.

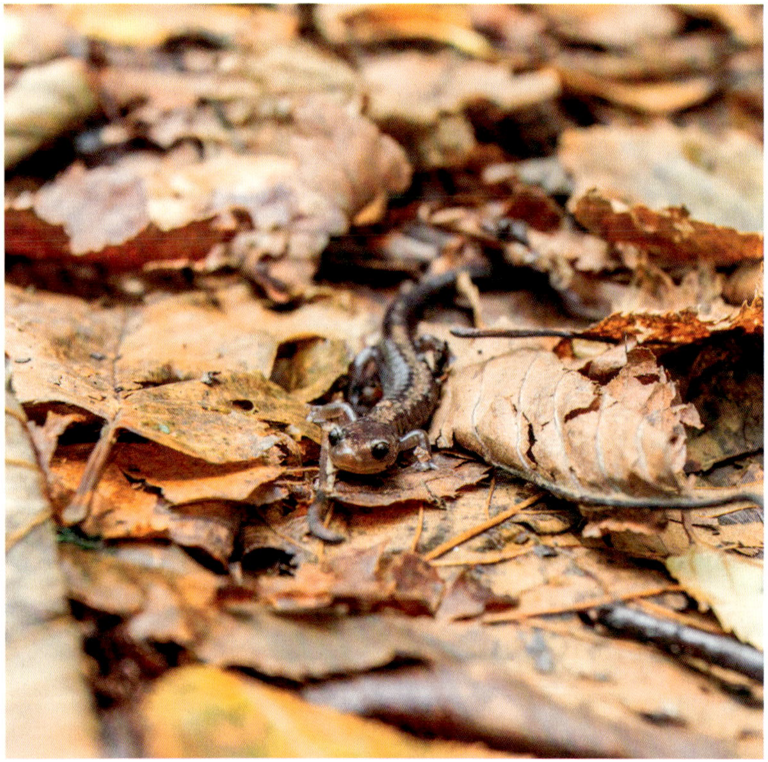

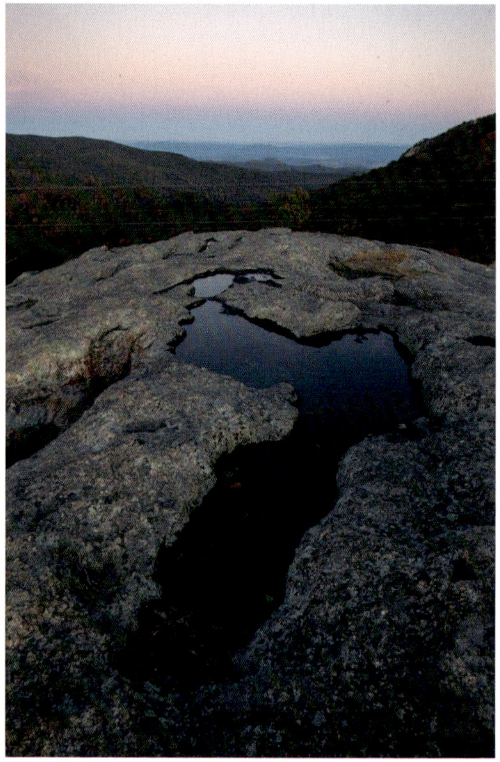

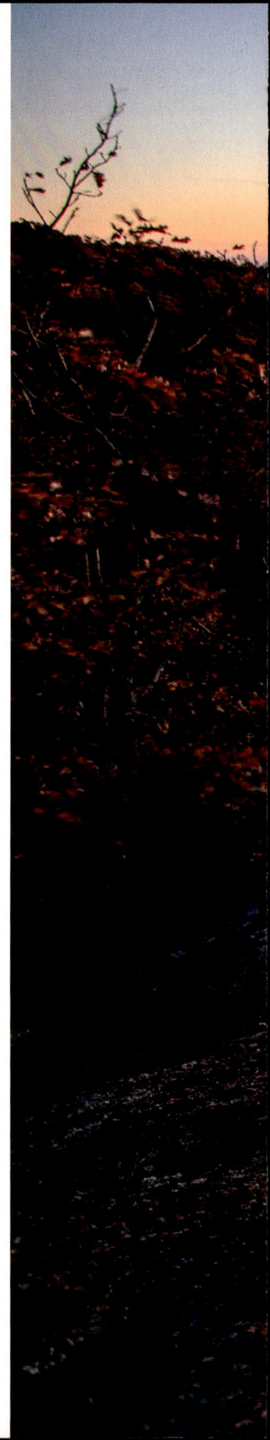

A Shenandoah salamander moves among the leaf clutter in preparation for winter hibernation.

Spy Rock, of Maintop Mountain, is an expansive granite dome that offers an incredible 360-degree view of the mountains and valleys that surround it. One can see other popular mountains of the Blue Ridge, such as Devils Knob, Trayfoot Mountain, the Three Ridges, and the Religious Range. George Washington National Forest, Virginia.

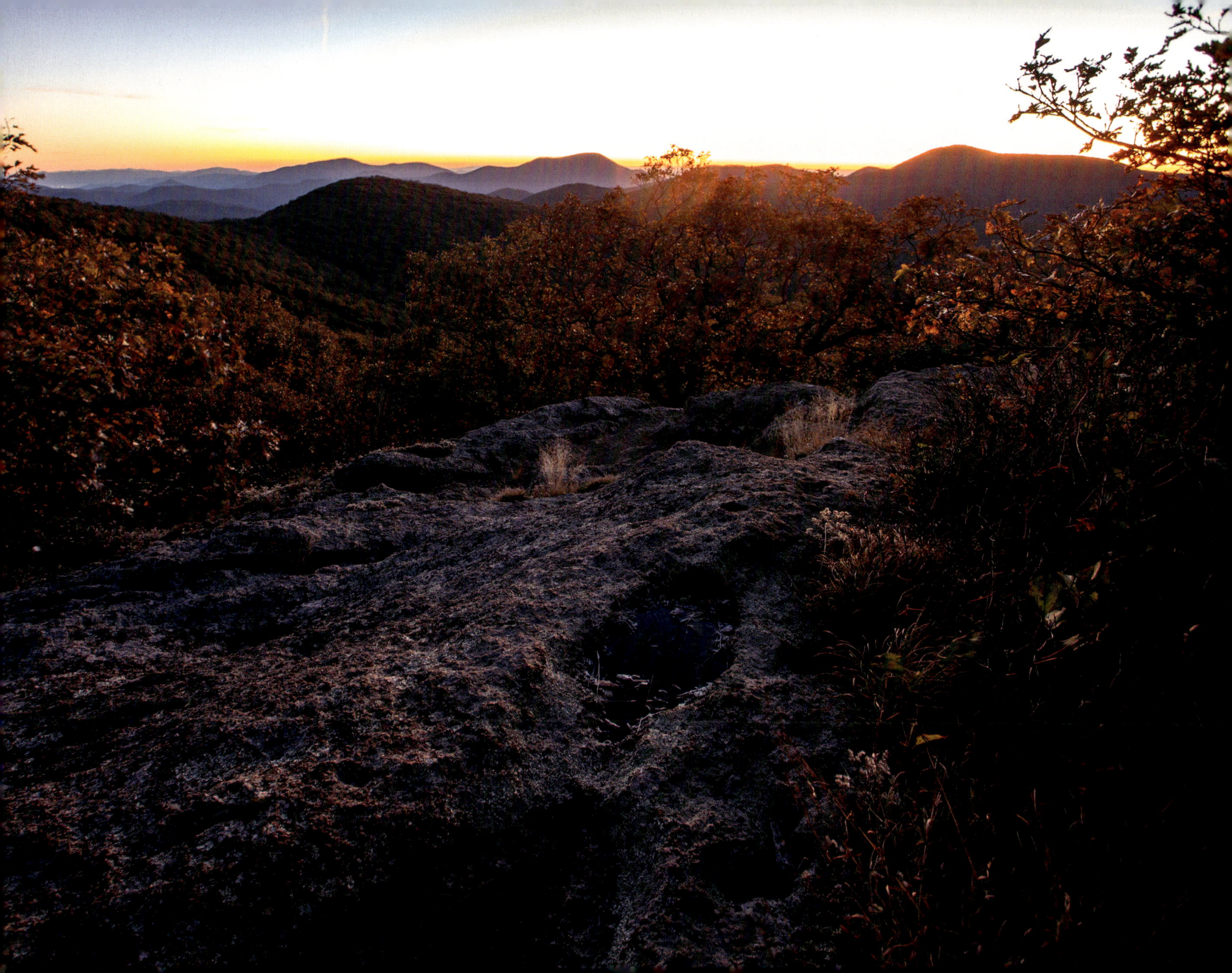

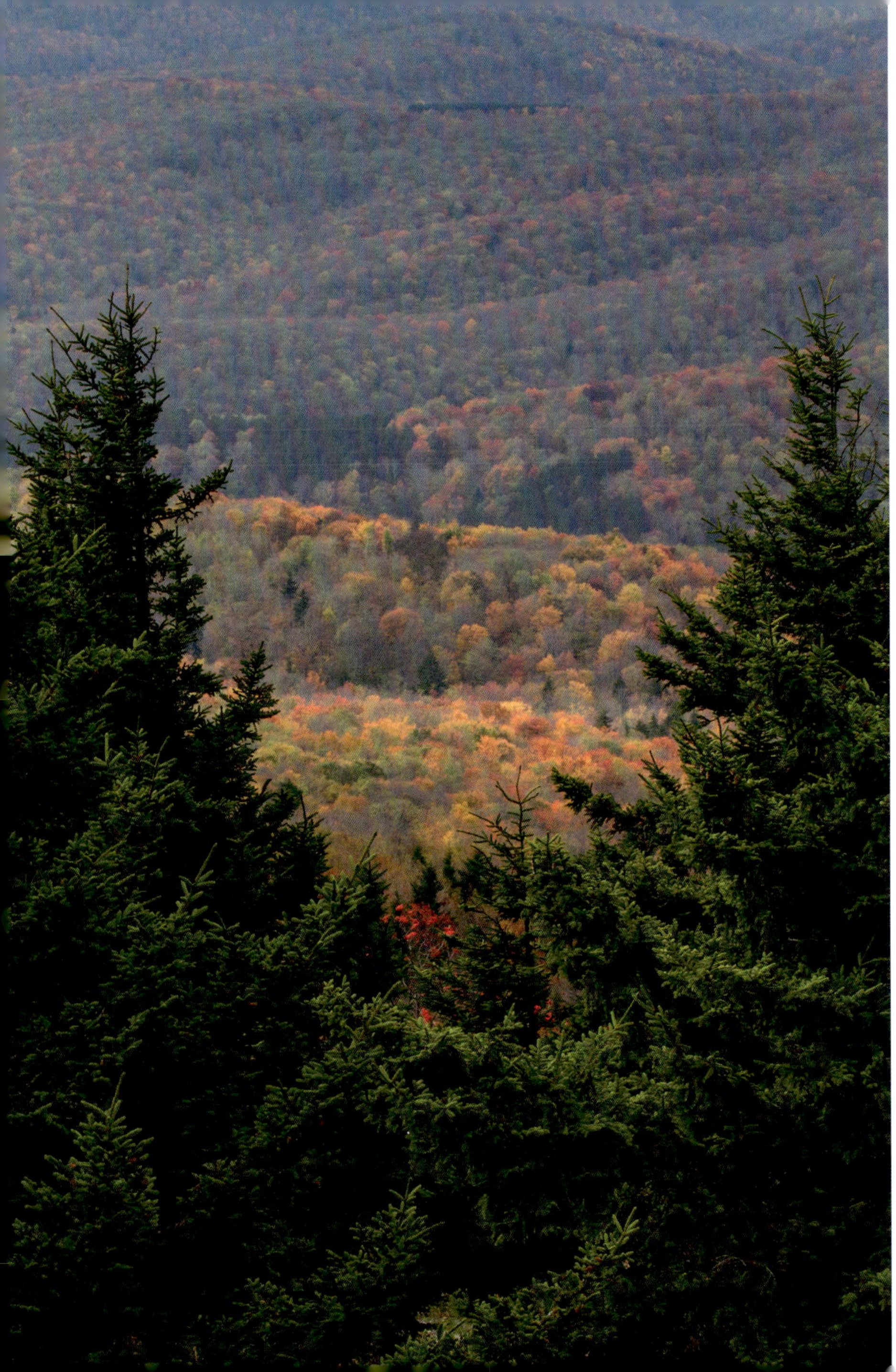

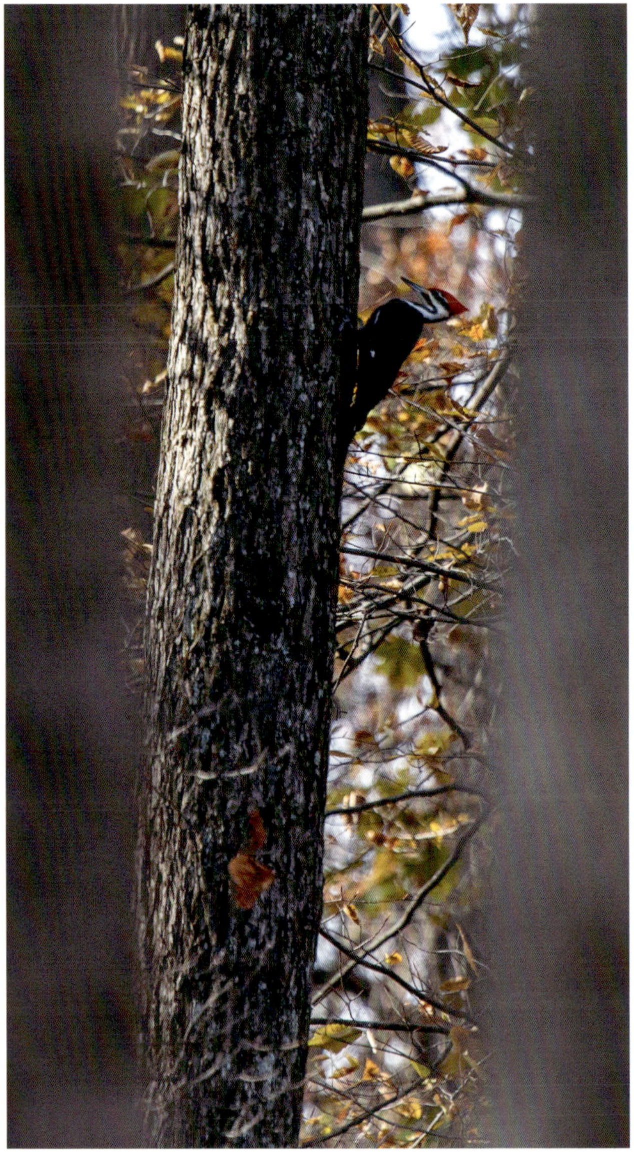

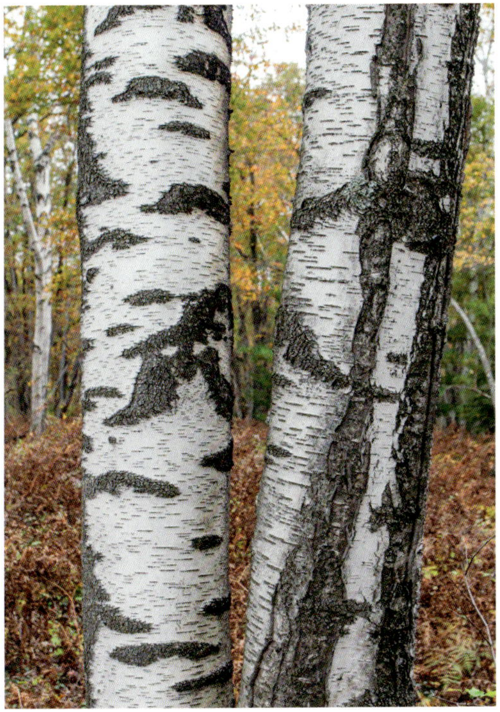

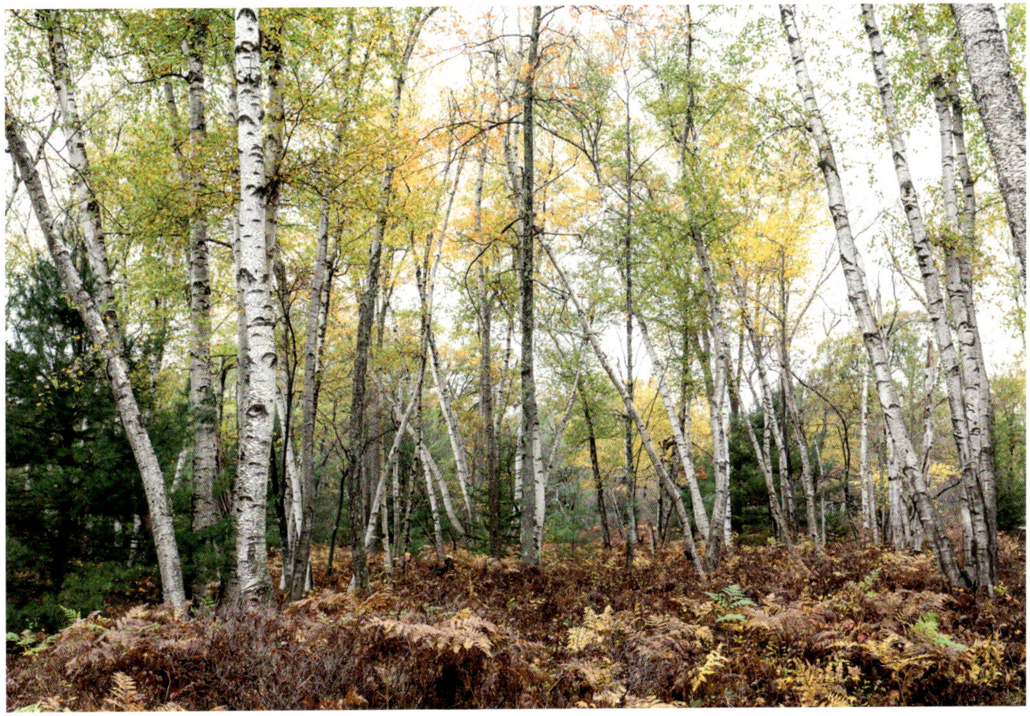

Colorful mountains below framed within the red spruces atop the summit of Spruce Knob

A pileated woodpecker. Cunningham State Park, Maryland.

Like a scene from upper New England, an expansive grove of white paper birches exist within the Marion Brooks Natural Area of the Moshannon State Forest in Elk County, Pennsylvania.

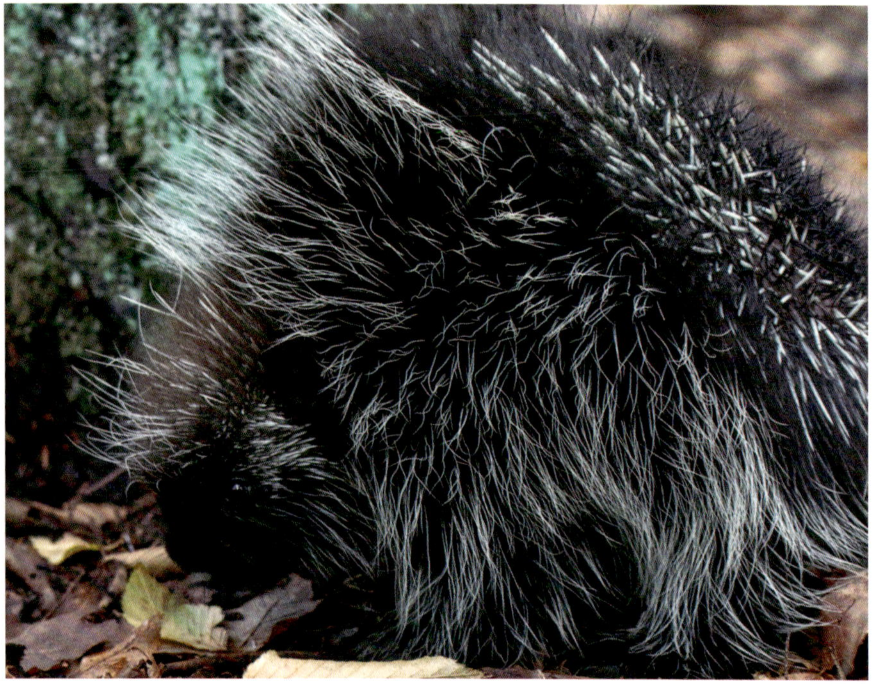

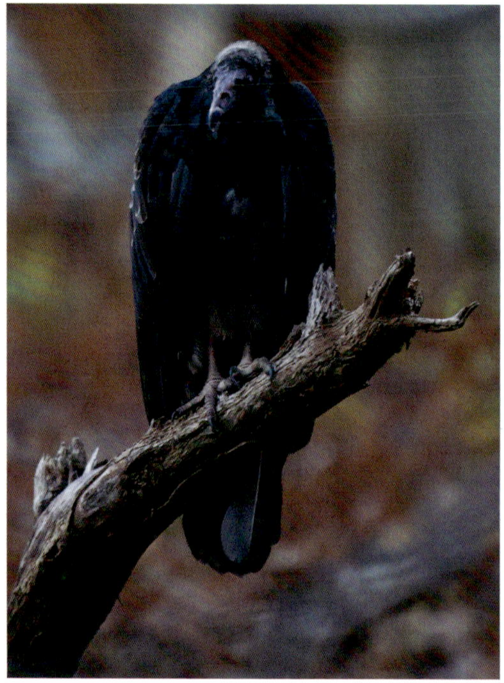

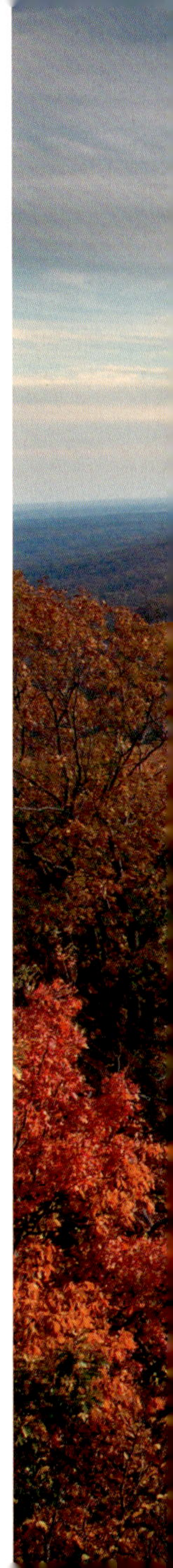

Many are surprised to learn that porcupines are found in the Central Appalachians. Populations in Maryland and Pennsylvania have grown, and some have been spotted in West Virginia and Virginia. Elk State Forest, Pennsylvania.

Like a statue, a turkey vulture perches motionless on a fallen branch. Shenandoah National Park.

By late October, the mountain views afforded by Chimney Top in Catoctin Mountain Park are breathtaking.

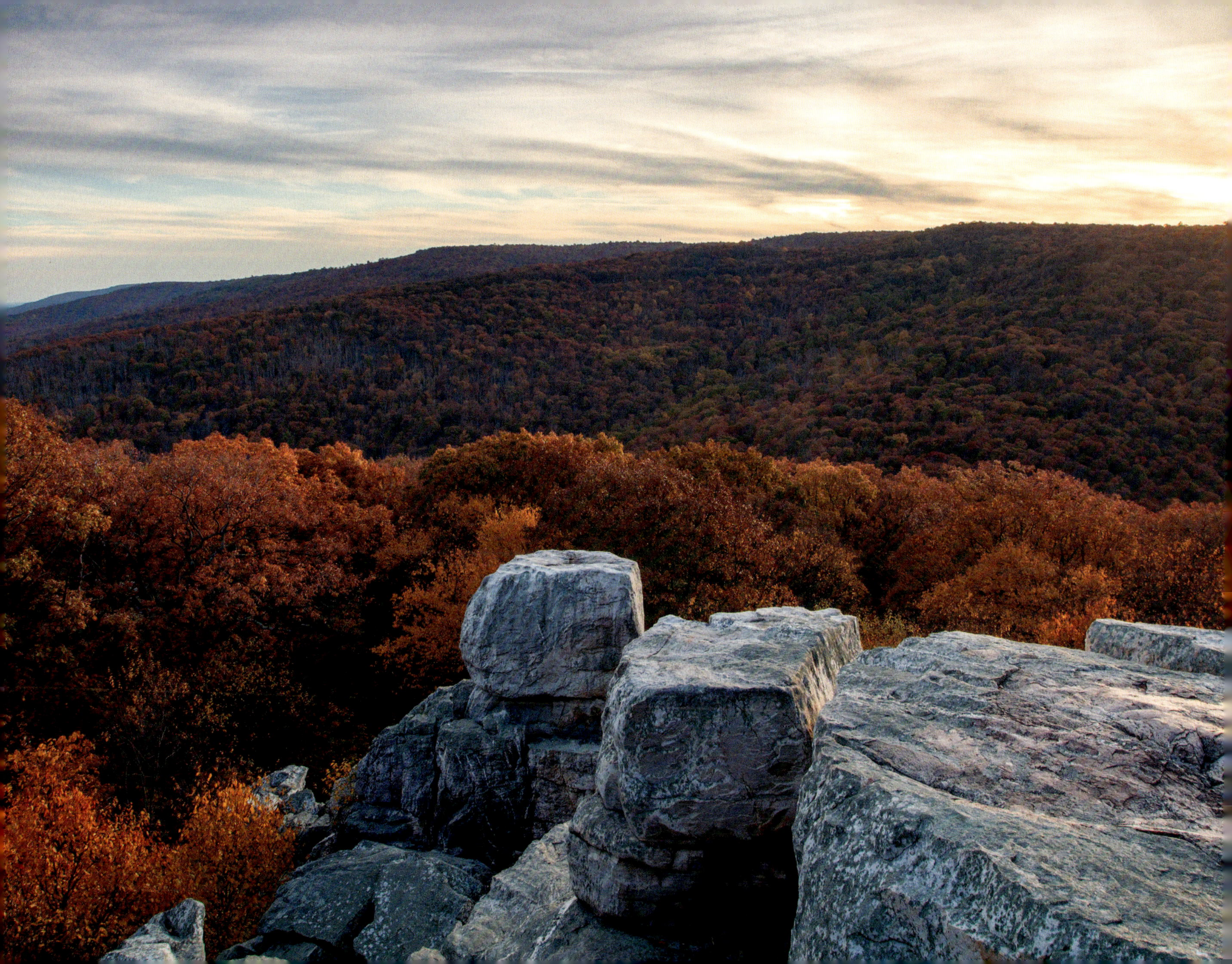

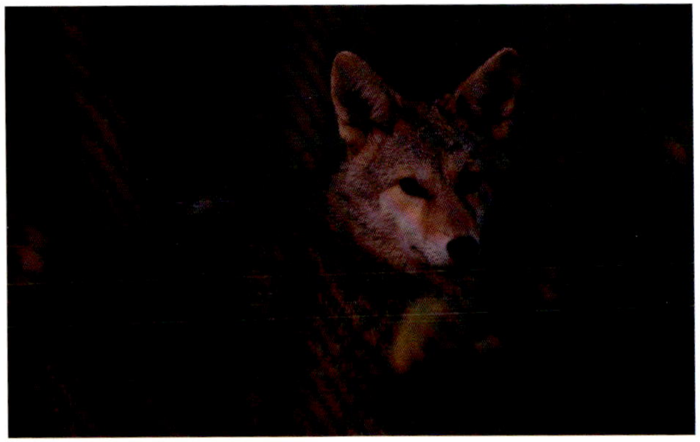

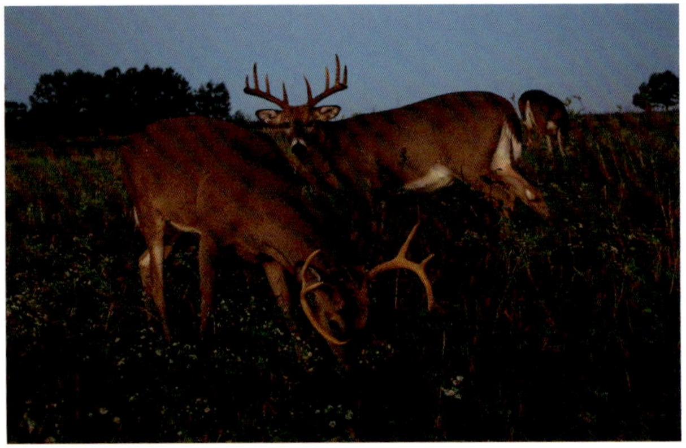

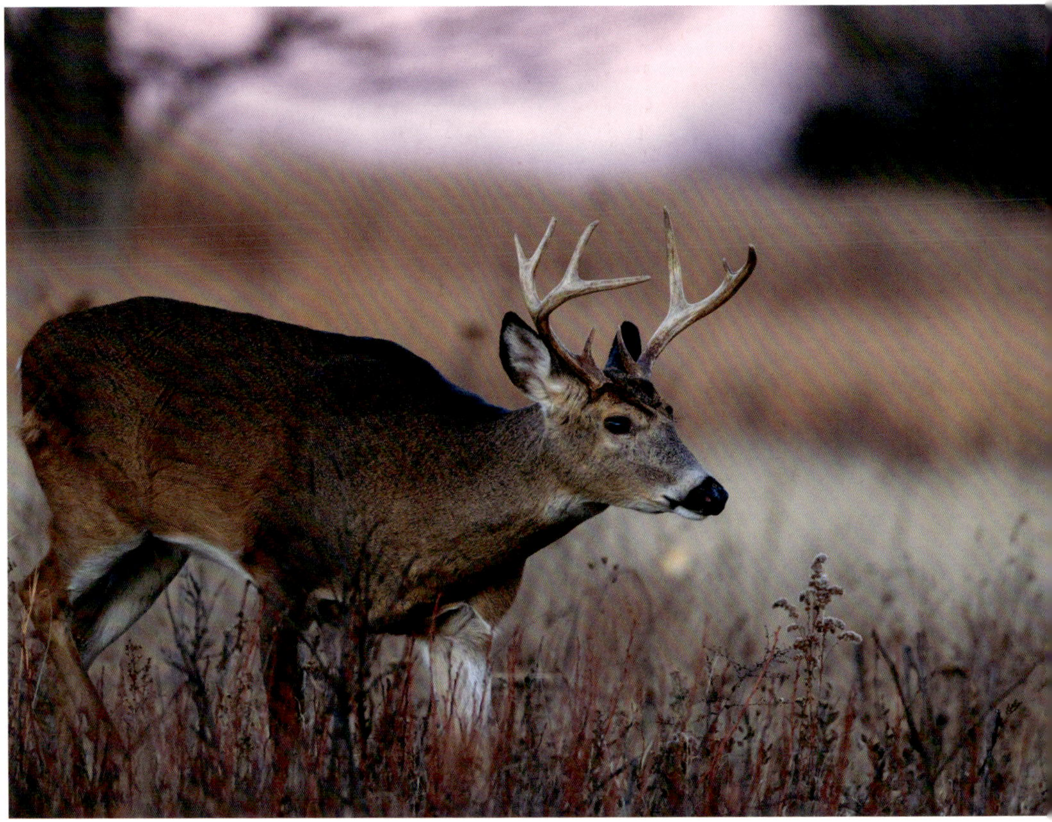

Eastern coyotes have expanded their range throughout the Central Appalachians. It is not uncommon to hear their eerie calls at night throughout the mountains. Rockingham County, Virginia.

At the beginning of the season, bachelor groups of white-tailed deer still travel together, but by November, when the rut is at peak activity, they are competitors.

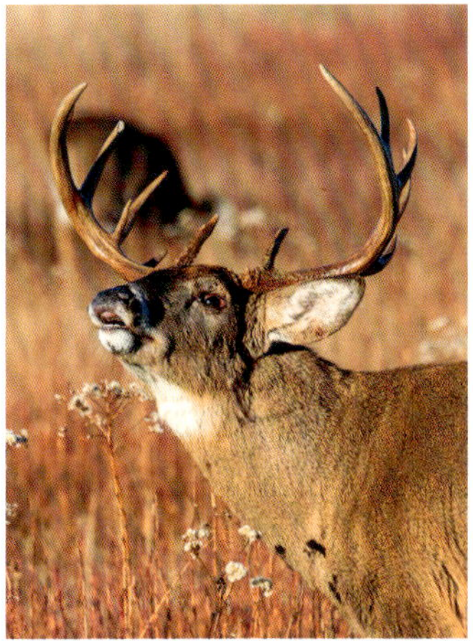

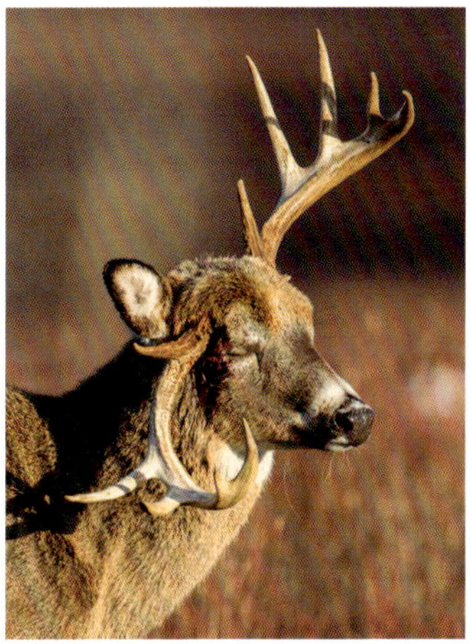

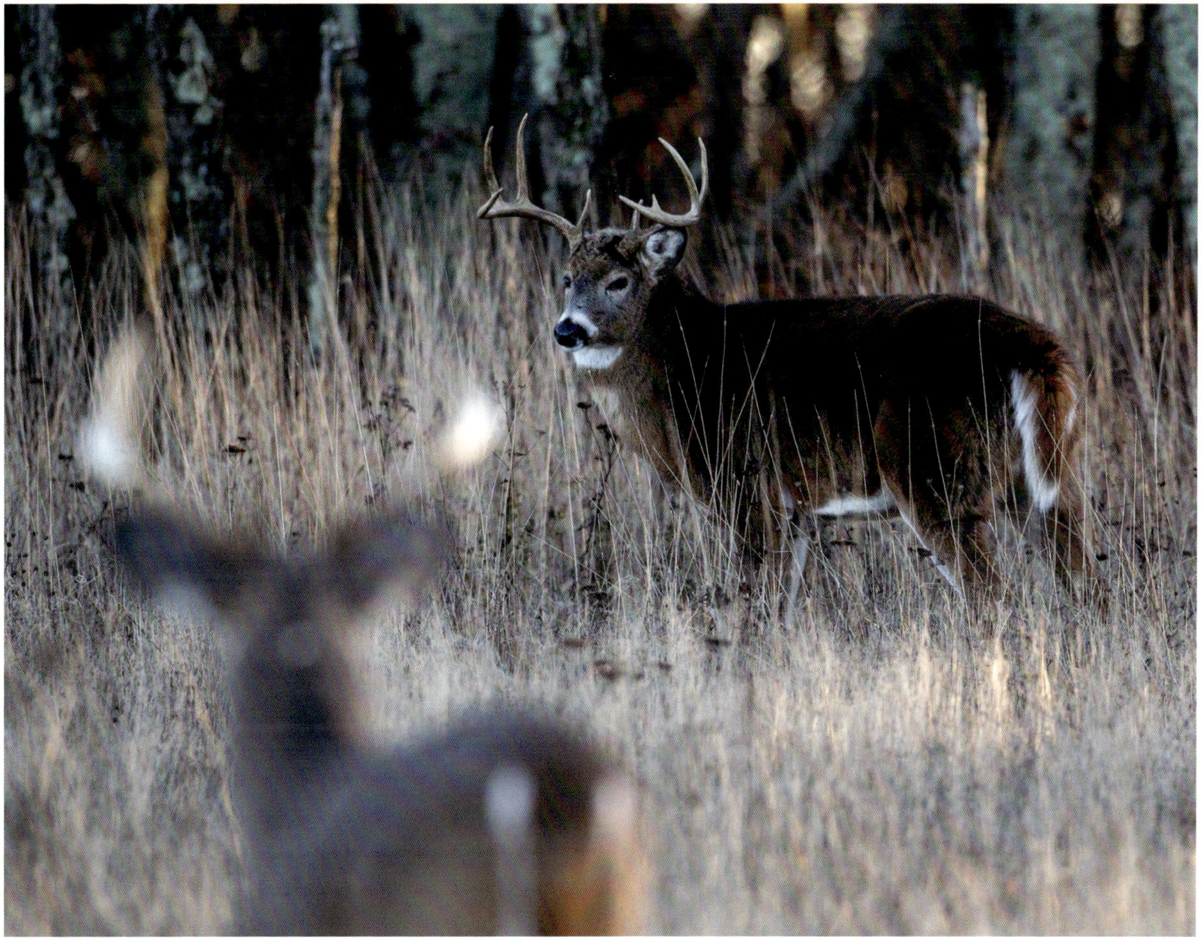

A large buck lip-curling. With an ability known as the Flehmen response, the buck is able to detect the scent of doe urine and determine if the doe is in estrus.

Battles among dominant males are uncommon, since the weaker animal will flee before combat can commence. But on rare occasions, these encounters can be incredibly violent.

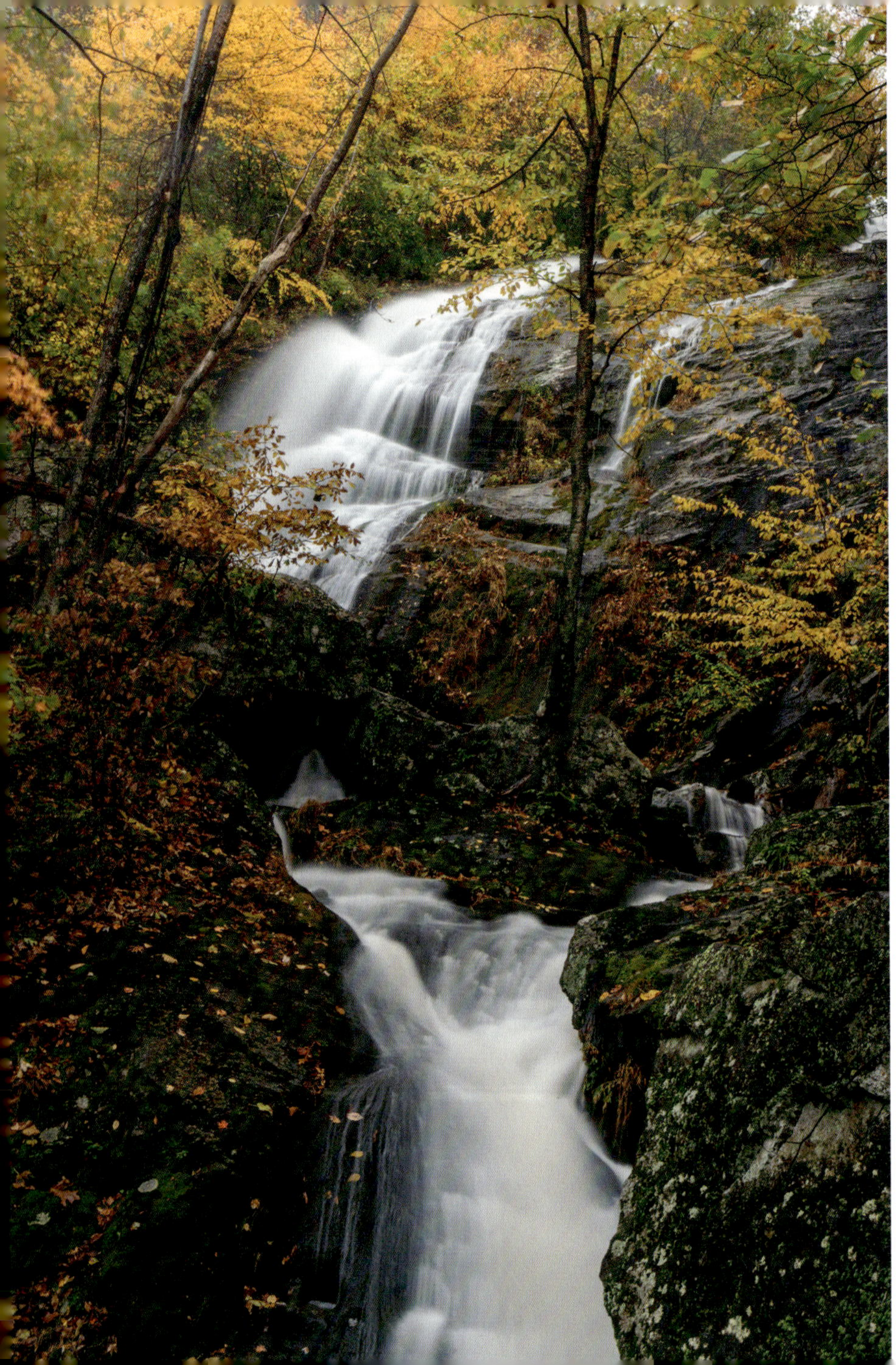

Crabtree Falls is Virginia's largest waterfall and features the highest vertical drop of any waterfall east of the Mississippi River. George Washington National Forest, Virginia.

Star trails above the forest on a clear autumn night. Shenandoah National Park, Virginia.

Autumn fog and leaves. Shenandoah National Park and the Blue Ridge Parkway.

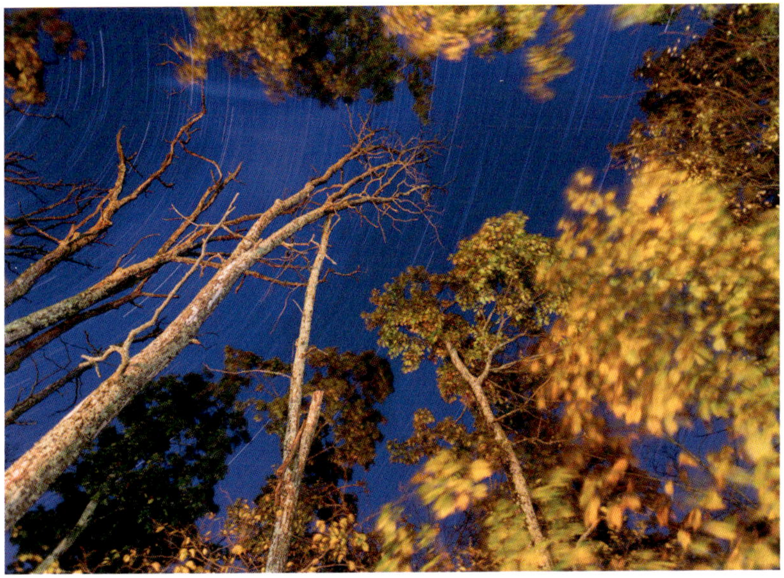

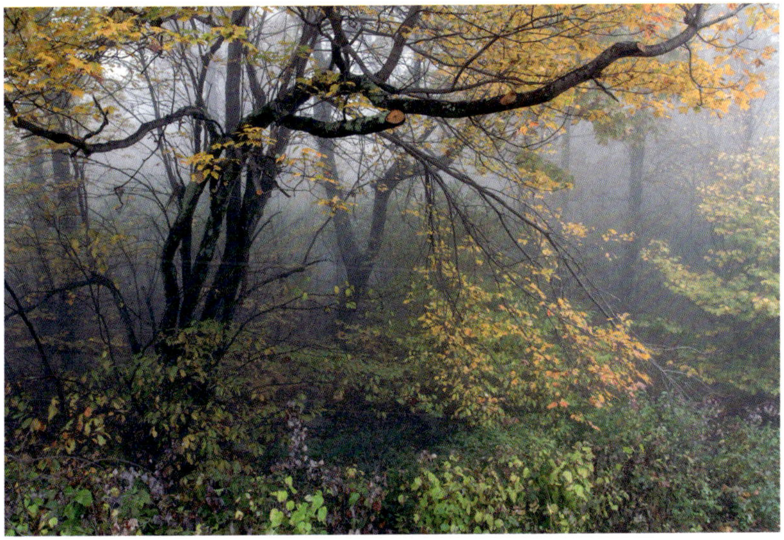

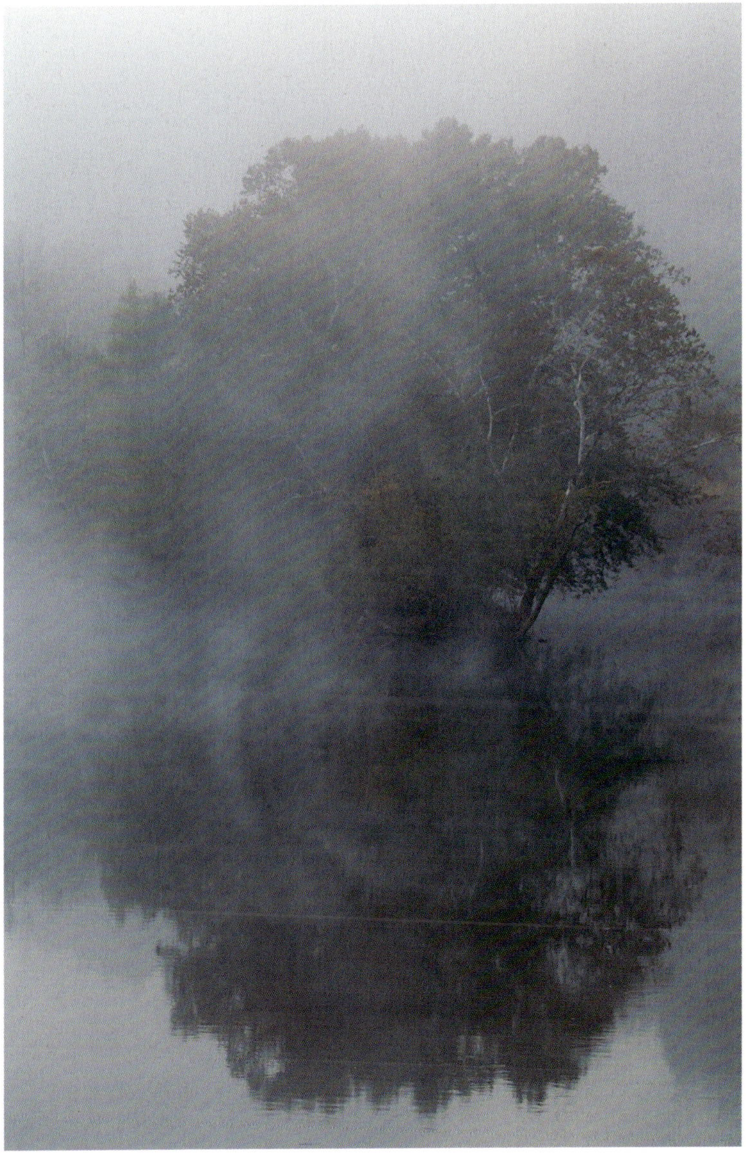

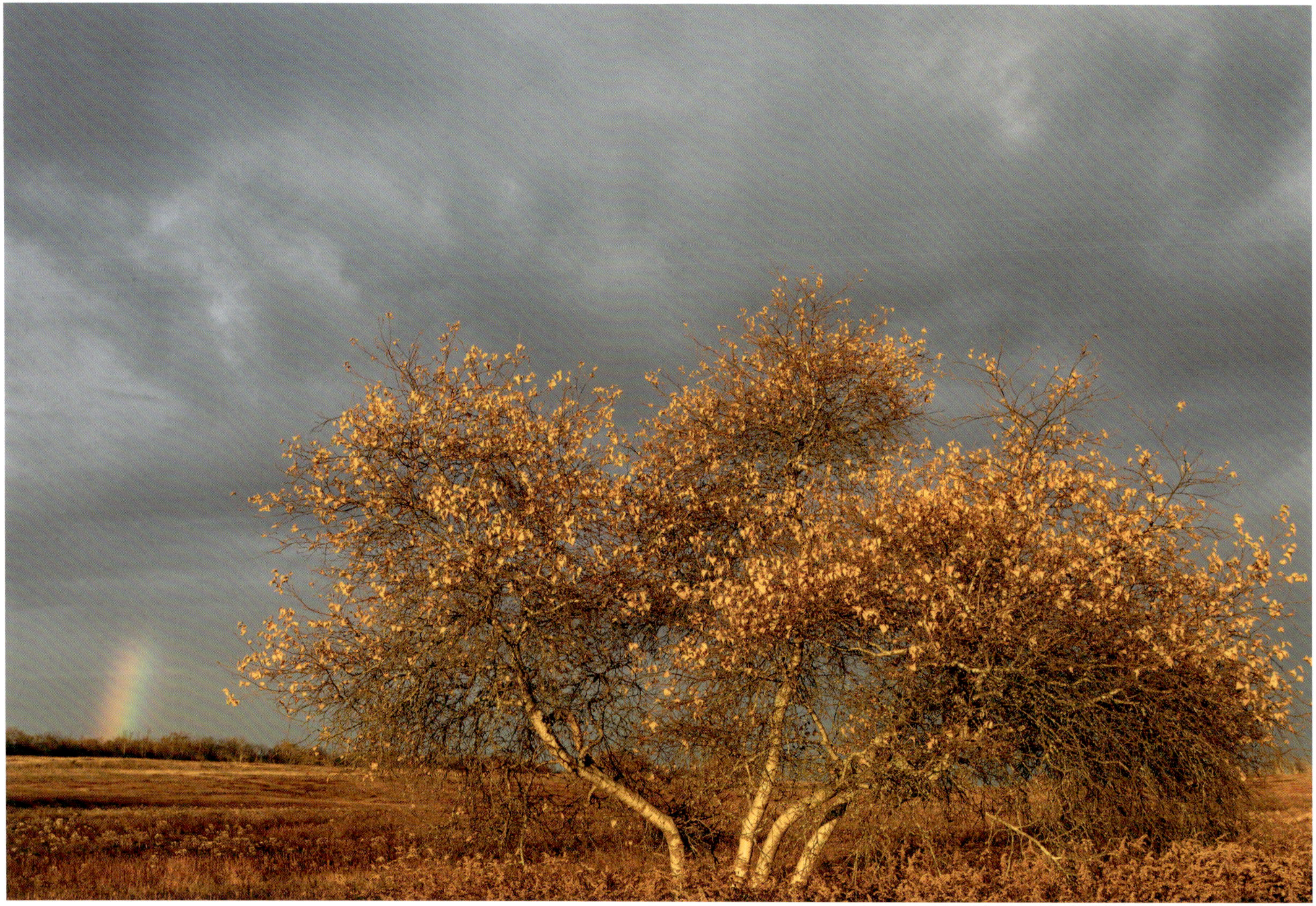

Following a storm, a rainbow breaks through the clouds near the birch.

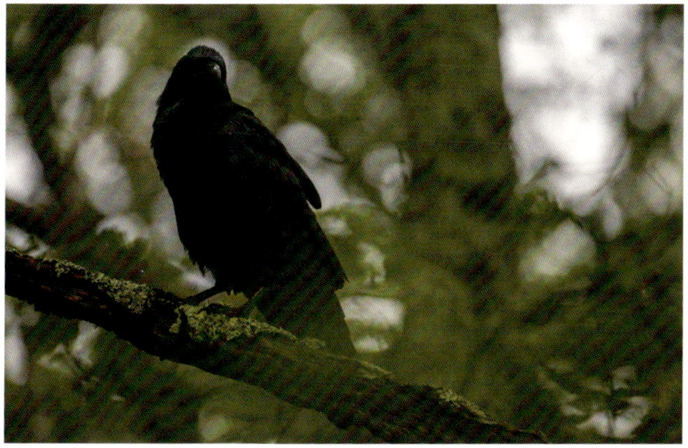

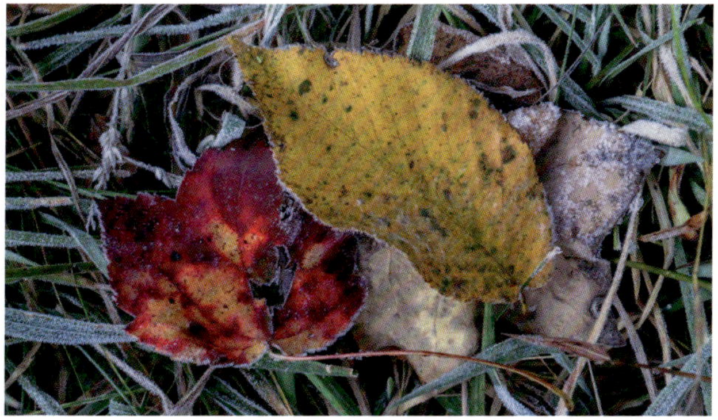

The call of the raven is a true symbol of the mountain wilderness. Savage River State Forest, Maryland.

Frost overtakes the autumn color as the transition to winter will soon be complete. George Washington National Forest, Virginia.

Autumnal twilight in the Blue Ridge

Restoring Larch at the Finzel Swamp

"You can't see anything, but watch your step."

On a chilly October afternoon, I am standing waist-deep in black water at the Finzel Swamp Preserve with conservation ecologist extraordinaire Deborah Landau, PhD. Entering the opaque water is like stepping into a large pool of black tea due to the high concentration of tannins in the fen. While there is nothing to see below, above the surface I am surrounded by a unique beauty unlike anywhere else in the Central Appalachians. As Deborah's arms disappear into the chilly darkness in search of the long black tube of a beaver baffle, I am mesmerized at just how foreign this place appears. It's almost as if I meandered into a different time period.

And with a little imagination, that is exactly what I did. Among the speckled alder thickets that surround us, I imagine a woolly mammoth emerging from the bordering forest. This wonderland, owned and operated by the Nature Conservancy (TNC), is the method one can use to visit the Pleistocene era. No time machine necessary.

The preserve protects a rare boreal fen that straddles the borders of Garrett and Allegany Counties in Maryland. While similar to a bog, which Finzel is often mistakenly called, it is a fen because it is less acidic and also fed by a steady source of groundwater. Finzel's source is the headwaters of the Savage River, which in turn feeds the Potomac River before eventually reaching the Chesapeake Bay.

Things look different here because, well, they *are* different. We are in the middle of a "frost pocket," where cold air is trapped by the surrounding hills, which in turn keeps species more at home in the northern latitudes comfortable. The resulting combination of temperature and altitude created a spectacular mountain peatland for a number of locally rare species. Most notable is larch, or tamarack, the deciduous conifer of the Far North. Here at Finzel, one of the southernmost disjunct populations of the tree still stands, far outside its present-day range. And it is because of larch that we are wading through the murky waters.

"Finzel was one of the first sites I worked on when I started at TNC because it was a

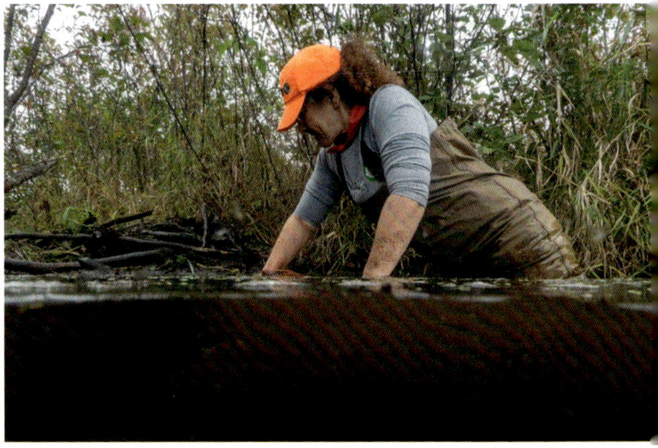

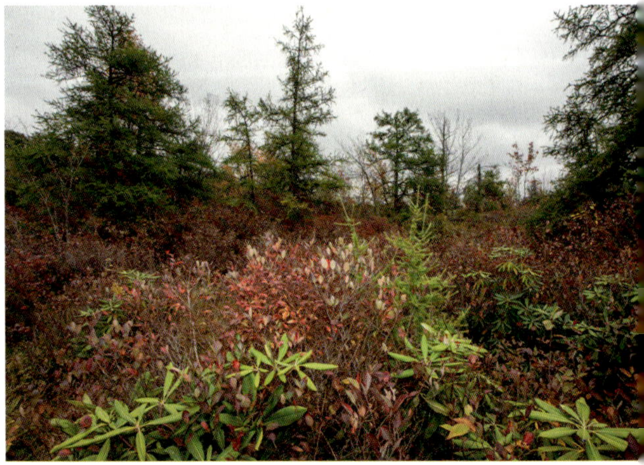

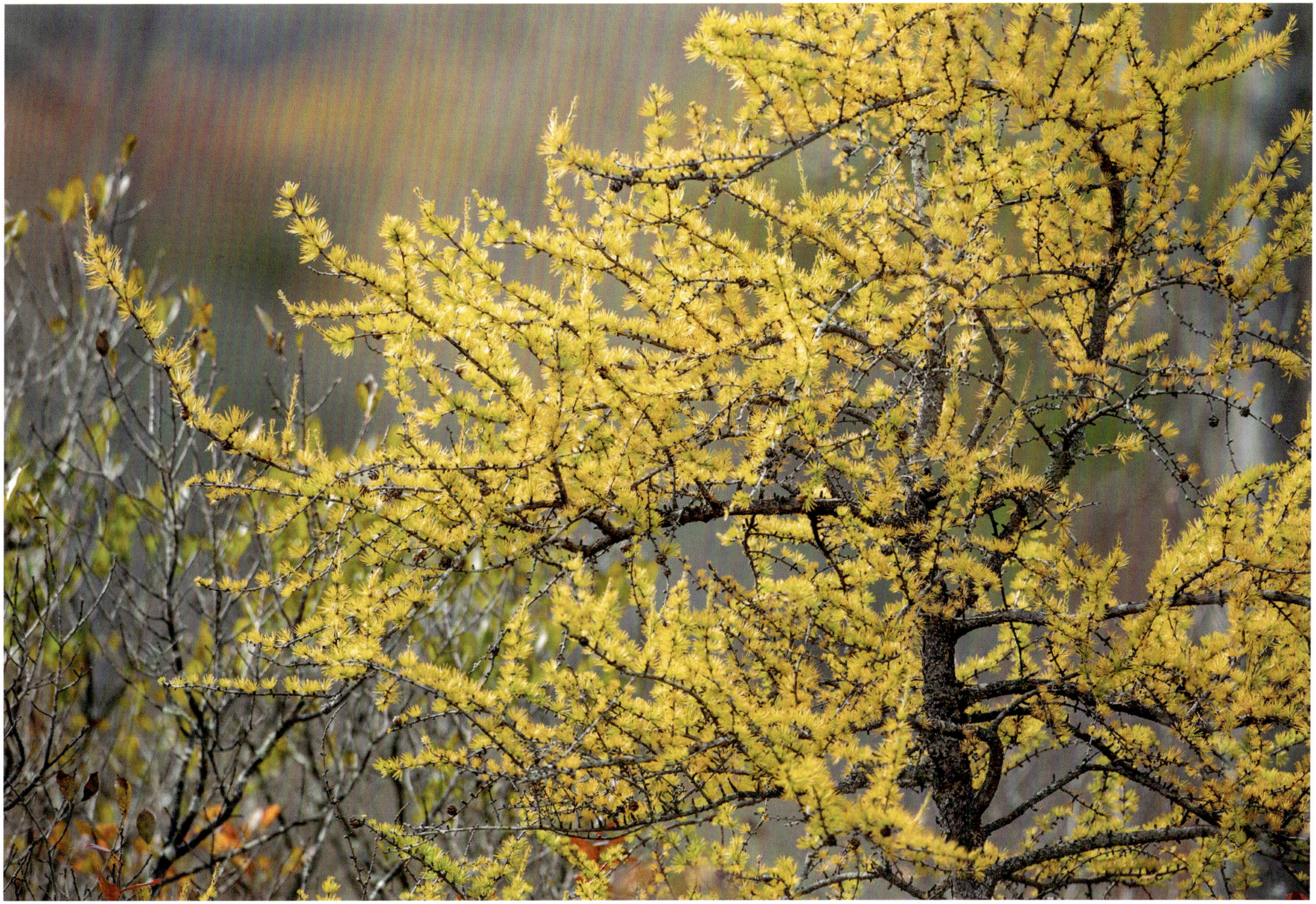

high-priority site for restoration, but the larch were slowly dying," says Landau. The natural hydrology of the preserve had changed, and it became clear that standing water was killing the larch, but repeated attempts at clearing the culverts under the road to remove beaver dams had been unsuccessful. Four other sites in western Maryland had lost larch due to changes in hydrology and climate, and it appeared Finzel was close to the same fate.

To reestablish water flow through the preserve, Deborah had successfully removed three sections of road and replaced it with bridge spans. However, after exploring the swamp, she found no evidence of regeneration. The shrubs were just too thick, even in areas that did not have standing water. Three openings were cleared under mature larch trees in the heart of winter, when the swamp is frozen over but can be very inhospitable.

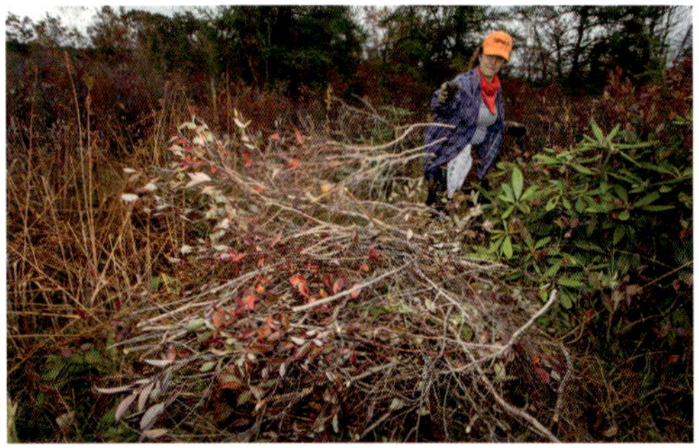

Her efforts paid off. Landau adds, "We started seeing regeneration of the larch about two years after we created the openings. Very exciting!" And, might I add, young larch trees are simply adorable looking.

Earlier in the morning, Deborah had led a team through a labyrinth of thick shrubs and deep swamp in an effort to maintain the clearings. Standing below towering old larch trees and surrounded by scarlet-hued blueberry bushes and golden bog fern, I forget I am still in Maryland. This primeval landscape is more to be expected in northern Maine or Canada. Once the team commences, the dense shrubs soon turn to large piles of trimmings. This must be done in order to give larch seeds space to germinate. Only where the clearings have been made has new larch grown.

In addition, Deborah regularly visits the swamp to do maintenance on beaver baffles, which help keep the water flowing in the swamp and are the reason we traversed through Finzel's caliginous water. Otherwise, the beaver dams cause the water to back up and flood the larch. If a dam is built under the bridge span, it must be removed. On this day, she had to do both. Beavers are an integral part of the region's natural history, but without these measures the larch would continue to decline.

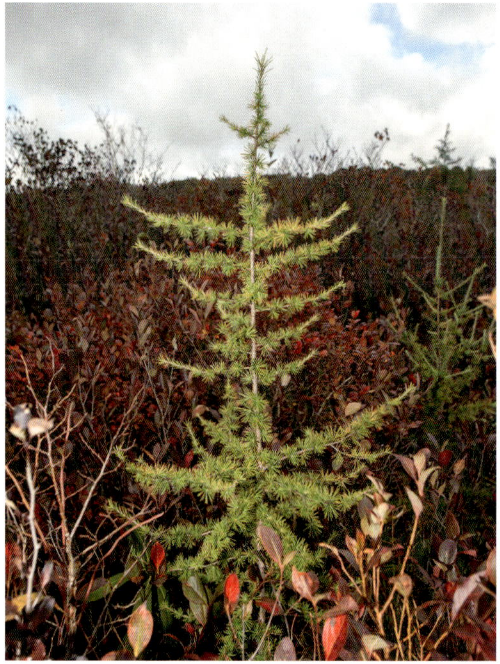

I returned with Deborah eight months later to check on the clearings. She scans the larch that surround us. The height of these trees barely reaches that of an average man; these are the "teenagers" of the preserve.

Holding on to a branch with a gentle touch like a mother looking over a child, Deborah meticulously checks the trees to ensure they are in good health. We even found a very young larch (barely a toddler) that I had somehow missed on our previous outing.

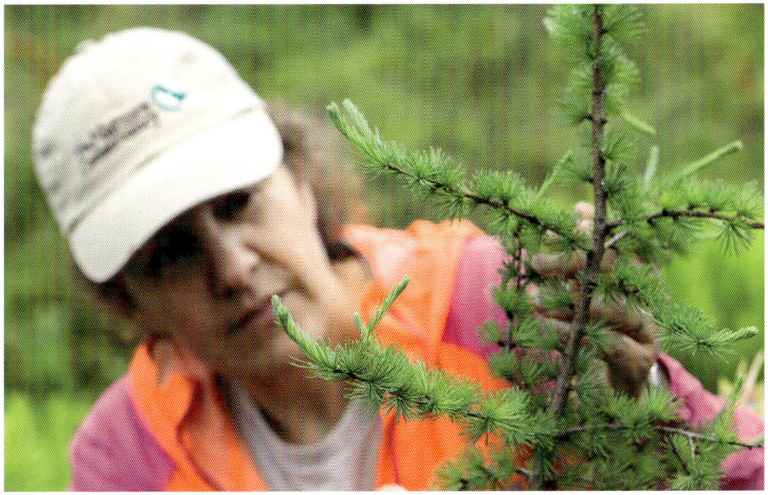

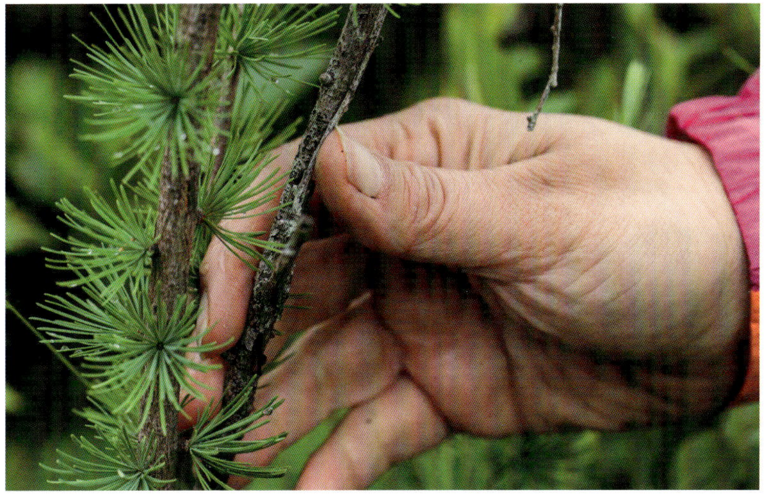

"Seeing the larch return to the openings we've made has made all the work we've done there worth it. Knowing that the trees are viable, and all they need is a little help from us to clear the way, gives me a deep sense of satisfaction that I'm one of their caretakers, and that we're doing the right thing," she adds.

This preserve is one of my favorite locations in the Central Appalachians. Subtle beauty imbues the landscape here, the sum of all the parts being extraordinary. When I was a child, I saw photographs of golden larch in faraway northern bogs. At the time, they conveyed a mysterious and wild landscape; yet here they are, native to my home state. To walk among these magnificent trees and experience the cyclical changes of the seasons is a unique experience.

For Deborah, as caretaker of this wondrous place, the connection to Finzel is intimate. After she won her battle with cancer, Finzel was one of the first refuges she returned to. It was during the winter, and the swamp was frozen over. When she reached the large culvert, the same culvert I had watched her clear of a dam, she saw a mink hopping along the ice. To see a mink is incredibly fortuitous, but the experience does not last long. They're agile and quick; every time I have encountered one, it lasted mere seconds. On this day, though, Deborah was able to watch this mink for a few minutes.

"It made me feel like this was my little 'welcome back' from the swamp, a magical peek just for me."

When she shared this wonderful story, it served for me as a gentle reminder of just how important it is to preserve these special places. In a society with an ever-increasing reliance on technology, it is easy to lose track of what the natural world provides for the human condition. Whether observing the larch's golden needles wafting in the wind each autumn or an exceptional encounter with a mink, it is because of dedicated people such as Deborah Landau that these places continue to exist—which is better for all of us.

WINTER

The first snows of winter will usually arrive in the highest elevations by November, even though the winter solstice is still weeks away.

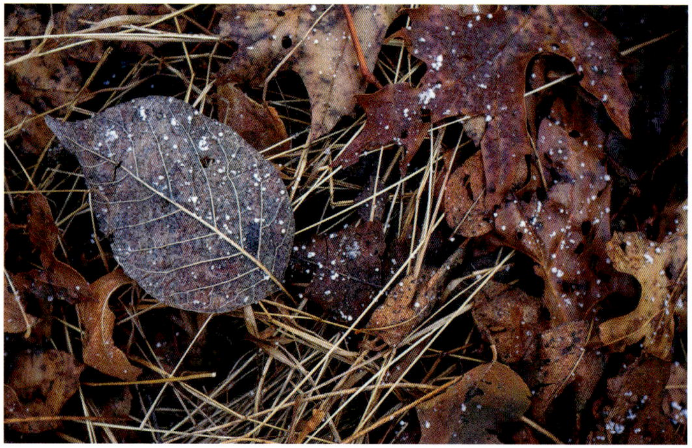

The last of the autumn leaves, a single basswood leaf loses its remaining grip on its branch and succumbs to the first snows of the season. Shenandoah National Park, Virginia.

Like a scene from a Christmas card, snow has transformed the mountains of the Blue Ridge into a winter wonderland. Shenandoah National Park, Virginia.

In the highest elevations, and in northern areas of the plateau, it is common to see the first snow in October.

The forests of the region are quiet and bare, but the cooler temperatures improve visibility. Large icicles have now shielded the exposed rock. Wildlife must constantly forage in the chilly, snow-covered mountains. Hoarfrost, the phenomenon that occurs when the dew drops become frozen on exposed surfaces, adds to the stunning, but harsh, environment.

The tracks of wildlife paint stories into the snow. A fisher follows a porcupine to the entrance of a den. A junco forages near a bare shrub. Fox, bobcat, and deer tread on the same game trail; one atop the other, the tracks become almost indistinguishable.

Bats hibernate in the region's caves and abandoned mines. Inside, a warmer and humid microclimate keeps the temperature comfortable.

In the crisp air and clear sky on a mountain summit, a raven wind surfs along a strong gust above a ridgeline. By late February, a skunk cabbage emerges. A remarkable early wildflower, it blooms when the ground is still frozen, and the buds can reach 70 degrees Fahrenheit, melting the snow around them. Other wildflowers will follow in a few weeks as the first neotropical songbirds return.

Another seasonal cycle is complete.

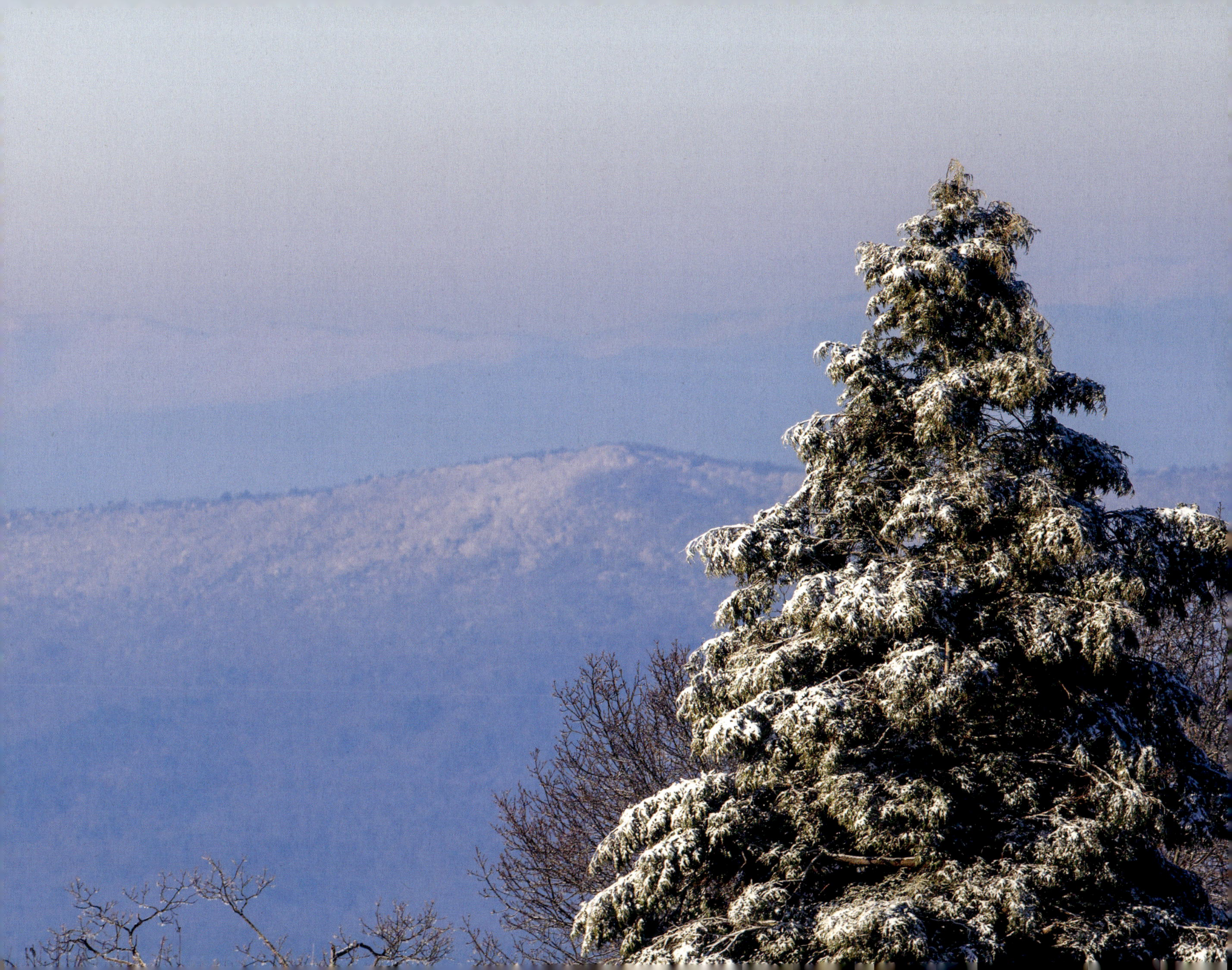

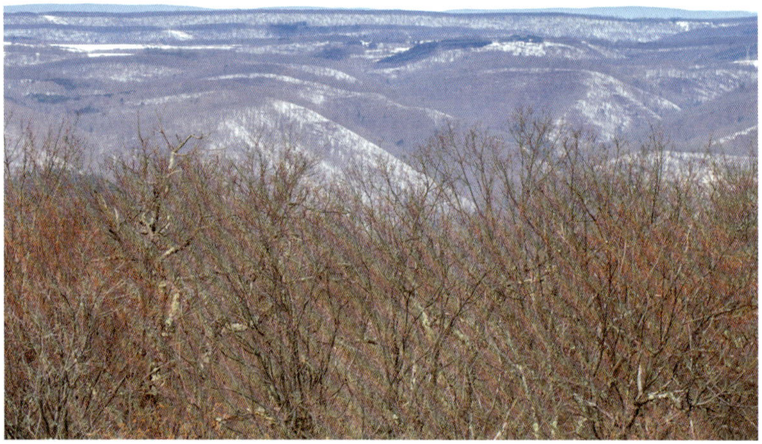

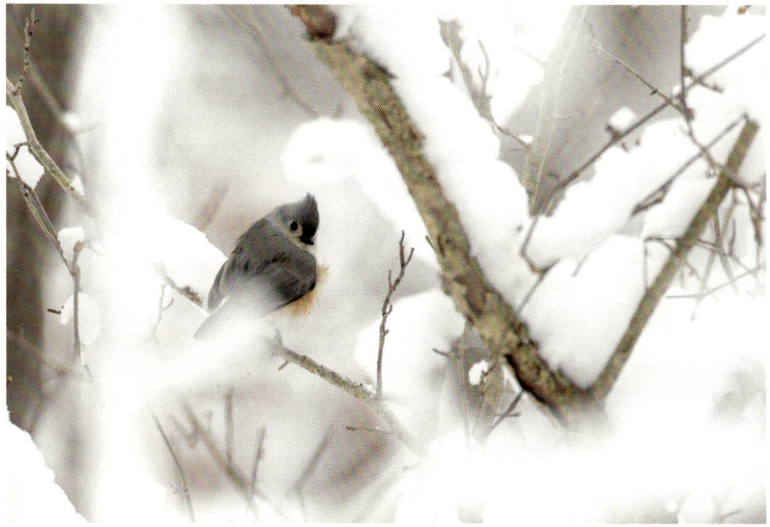

Bare trees and snow-covered mountains high into the plateau. Savage River State Forest, Maryland.

A tufted titmouse does not migrate and is active during the winter, hoarding food to see itself through the winter. Dans Mountain State Park, Maryland.

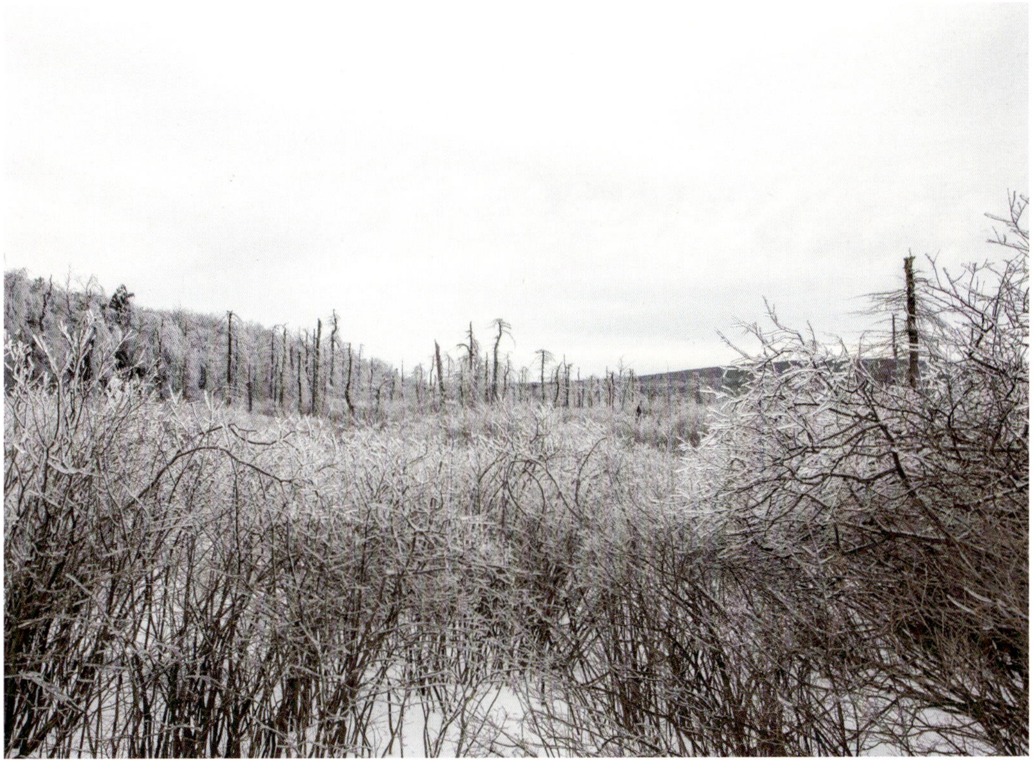

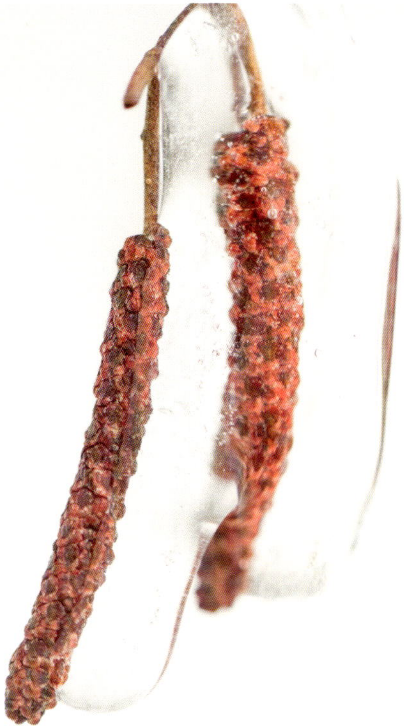

An ice storm comes swiftly and covers the landscape of the Finzel Swamp, adding a quiet and simple beauty to the ornamental red catkins of a speckled alder tree.

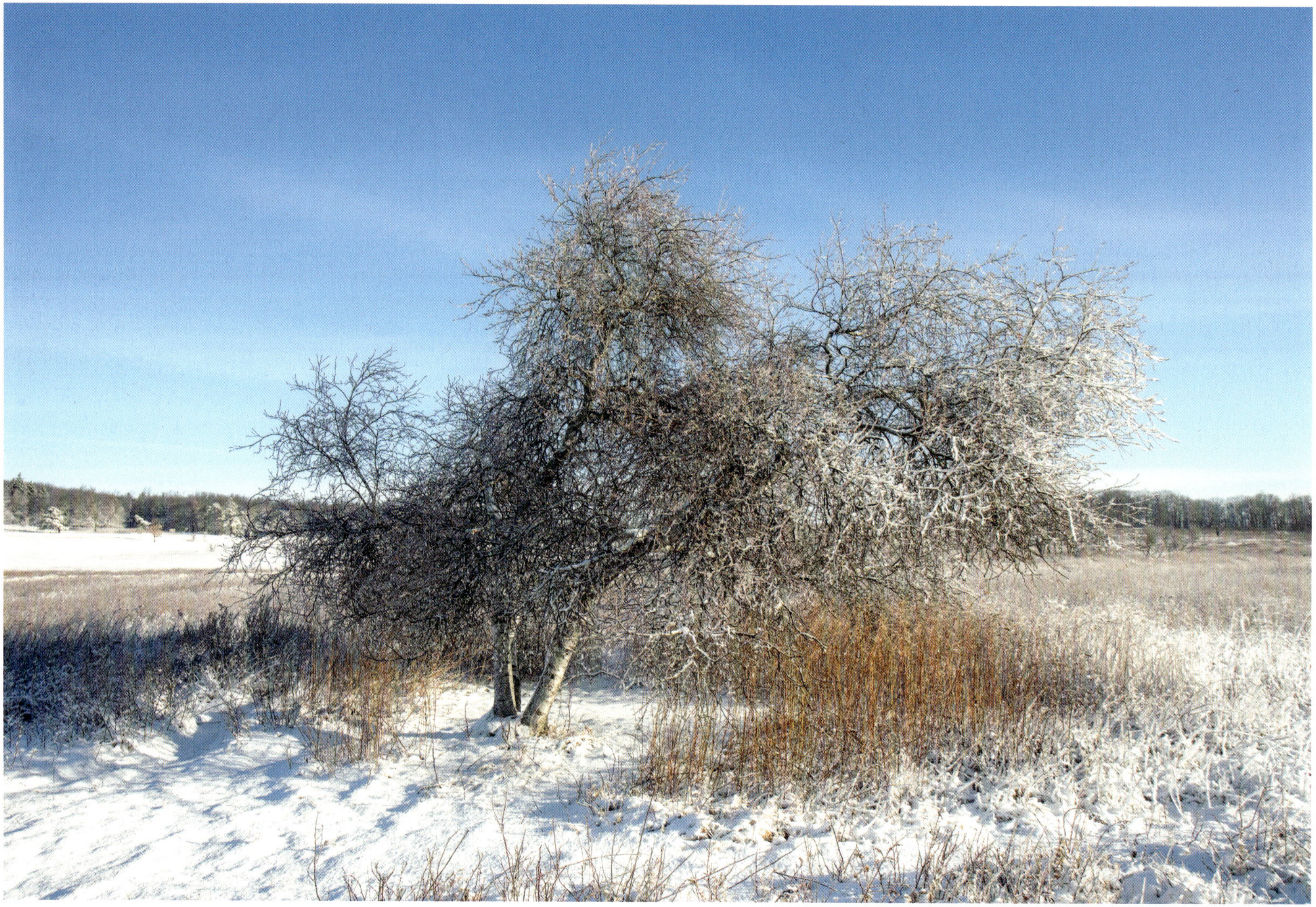

For the gray birch, winter is the time for the female catkins to drop their seeds, providing a food source for songbirds.

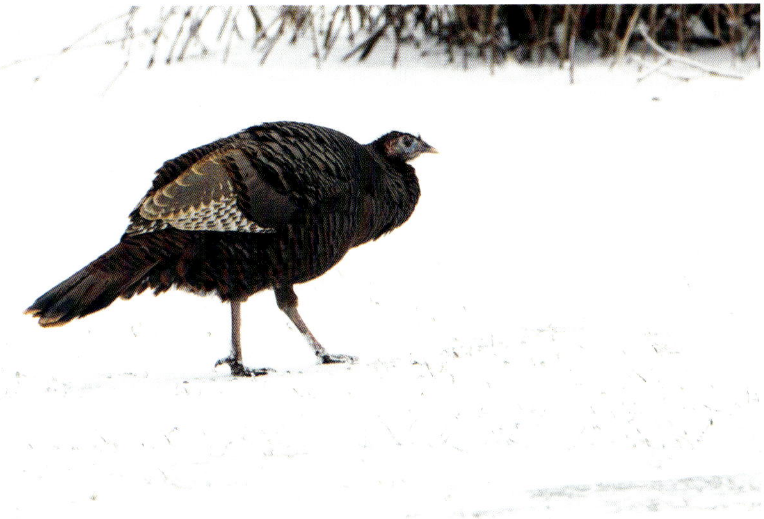

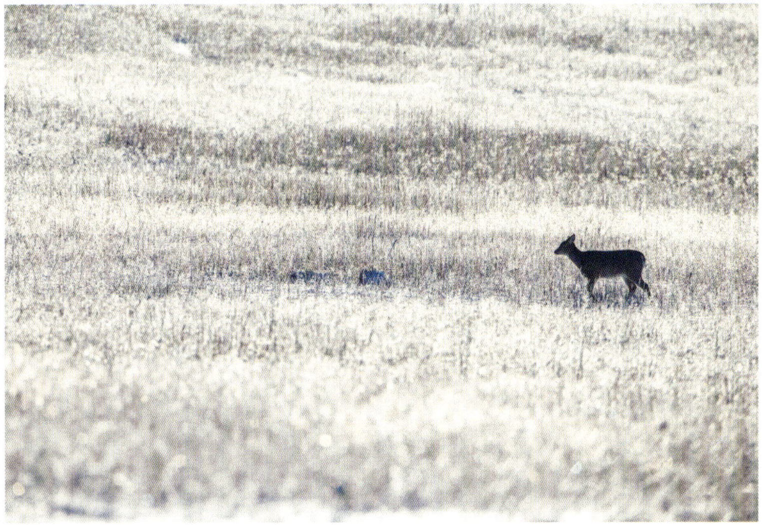

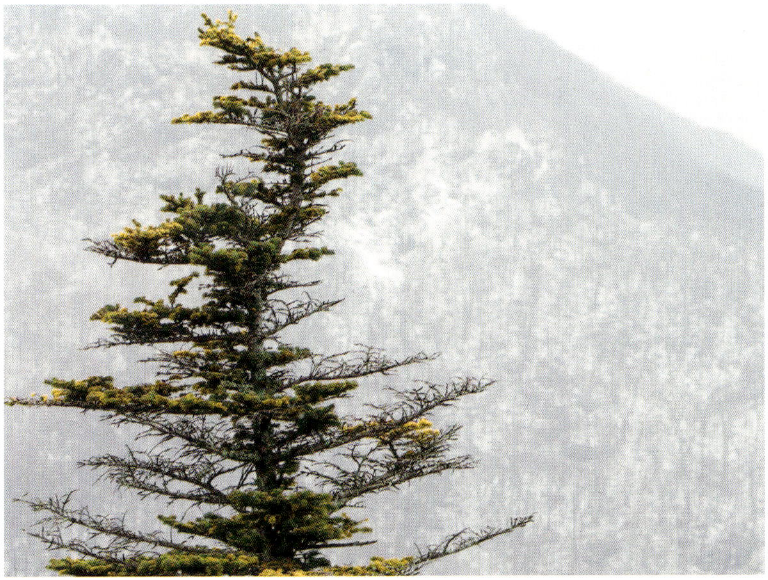

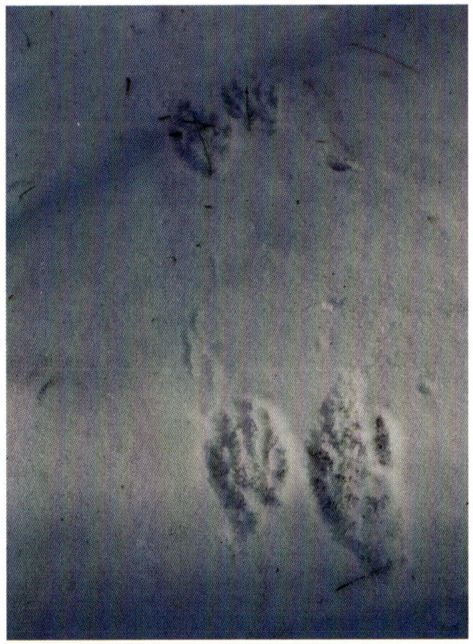

Winter is a harsh time for many species, such as wild turkey and white-tailed deer, having to balance foraging and moving on so as not to waste energy reserves.

Balsam fir in winter

The tracks of a fisher, a wilderness predator that thrives in the winter landscape. Elusive and crepuscular, these creatures leave tracks that offer a glimpse into their mysterious lives.

Appalachian Camera Trapping

Camera trapping is an unobtrusive technique in which a camera is connected to a motion sensor that is activated when an animal crosses its infrared beam. This technique allows the photographer to capture images of nocturnal and elusive animals in the context of their natural environment, which would otherwise be impossible. Popularized by consumer "trail cameras," one may utilize these small and inexpensive cameras to glimpse into the mysterious daily routines of wildlife. Others, like myself, use DSLR cameras to create a custom "trap" setup where you can control the lighting and composition. The sole reason I began this method was to capture an image of a fisher.

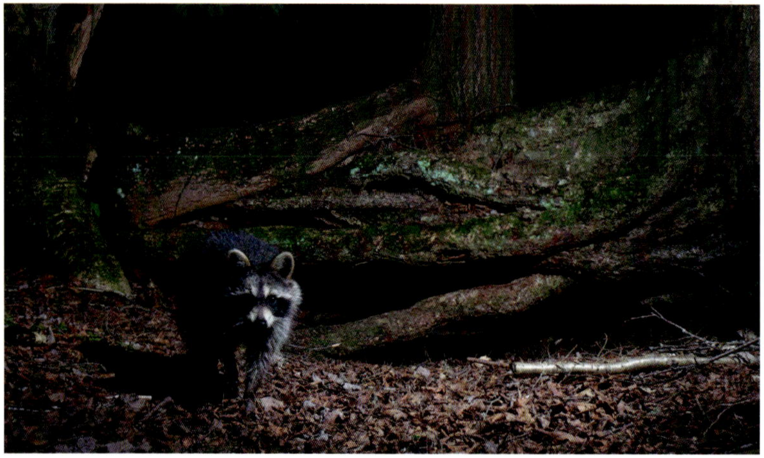

The fisher is a medium-sized weasel, mostly crepuscular, that moves through the forest like a phantom—which is fitting because, in the not-so-distant past, they were nothing more than the spirit of the Central Appalachians. Unregulated trapping and habitat loss were the causes of their extirpation, but in 1969 the species was successfully reintroduced to West Virginia when twenty-three animals were released from New Hampshire. In time, fishers expanded their reclaimed range into western Maryland, northern Virginia, and southwestern Pennsylvania. Additional reintroductions in the Catskill Mountains in New York as well as in northern Pennsylvania resulted in a conservation success story. The absent phantom had returned. An impressive animal, fishers are perhaps the only species that specializes in hunting porcupines; the sharp quills having no ill effect on them.

To capture the fisher, I use cameras inserted into protective housings. This is done to protect the cameras from the elements and in case a bear decides to make one into a chew toy. Connected to each camera is a passive infrared sensor that is set to trigger the camera when motion is detected. I then use two or three strobes for lighting the scene. Setting up the gear is the easy part. Finding where to place it? Not so much. To successfully camera-trap, one must think in equal parts like the animal

Above: A raccoon

Right: A fisher

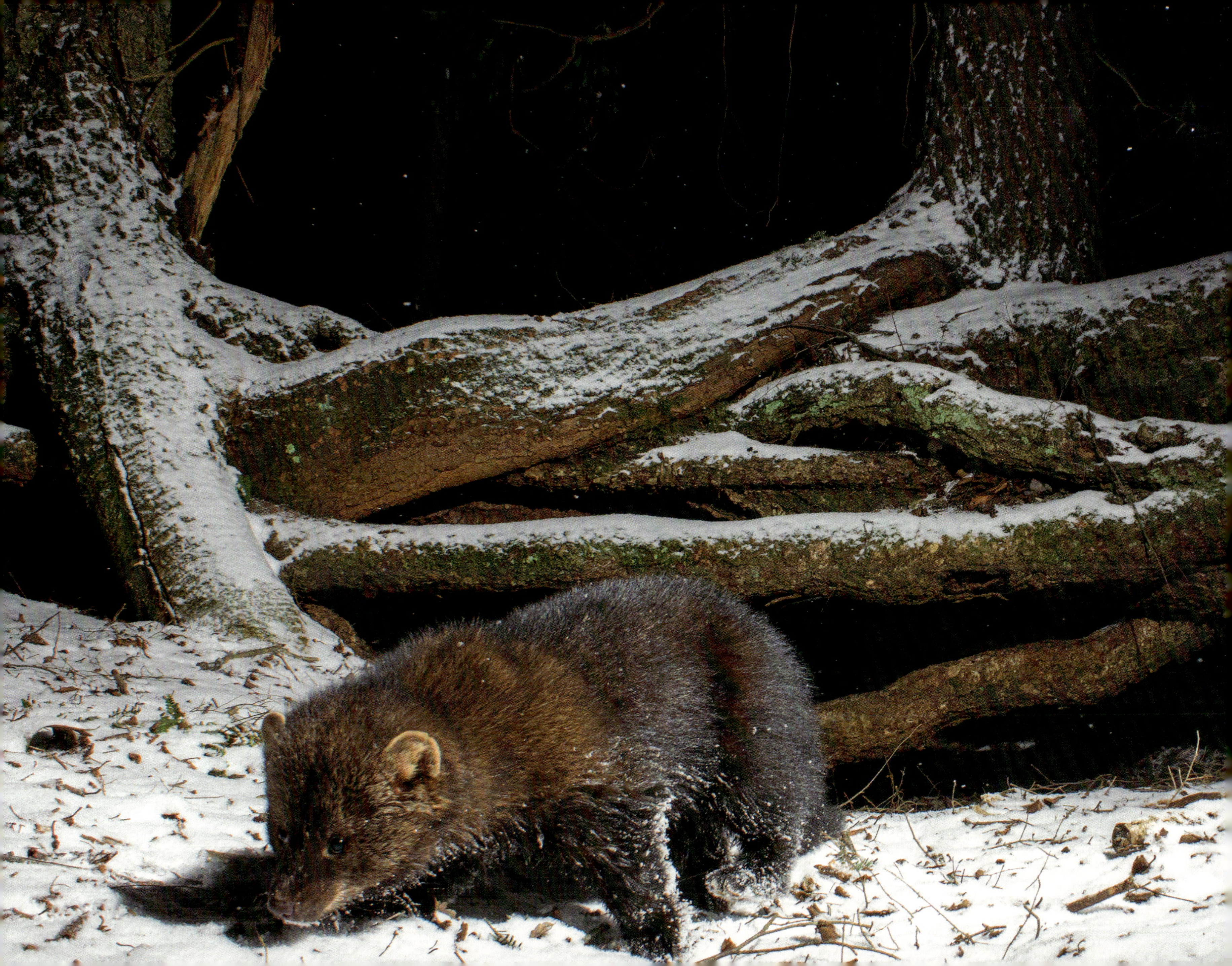

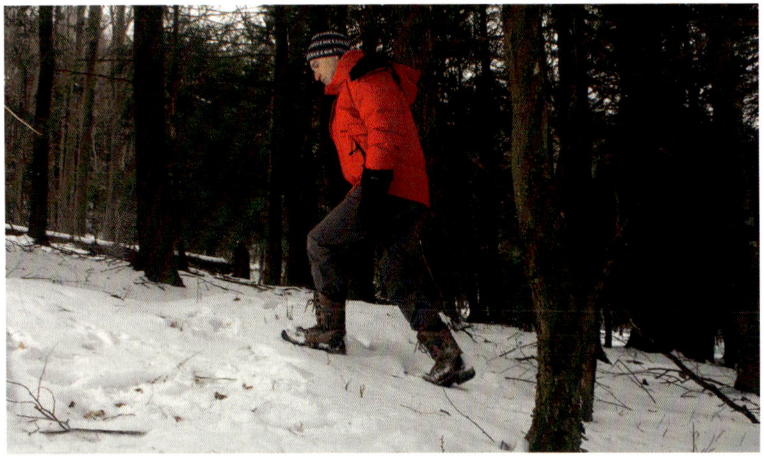 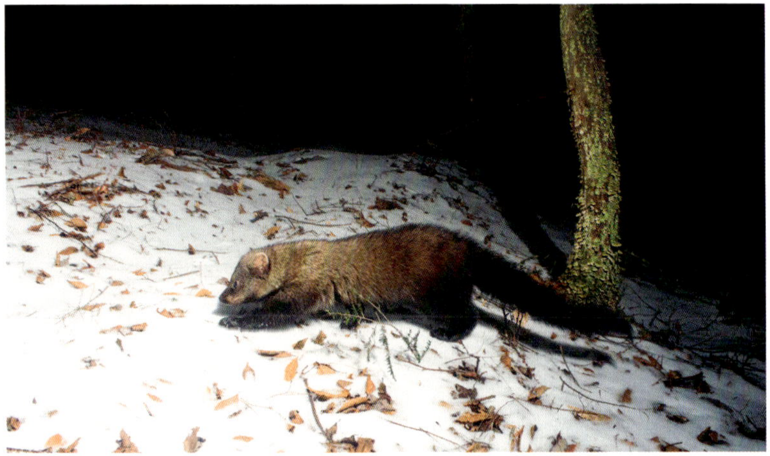

one hopes to capture and like an old mountain trapper of yore. Tracks, game trails, scat, and available food sources provide the clues as to where to place the camera. A paucity of actual wildlife photographs results from this method. My most common subject? Leaves in the wind.

For three winters, I tried to capture images of fishers on my camera traps. With a home range of up to 10 square miles that fishers continually move through, my chance of success was low. When you factor that with inevitable equipment failures, dead batteries, and storms (or playful raccoons) resulting in a knocked-over camera, even lower. But each time I hiked into the wilderness to check a camera, my feeling was akin to that of a child on Christmas morning: I was full of excitement.

My good friend Rich Raesly, professor of biology at Frostburg State University, has become a constant companion in this endeavor. It was with him that I finally captured my only images of fishers. When Rich, a seasoned adventurer, identified fresh fisher tracks, we played the part of wildlife detectives in trying to figure out when and where the fisher would appear next. Evidence of a red squirrel, a common prey source, was also present. As it turned out, Rich is pretty good in this role, because a fisher triggered the motion sensor ten days after we first set the camera.

Deer, bobcats, opossum, raccoon, and fox all have triggered the sensor. In summer, even curious black bears appear. Bite marks and off-center cameras are evidence of their inquisitive nature. One bear was so dangerously curious it destroyed an entire set, stripping my camera housing from a tripod and tossing the flash units out of trees. While the housings were destroyed, miraculously the camera and flashes survived.

At least leaves in the wind never did that.

A fisher and Rich Raesly captured by the camera trap

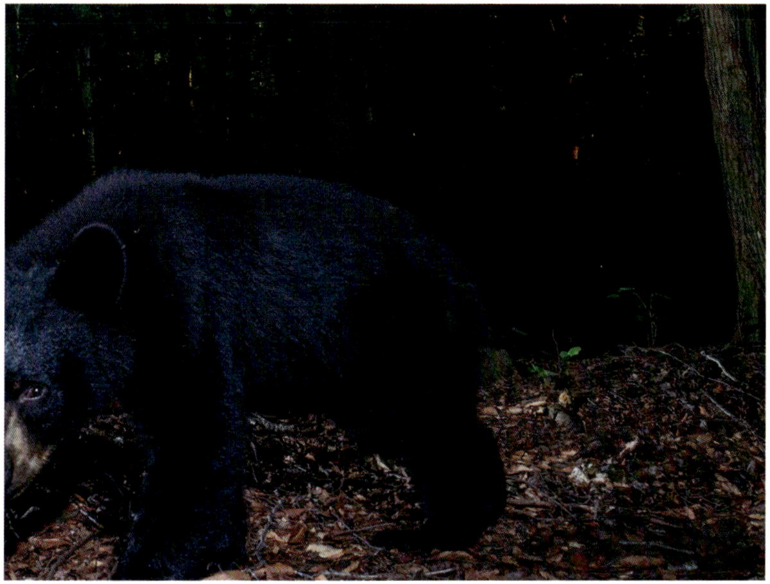

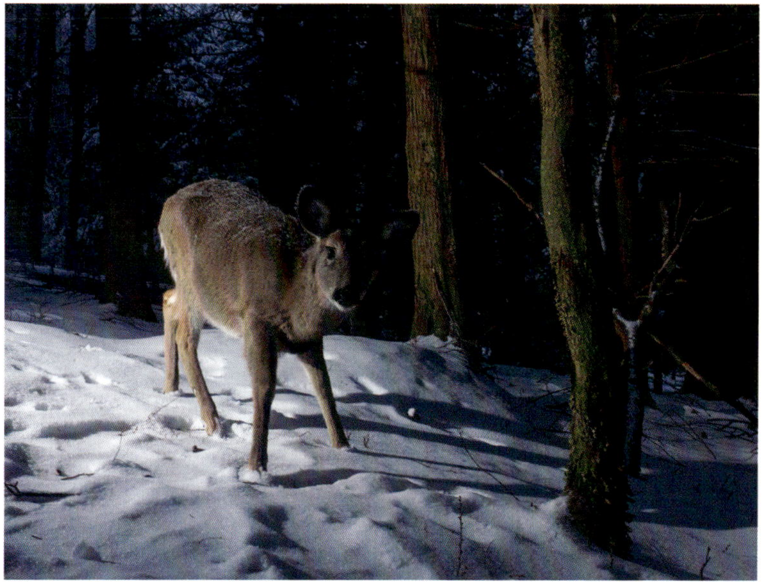

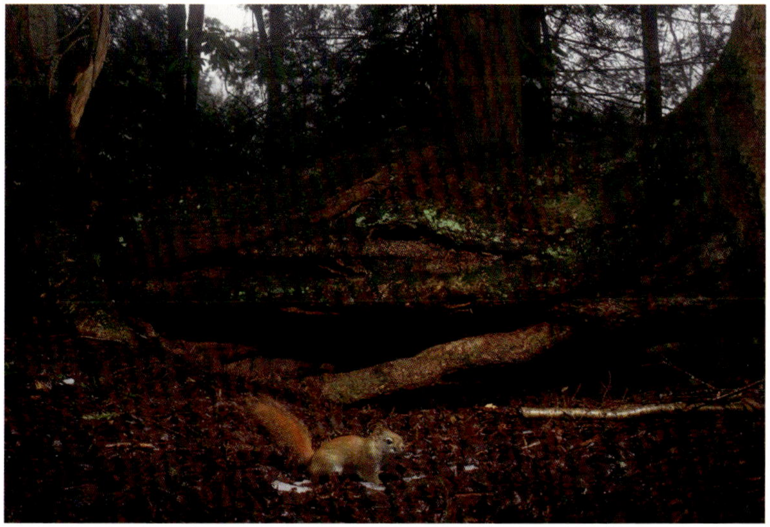

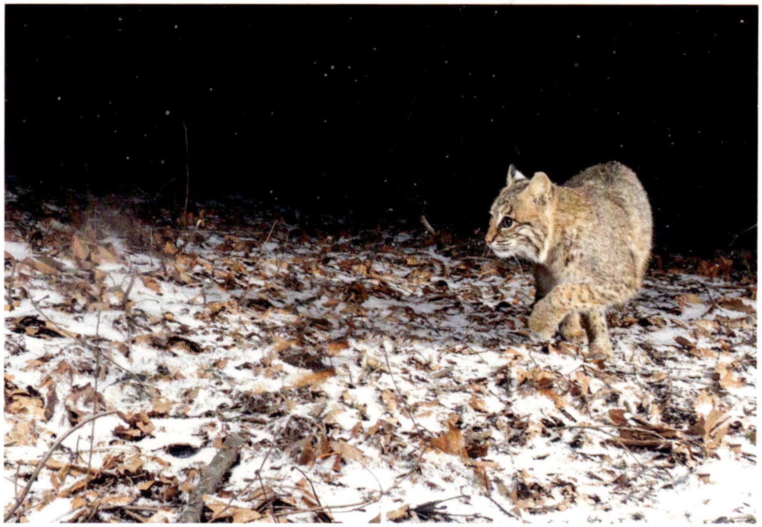

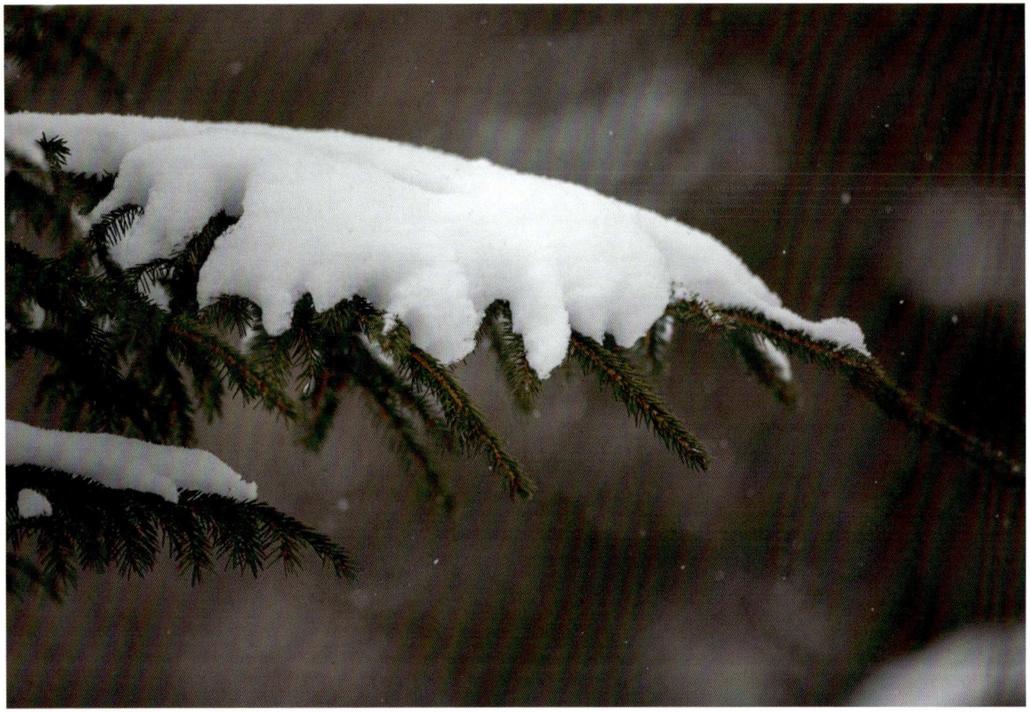

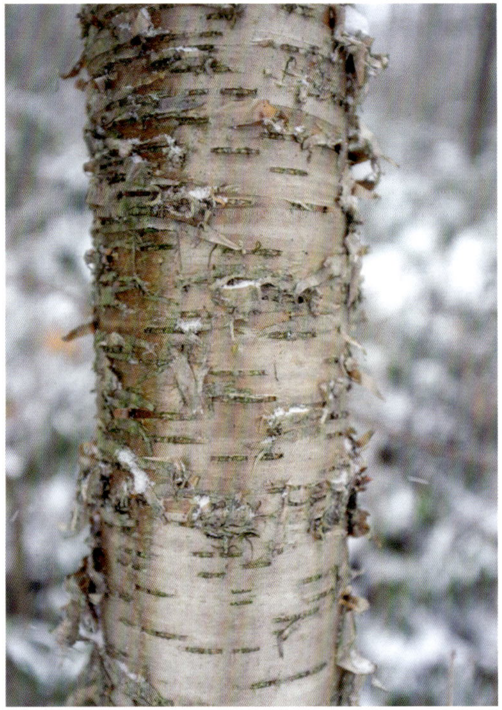

The boughs of red spruce and the bark of yellow birch provide a striking contrast to falling snow.

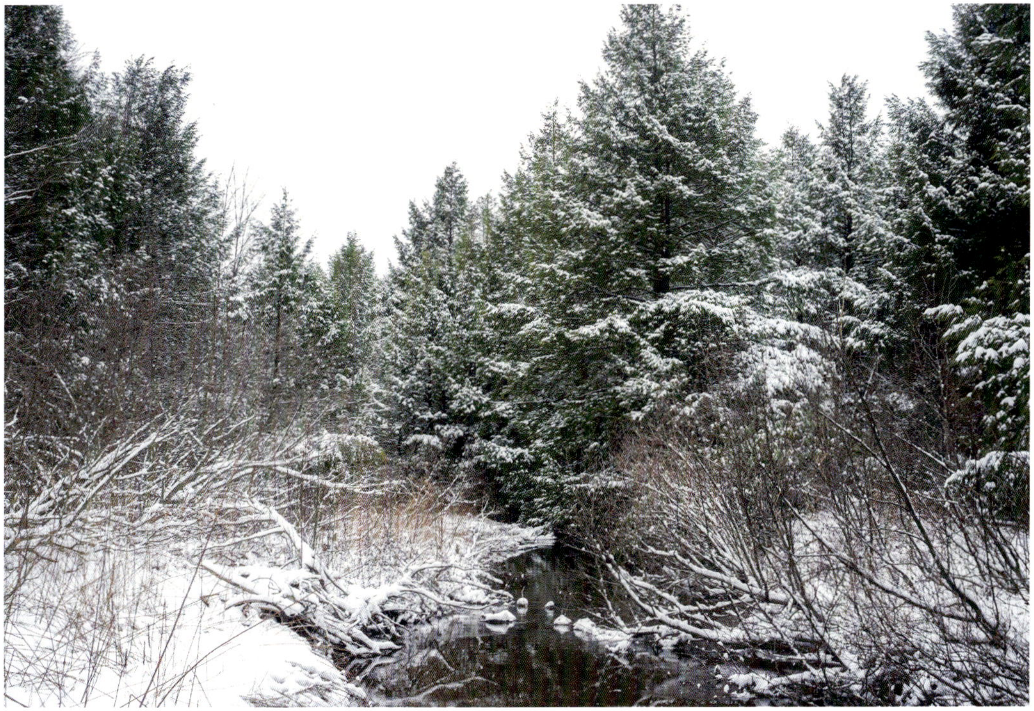

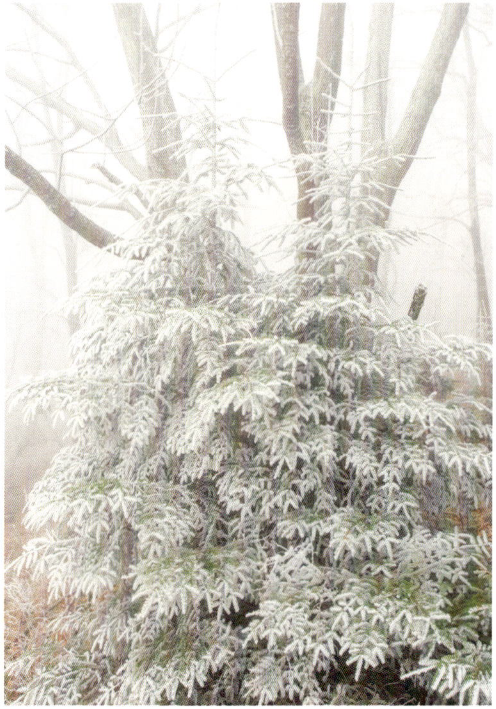

Conifers, such as these hemlocks (*left*) and red spruce (*right*) are well adapted to the season, providing shelter to numerous species. Maryland and West Virginia.

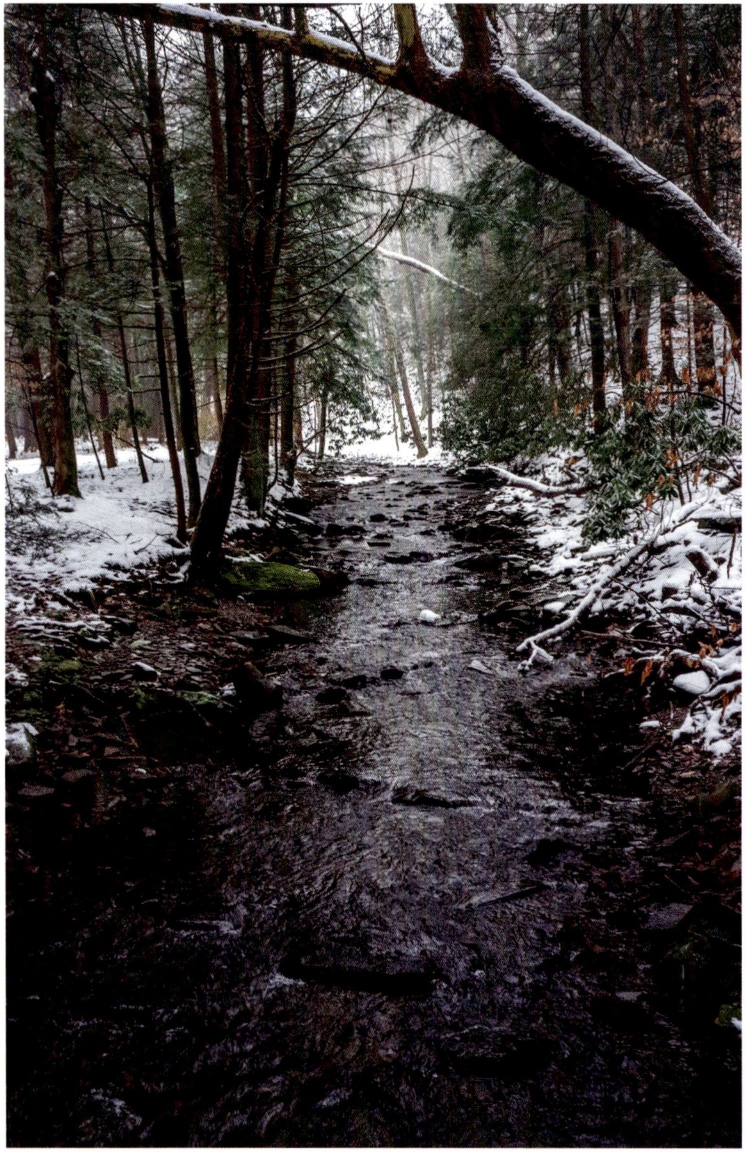

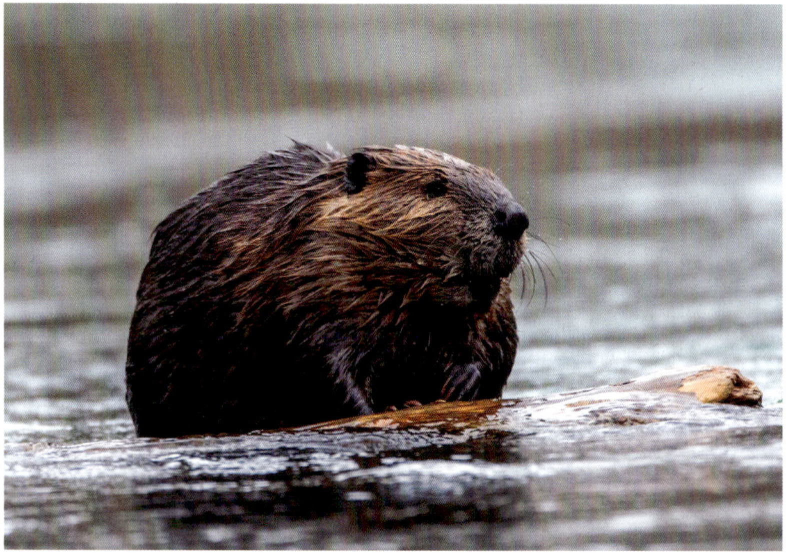

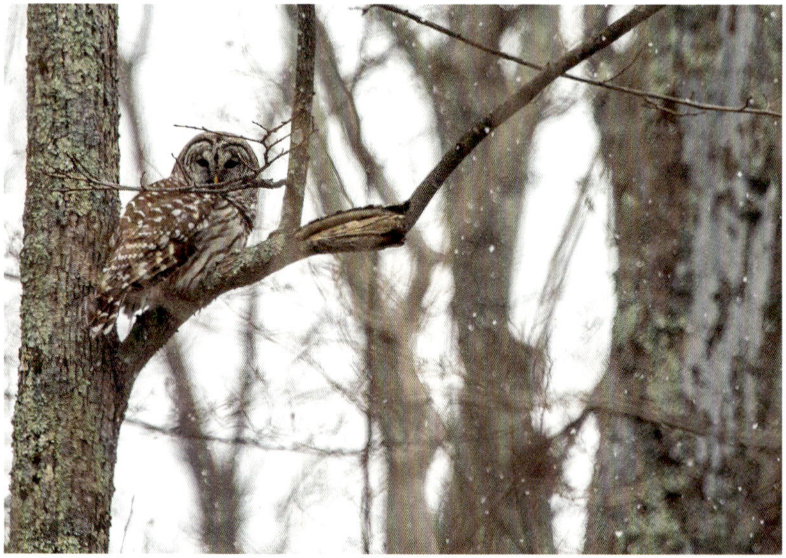

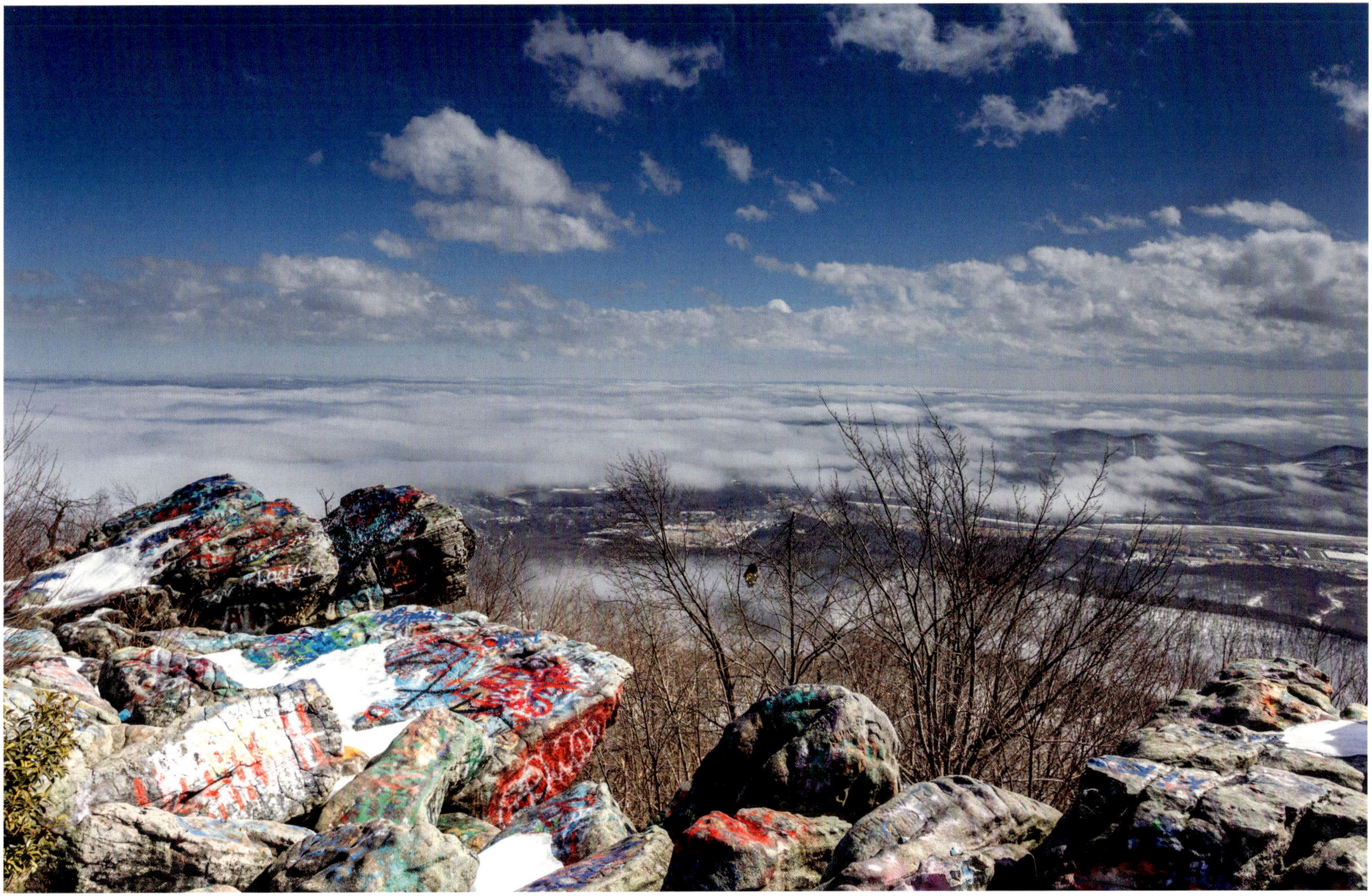

The Savage River still flows even in the heart of winter, during which period a resident beaver must still maintain its den. Big Run State Park, Maryland.

A barred owl searches for its next meal in the storm.

The graffiti-covered summit of Dans Mountain. Dans Mountain State Park, Maryland.

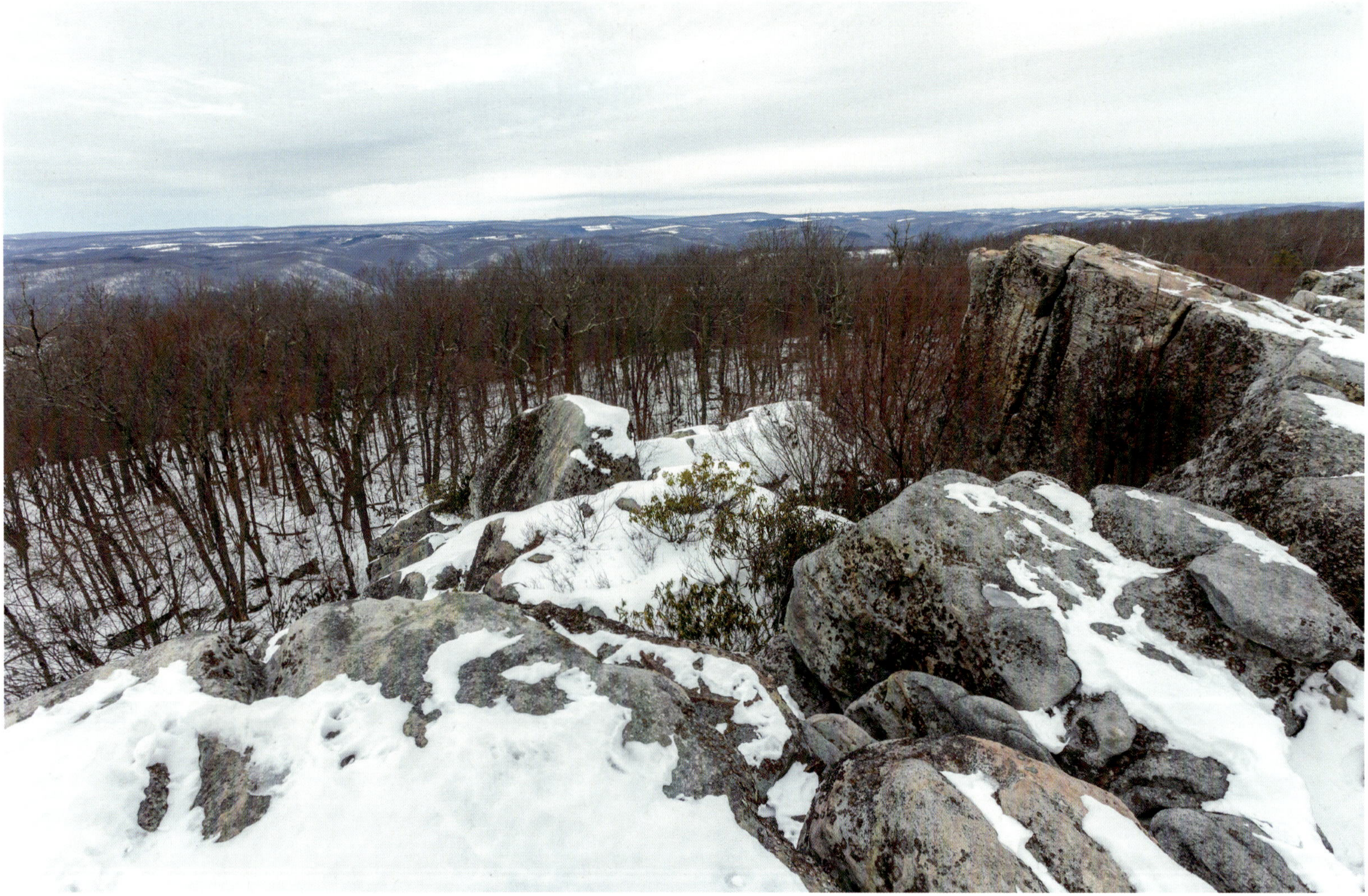

A winter scene on High Rock. Savage River State Forest, Maryland.

Cave rock in Maryland. *Adjacent page, top left*

The exquisite forms of ice, generated in a season, and the forms of a cave flowstone, generated by the eons. Virginia and Maryland.

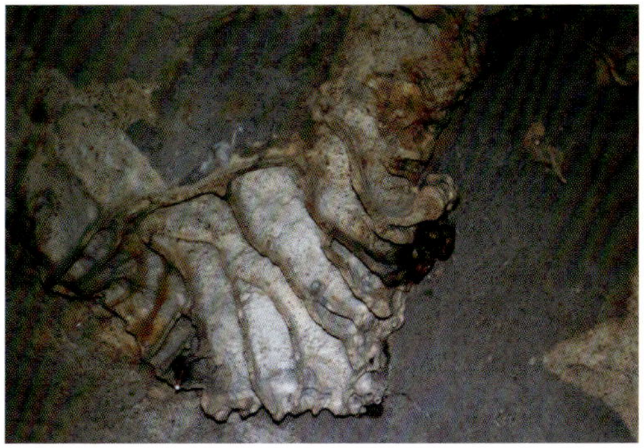

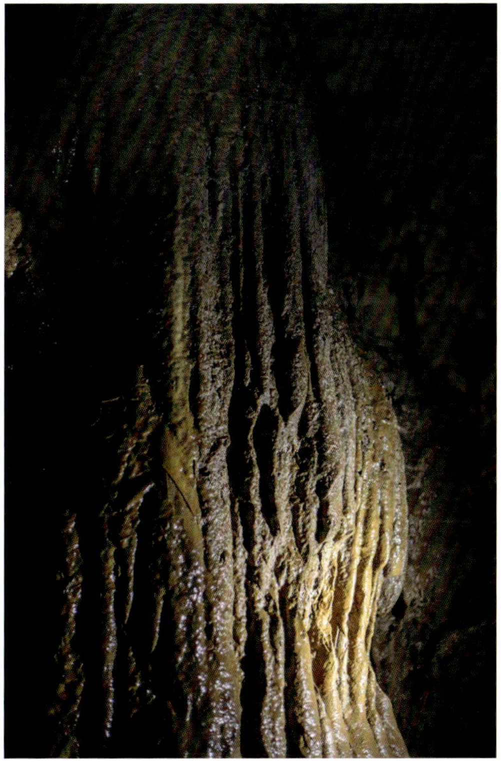

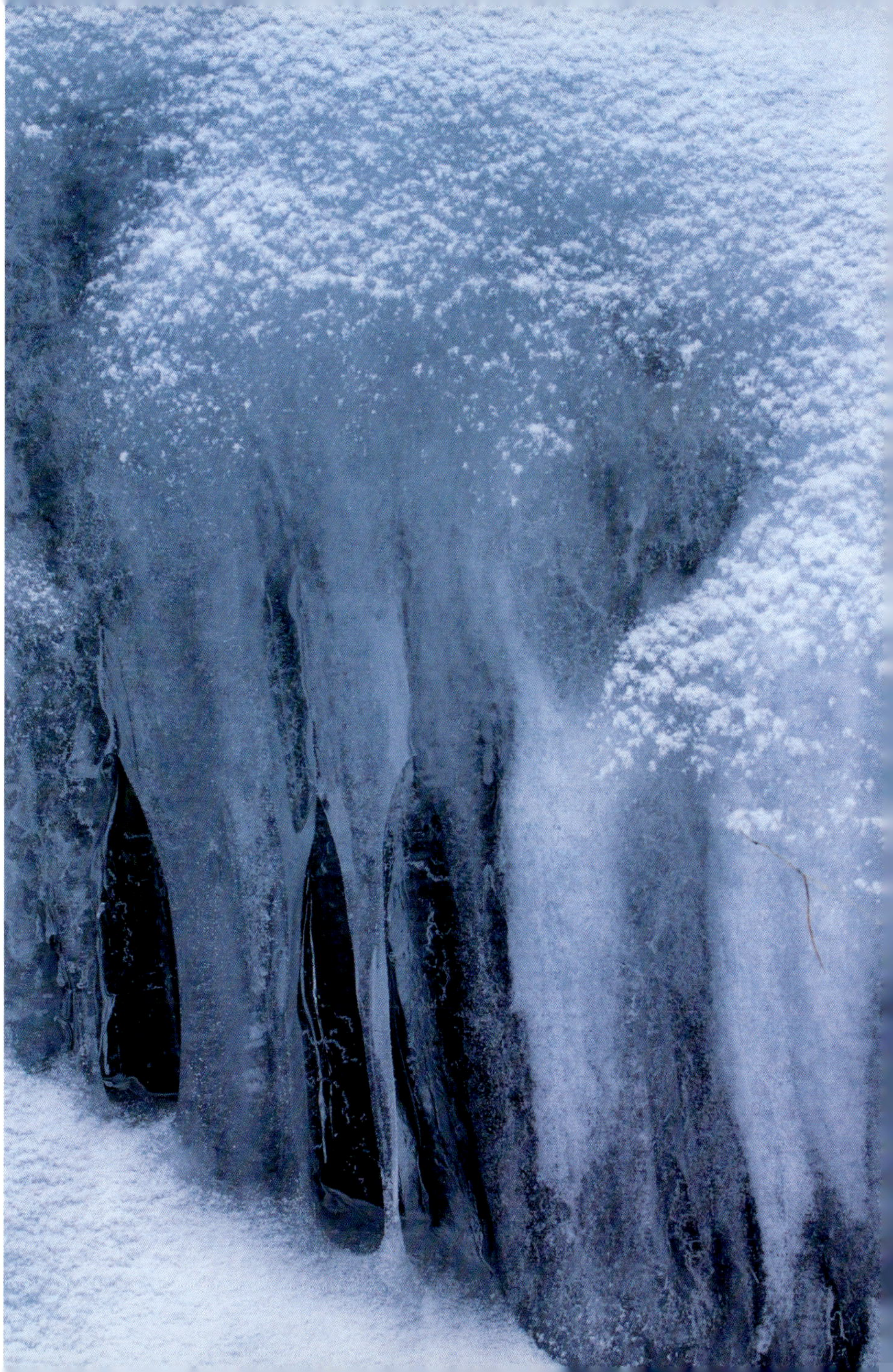

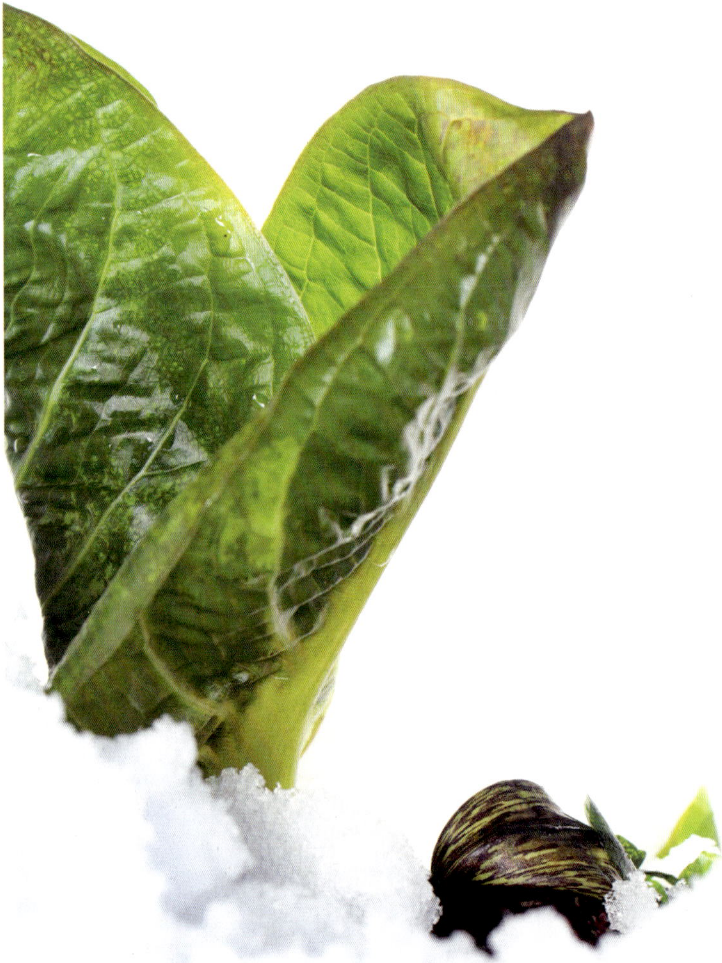

But by February, skunk cabbage appears in wetter areas in some regions, and by early March the rare snow trillium is almost ready to bloom.

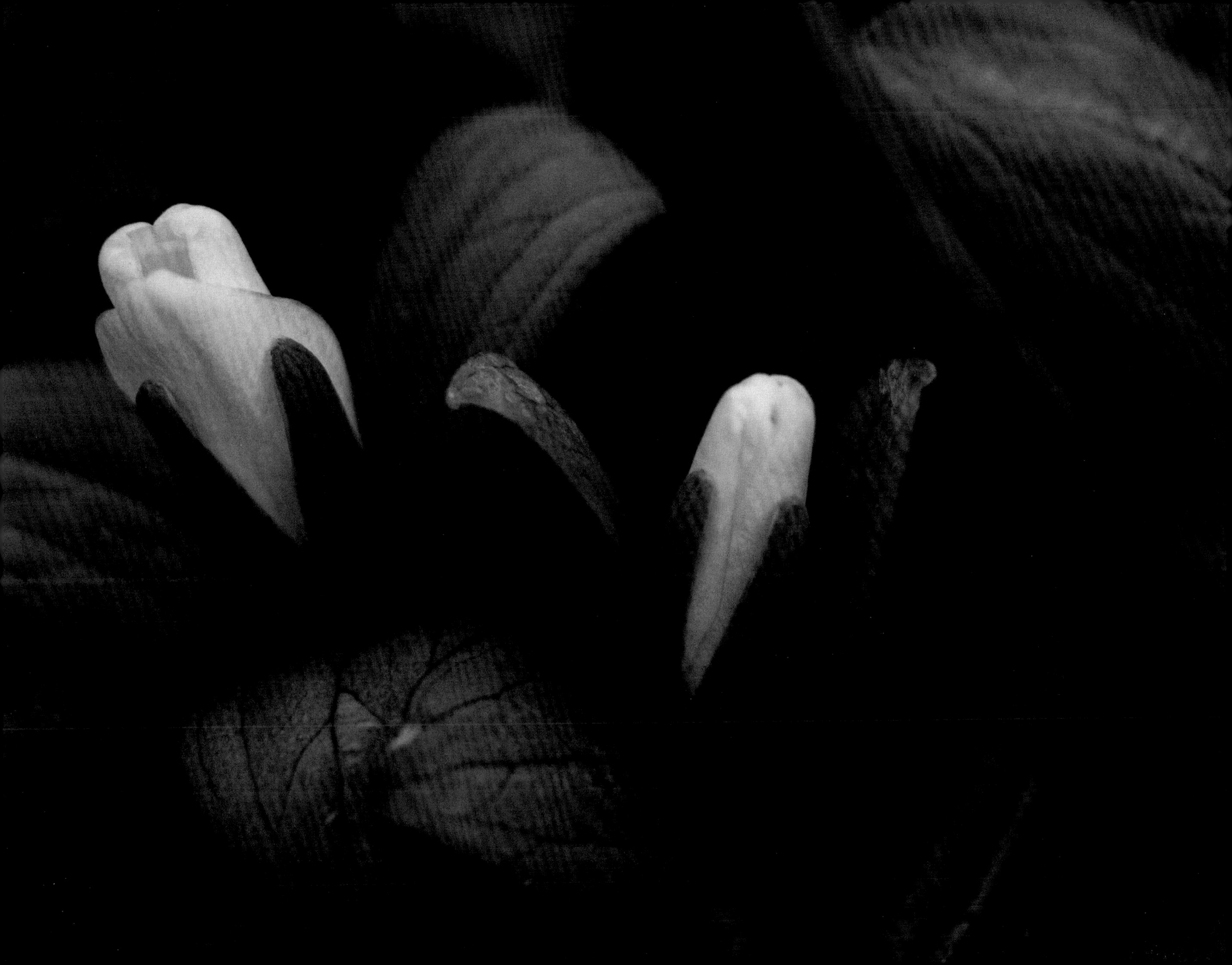

Saving Bats with the "Bat Man"

Sliding through tight rocky crevices, Dan Feller moves with relative ease as he twists his body at every turn like a contortionist. Rappeling by rope downward into the dark labyrinth and crowned with a headlamp-covered helmet, he's enthusiastic about this difficult sport. Feller is both a talented caver and biologist.

He is also Maryland's resident "bat man." Today, he dons his white Tyvek coverall suit to survey bats in hibernacula—their winter dwellings.

Feller's fascination with the world of caves began when he was a young boy. "It was a sign for Meramec Caverns that I saw on vacation with my parents, and I cried when they told me we couldn't stop," he says. "I had awesome parents; they turned around, and we went in. I still have the stamped coin today from that visit." The almost otherworldly fauna of caves, such as woodrats and blind invertebrates with no pigment, led him to biology. "My primary interests as an ecologist were cave-adapted invertebrates and small mammals, so bats joined those two niches." His enthusiasm for caves is contagious. Inside, he finds two rare, incredibly small, blind cave isopods and shares their natural history with the ease of recounting a great novel.

The subject of Feller's survey is cave bats. These bats, such as the big brown bat and tricolored bat, hibernate in caves and mines during the winter months due to the mild temperatures and high humidity inside. It is within these ancient formations that the bats are most vulnerable to the threat of white-nose syndrome (WNS), a disease caused by a fungus (*Pseudogymnoascus destructans* or simply *Pd*) introduced from Europe and first discovered in a cave near Albany, New York, in 2006. Its spread has been very rapid. WNS is now confirmed in thirty-three states and seven Canadian provinces.

It's a nasty disease that first attacks the skin of bats when they are hibernating in caves. As the fungus progresses, the bats become more active and engage in odd behaviors such as flying outside during daylight hours in the midst of winter, when they should stay in the hibernacula. This causes the depletion of stored fat they need to survive the season. Perhaps the most

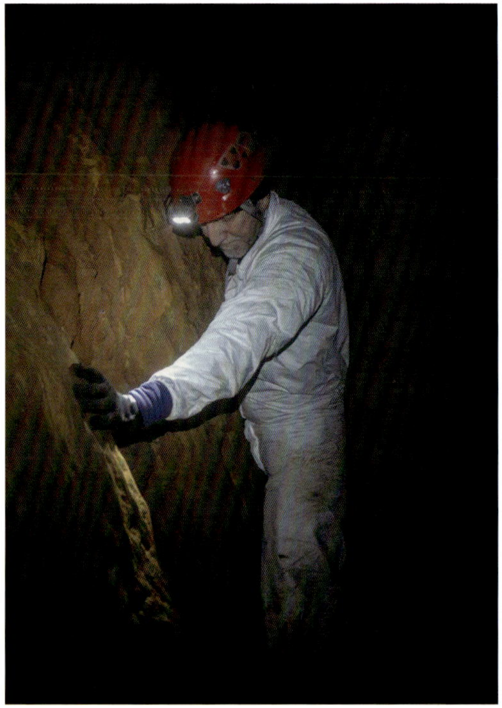

infamous symptom is the white powder or fuzzy-looking substance that is found growing on the skin and muzzle of affected bats, evidence of the fungus visible to the naked eye. This awful disease is now responsible for the deaths of more than 5.7 million bats in the eastern United States,

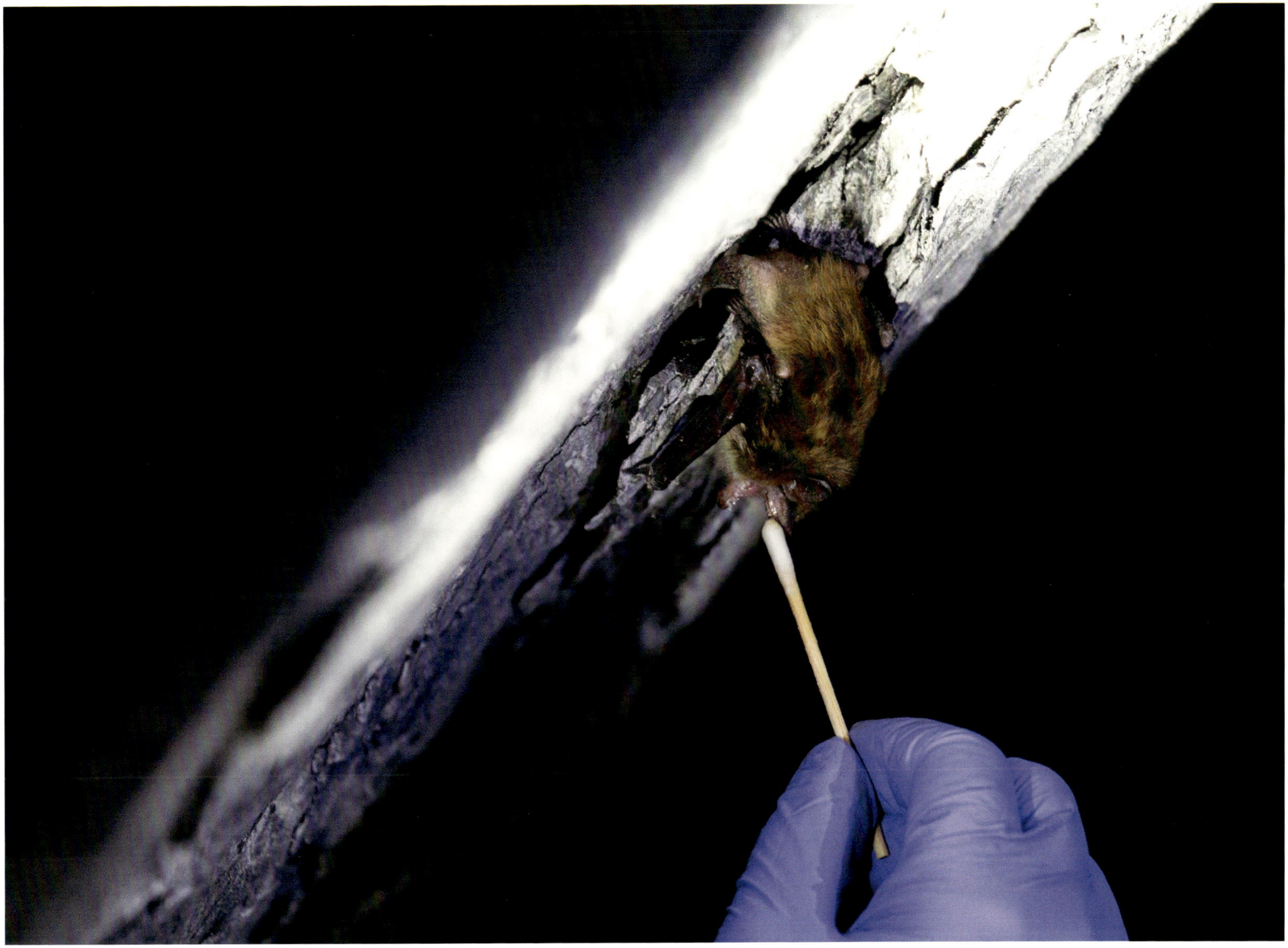

and there is no cure. Caves that once hosted thousands of bats are now empty.

A thirty-year-plus veteran of the Maryland Department of Natural Resources, Feller has witnessed the devastation. "Hibernacula were littered with dead or dying bats often covered with the white fungal hyphae characteristic of WNS," he says of the early years after its arrival. Feller initiated the state's annual bat hibernacula monitoring program and was able to confirm the extent of the disease's impact.

White-nose syndrome "has wreaked havoc among cave hibernating bats, reducing populations by more than 90 percent for most species. Many of our most common species, such as the little brown bat, the tricolored bat, and the northern long-eared bat, have become remarkably rare. The latter is now listed as threatened in Maryland and nationally as a result of WNS," he laments. Other caves where WNS has been documented experienced 100 percent mortality in hibernating bats. Bats are the only species susceptible to this disease, and it causes no ill effect to humans.

So how did it spread, and spread so rapidly? Its introduction to New York was most likely from humans who had been caving in Europe. The fungus is easily spread if left on clothing and gear that has not been

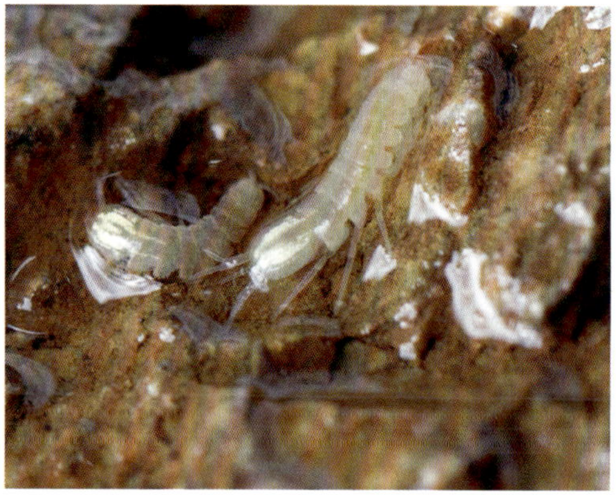

decontaminated. In 2016, the fungus appeared in Washington State. The next-closest site contaminated with the fungus, though, was in the Midwest. The most plausible explanation is that it arrived via humans entering the cave who had unknowingly carried the fungus on their gear (this is why Feller changes coveralls between caves). Fast-forward to the present: it is believed that the fungus is spread mostly from bat-to-bat contact. Vulnerable species of cave bats hibernate together in close clusters, from only a few to very massive numbers, essentially acting as a petri dish for the fungus to grow.

It's an unfortunate reality for such remarkable animals. Aside from the bat's uniqueness as the only mammal capable of flight, bats also echolocate like cetaceans,

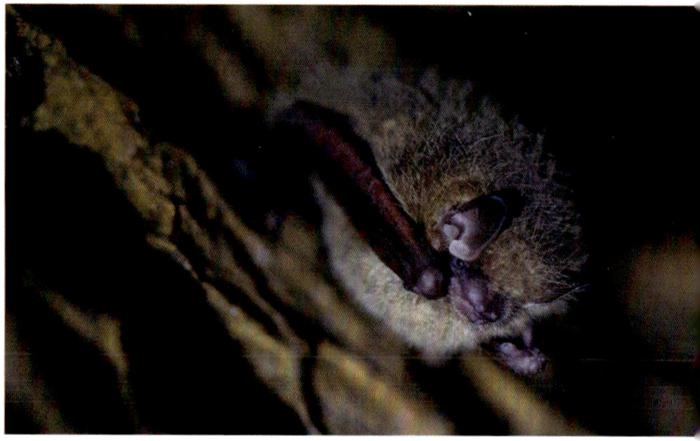

making them adept at navigating and hunting insects in total darkness. They also serve multiple notable ecological niches. Some bats pollinate plants, while some fruit-eating bats are responsible for seed dispersal. They also eat a lot of insects—so much so that it is estimated they provide billions of dollars' worth of insect-control service to farmers in the US alone. Bats are essentially nature's perfect insect-control team.

Feller is involved with a continental-scale study investigating WNS transmission and surveillance to better understand the disease and combat its spread. Weaving through the cave in an almost poetic fashion, he meticulously scans a wall. If he locates bats, he makes note of the species, the cluster size, and how close together they are. He also records details of the cave's microclimate, since there are multiple studies underway that are investigating links between the pathogen's spread and hibernaculum microclimates.

On this cave visit, he locates a bat within reaching distance, giving him the opportunity to swab it. He does not find any evidence of the infamous white fuzz—that is, the obvious superficial sign—on this bat. But that doesn't mean the fungus is not present. "In recent years, remaining diminished bat populations in hibernation rarely show obvious signs of the fungus, as it's taken on a more diminutive form but may still be observed fluorescing under UV light. The fungus does not disappear even though all the bats at a site have perished; it continues to persist in the cave environment on other organic material," he adds.

This big brown bat briefly awakens from its state of torpor and emits a slight hiss. Once Feller completes a swab of its muzzle, he places the sample in a vial and records his notes. The bat is back in hibernation within minutes of the interruption.

A growing number of biologists and scientists are working on combating the WNS invasion. The Northeast Bat Working Group, made up of representatives from federal and state agencies as well as biologists affiliated with universities, conservation organizations, and even private consultants, gathers to make presentations, share current data, and discuss methodologies and management practices to better protect bats. Current research into potential treatments and curbing the growth of the fungus was presented. Researchers from Temple University, Lock Haven University, and the Pennsylvania Game Commission observed that the growth of the fungus is slowed on bat wings at colder sites in a hibernaculum. Because this infliction knows no borders, it is through collaboration and the sharing of methods that researchers hope to gain traction on the disease.

If there is a silver lining regarding the plight of hibernating bats, it is that there appears to be a growing appreciation for these fascinating mammals. "WNS has created a new focus on bats, with research documenting their importance and the public gaining interest and acceptance of the once-maligned creatures," says Feller. This includes compliance in not entering protected caves, of which there are many in Maryland. "Conservation efforts have stepped up management efforts such as gating caves to reduce disturbance during hibernation and creating more-bat-friendly forest management practices," he adds.

In the chill of the fresh air, I walk with Feller down a path as our eyes readjust to the daylight. He points to a small rocky opening at foot level that leads to a cave. "That one requires belly crawling for about 150 feet before you can stand," he says. "We don't have to do that one today."

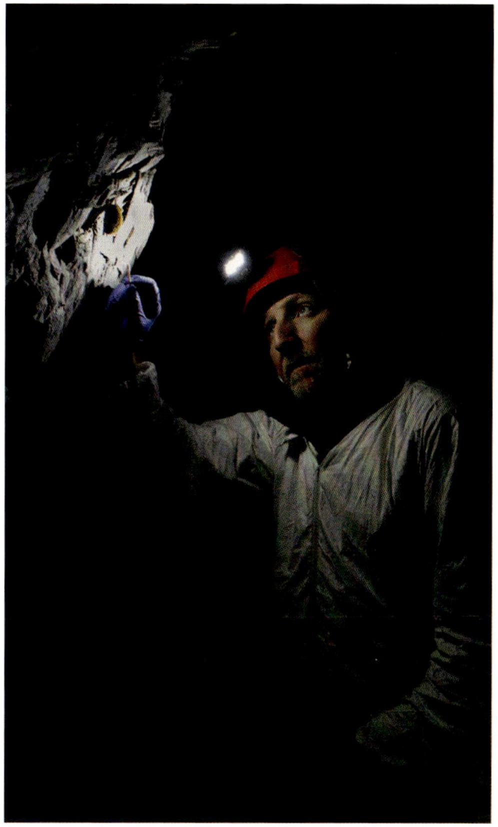

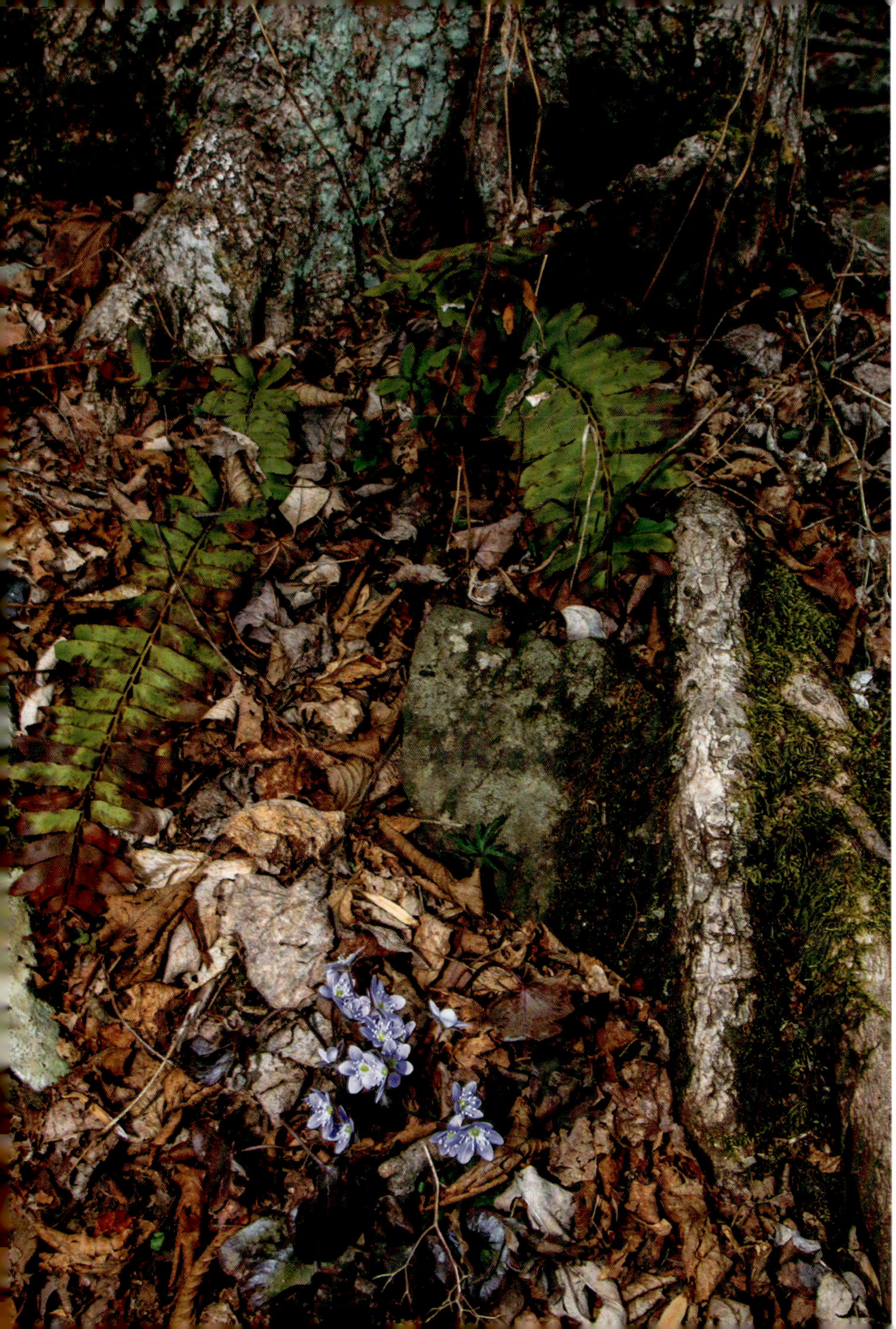

As they have for ages, round-lobed hepaticas color the winter landscape, signaling that spring will soon commence.

AFTERWORD

Christmas Day. Miami, Florida

Far from the mountains, I write this entry from the poolside, surrounded by lush tropical foliage at a family residence.

While this may seem like an odd location, given the subject matter of this book, this was the first time that I had been able to share my work from this project. And as I'd hoped, my family was surprised to learn of all the natural wonders of the Central Appalachians and of how that water helps feed North America's largest estuary. There were informal discussions of family trips to the mountains for autumn leaf peeping and summer camping.

And that is what I hope this project serves as an impetus for: responsible travel and care of the Central Appalachians. Many areas within the Central Appalachians are poverty stricken. Supporting the local communities when venturing into this Eden is as important as caring for the environment.

The Central Appalachians (and the greater Appalachian chain) are a globally significant biological hotspot. The diversity of species found here is simply magnificent, especially for a temperate environment. Many residents of the Chesapeake Bay watershed depend on the Appalachians for their water supply.

But there are issues the region faces: a warming climate, invasive species, and energy development. How does one balance the need for environmental preservation with the needs of a growing population? While there are no easy answers, and this is not the overarching subject of this book, it is my hope that one thinks of their environmental impact and how, if each of us does something more sustainable, it would be to the benefit of the mountains.

To better illustrate this, consider the Shenandoah salamander. While endangered, the amphibian fortunately exists within the protective boundaries of a national park. However, outside influences contributing to a warming climate know no boundaries. An experiment forecasting how the salamander would do in probable future climate scenarios due to climate change were poor (Dallalio, Brand, and Grant 2017, 193). Paradoxically, for other species, it has been predicted that the Central Appalachians will be a stronghold for plants and animals fleeing the effects of climate change.

Former surface mines are being reclaimed in an expensive process to mitigate the negative effects of acid mine drainage and prevent land subsidence, a gradual or sudden sinking of the earth's surface. Organizations such as the Nature Conservancy and the American Forest Foundation are partnering with private landowners by providing incentive programs for allowing their forests to mature as well as for carbon sequestration.

Supporting local Appalachian communities by purchasing their products and services, supporting the organizations that are trying to protect this region by donations and volunteering, and experiencing the gifts of nature that the Central Appalachians give us are some things I hope you may consider.

This has been a wonderful chapter in my life, the sum of many experiences in the mountains. Where the health of the mountains, and the Bay itself, go in the years that follow depends on us. Thank you for taking this journey with me.

With gratitude.

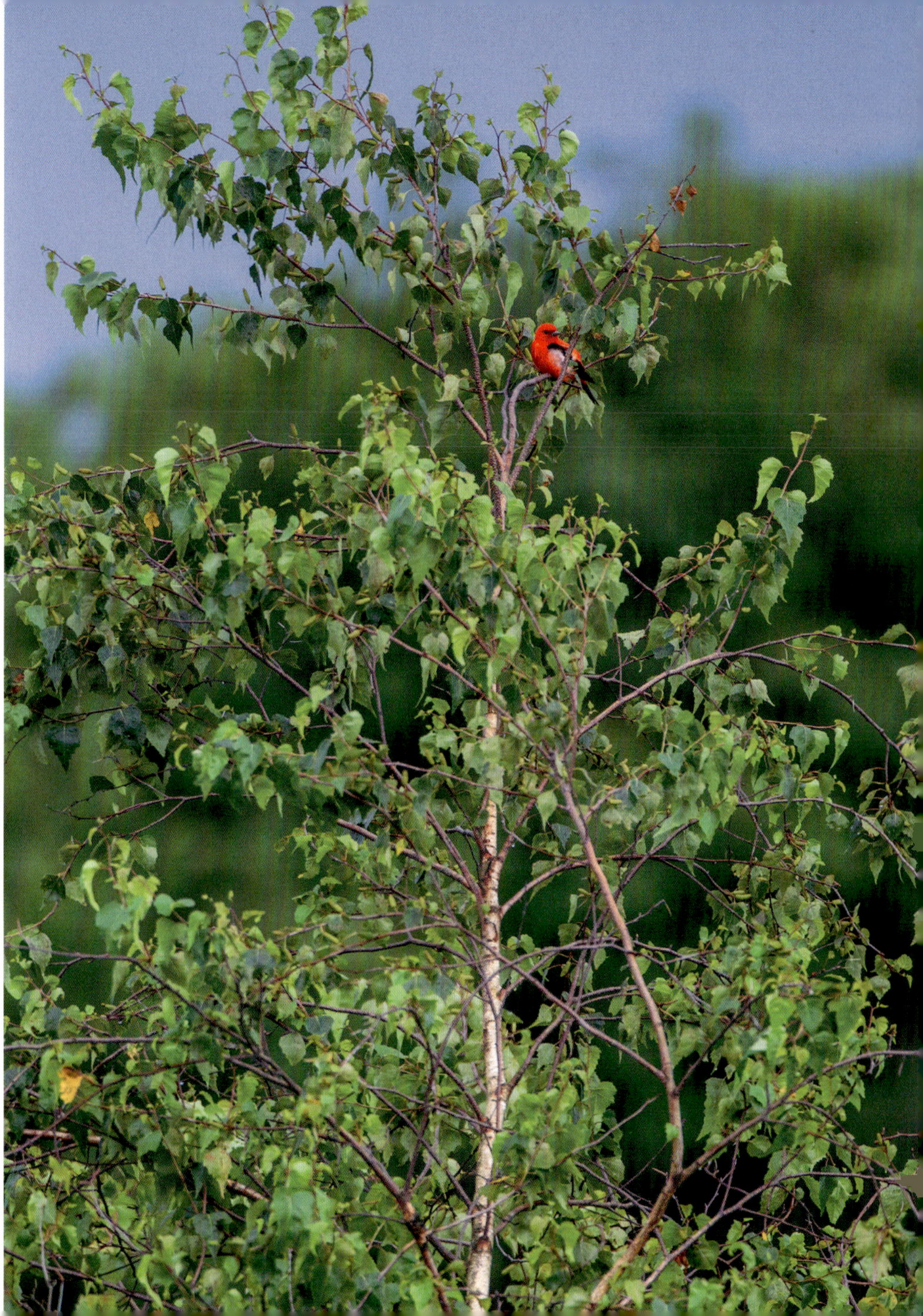

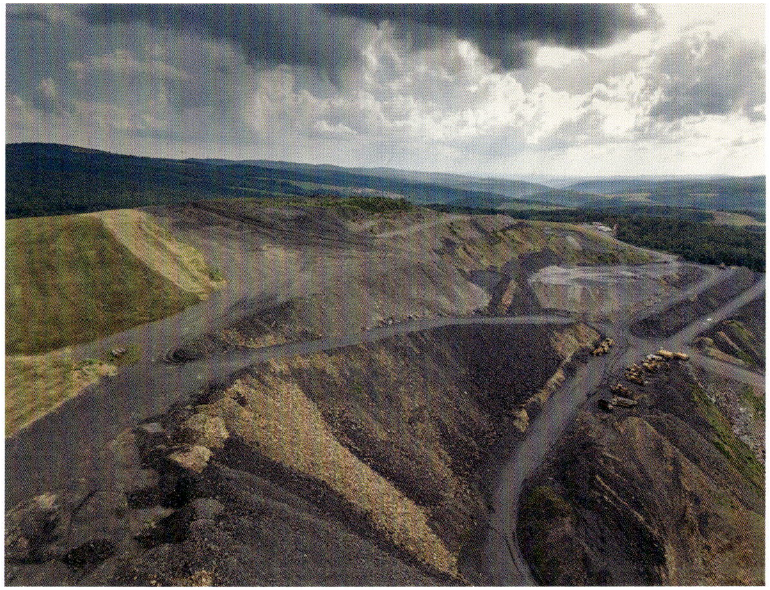

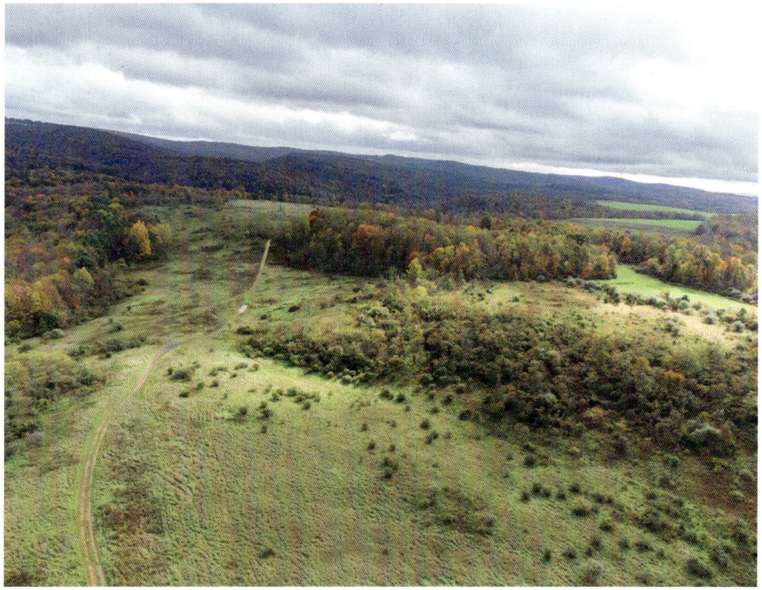

Like a jewel, the scarlet tanager is hard to miss in the lush Appalachian forest.

An active surface mine in Garrett County, Maryland, and a reclaimed mine in Allegany County, Maryland

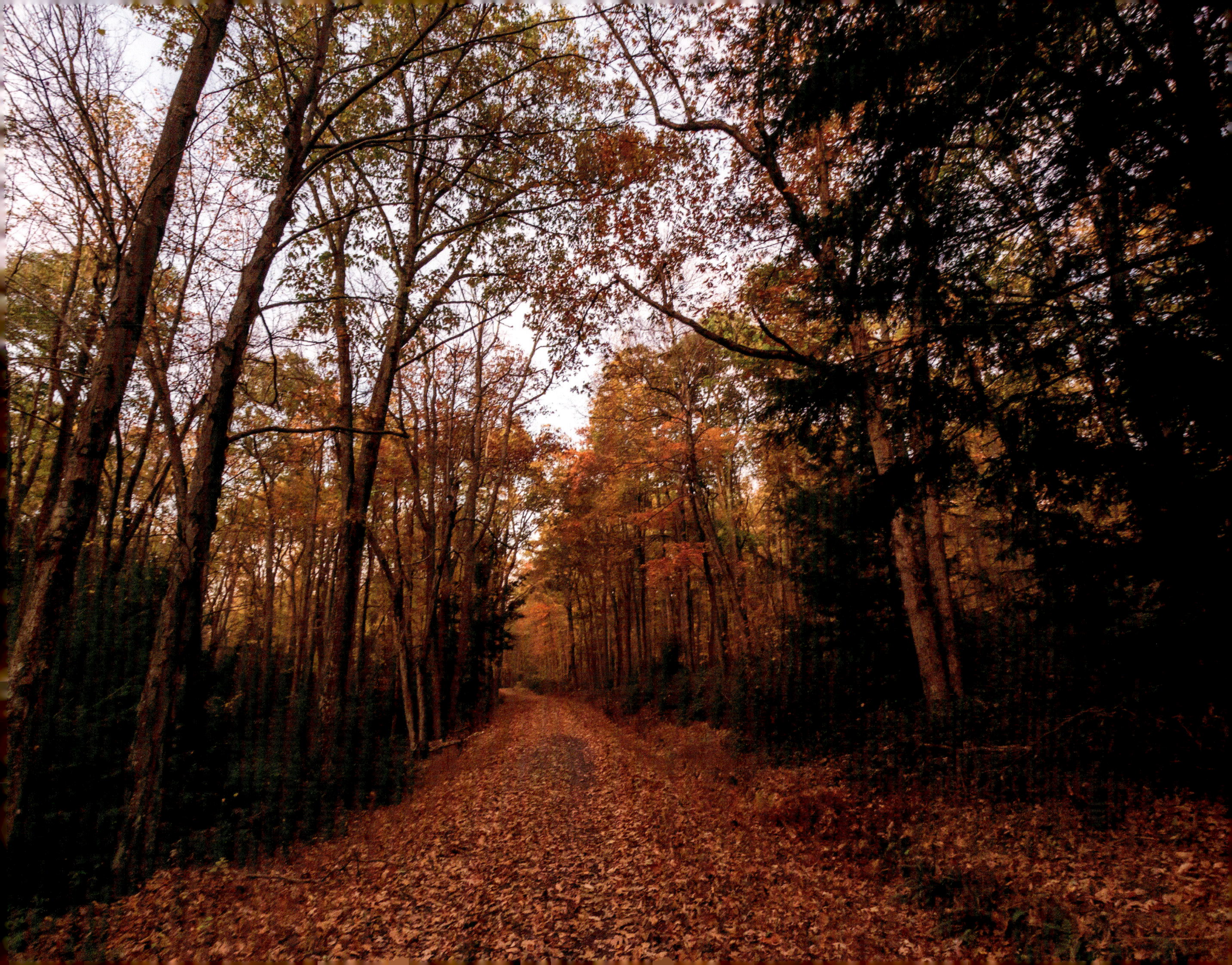

FAVORITE LOCATIONS

While this is not a complete list of parks and preserves, since there are many, the list below highlights my favorite public areas that I visited multiple times while working on this project.

MARYLAND:

Savage River State Forest
Big Run State Park
Finzel Swamp Preserve
Dans Mountain Wildlife Management Area
Green Ridge State Forest
Sideling Hill Creek Preserve
Cunningham Falls State Park
Gambrill State Park
Catoctin Mountain Park
Frederick Municipal Watershed
Appalachian National Scenic Trail (the "AT")

VIRGINIA:

Shenandoah National Park
George Washington National Forest
Saint Mary's Wilderness (within the GWNF)
Sky Meadows State Park
Appalachian National Scenic Trail
Note: The Virginia section of the AT has some of my favorite hiking in all of the Central Appalachians.

WEST VIRGINIA:

Bear Rocks Preserve / Dolly Sods Wilderness Area
Spruce Knob–Seneca Rocks National Recreation Area
North Fork Mountain (access via the Monongahela National Forest)
Pike Knob Preserve
Appalachian National Scenic Trail (a very small section of the trail passes through the state)
The historic town of Harpers Ferry

PENNSYLVANIA:

Hyner View State Park
Quehanna Wild Area
Marion Brooks Natural Area
Elk State Forest
Moshannon State Forest
Michaux State Forest
Bear Meadows Natural Area

Outside the boundaries of the watershed, but still within the Central Appalachians, are other incredible areas worth your visit, such as the Cranesville Swamp on the Maryland / West Virginia border, the Roaring Plains (in the Monongahela National Forest), the Canaan Valley of West Virginia, Swallow Falls State Park in Maryland, the southern Blue Ridge Parkway, Mount Rogers Recreation Area in Southern Virginia, and the Allegheny National Forest of Pennsylvania.

Predawn on a forest road in Rothrock State Forest, Pennsylvania

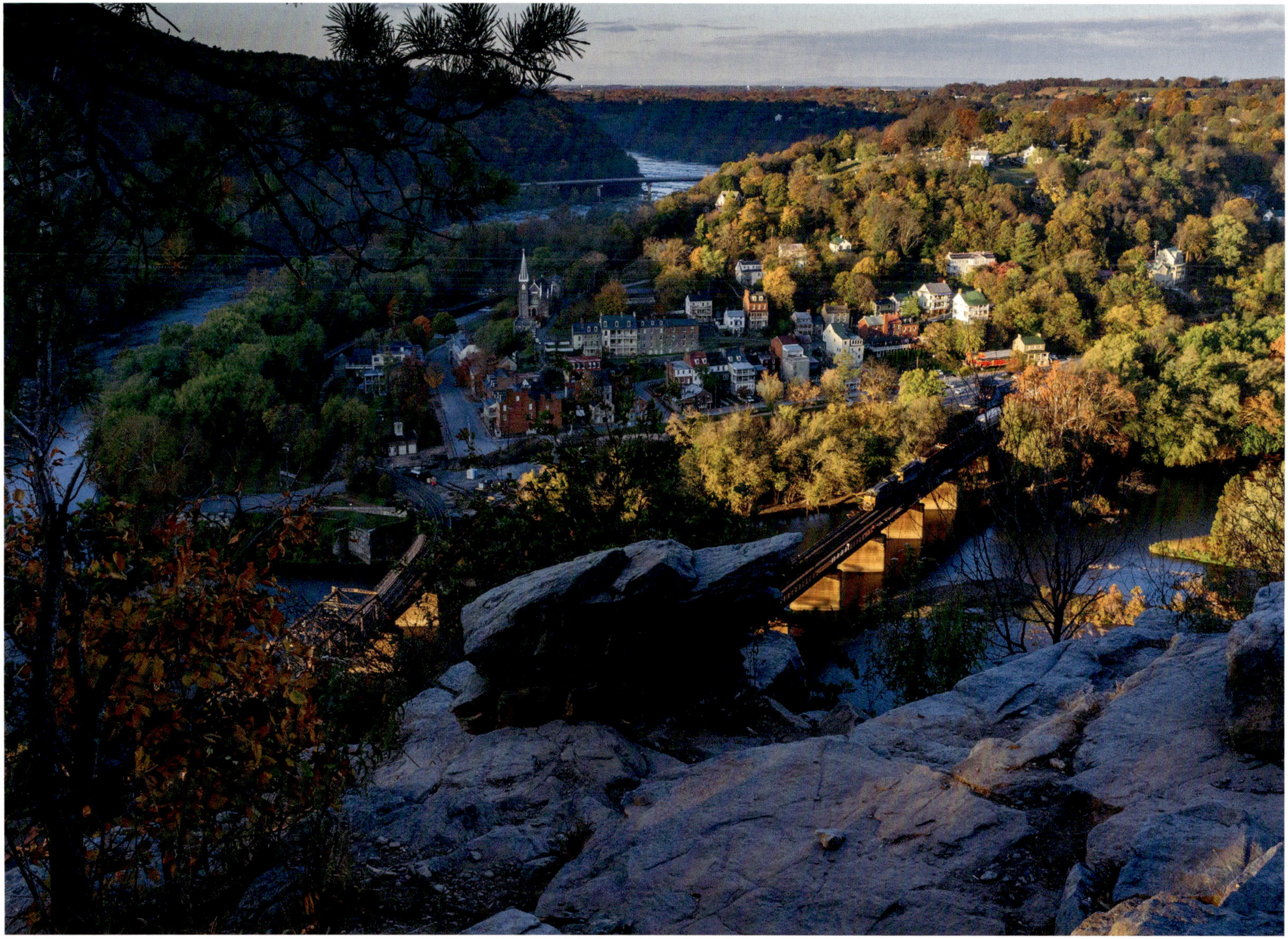

PHOTOGRAPHIC NOTES

I do not obsess over gear or upgrade to the newest and fanciest system each year. While I have amassed quite a collection over the years for this line of work, I carry only so much with me when in the field. Unless on a commercial shoot, carrying loads of gear slows me down. Carrying minimal equipment allows me to move and home in on what I am trying to accomplish.

I am agnostic on the subject of camera brands. The simple truth is that all the brands are amazing, including smartphone cameras. The best camera is the one you have on you. What is most important is having a personal vision and then executing that vision through knowledge of that gear to unlock one's creative potential.

My "traditional" cameras are Canon full frame DSLRs; the 5D Mark IV and 6D models were used most often. I use a variety of different lenses, since each has a special purpose. These include the Canon 500 mm f/4 L (sometimes with a Canon 1.4 tele-extender version III, making it a 700 mm f/5.6 lens), a 100–400 mm L, 100 mm macro L, a 50 mm f/1.8, and the 17–40 mm f/4 L. To decrease the minimum focusing distance for some lenses, I attach Kenko extension tubes. With this system I also use 3-stop hard-edge and 2-stop soft-edge Singh Ray neutral-density filters to capture high-contrast landscape scenes. Occasionally, I will use a circular polarizer filter to reduce reflections.

My "toy" camera, the Fuji X100, was around my neck fairly frequently. It is light and well built, and I love that it is a rangefinder camera. It features a fixed 35 mm lens for that classic photojournalism look. I use it for landscapes, both intimate and

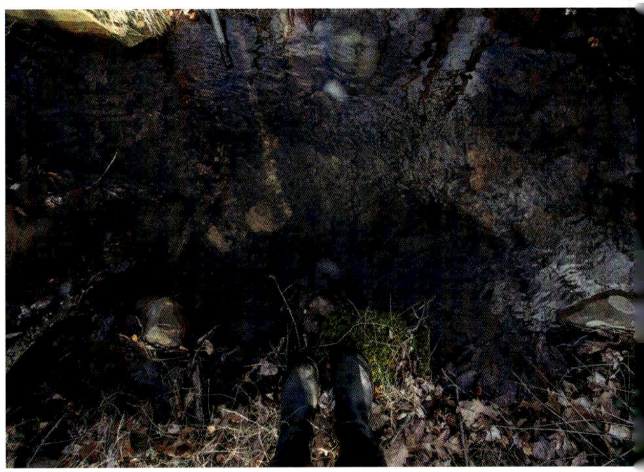

Looking below to the historic town of Harpers Ferry, West Virginia, from the Maryland Heights overlook, which is a short hike from the Appalachian Trail

The author standing before a vernal pool

wide, and for capturing people in the field. An added bonus is the embedded Fujifilm color profiles that can be employed for different scenes. For nature, I prefer Velvia for the way it renders greens and blues (just like the classic film type), but for humans I prefer Provia for the way it renders skin tones.

Underwater photography is an inherently equipment-driven (and expensive) pursuit. Shooting macro or wide angle requires separate ports to work with different lenses, and you can't switch ports underwater. But times have changed, and advancements in camera technology have increased so much that for my underwater work I downsized to a compact system. The Sony RX100 series of cameras are incredible and offer DSLR-like quality in a very small package. For this project I used the VI model. The camera is kept safe and dry in an Ikelite housing. I have a dry dome port to photograph at the camera's widest setting, 24 mm, as well as an external "wet" lens made by INON for semi-fish-eye images. I also use Ikelite DS51 strobes for lighting. This compact system makes traveling and carrying gear in the sometimes-rough Appalachian waters much easier (but not completely!).

For aerial photographs, both a DJI Mavic Air drone and DJI Mini 2 drone were used.

For camera trapping I use Canon Rebel DSLRs and Nikon SB-28 flashes, both housed in homemade protective cases. The Nikon flashes have an incredible standby mode that conserves battery power but saves just enough of a charge that they flash on the first motion trigger even after being "asleep" for days or weeks. I use passive infrared sensors and Camtraptions-brand wireless triggers and receivers. I also use countless amounts of duct tape, zip ties, clamps, and tripods to set all the gear in place. With camera trapping, things will often go wrong, so a sense of humor is going to be the best asset to bring with you into the field.

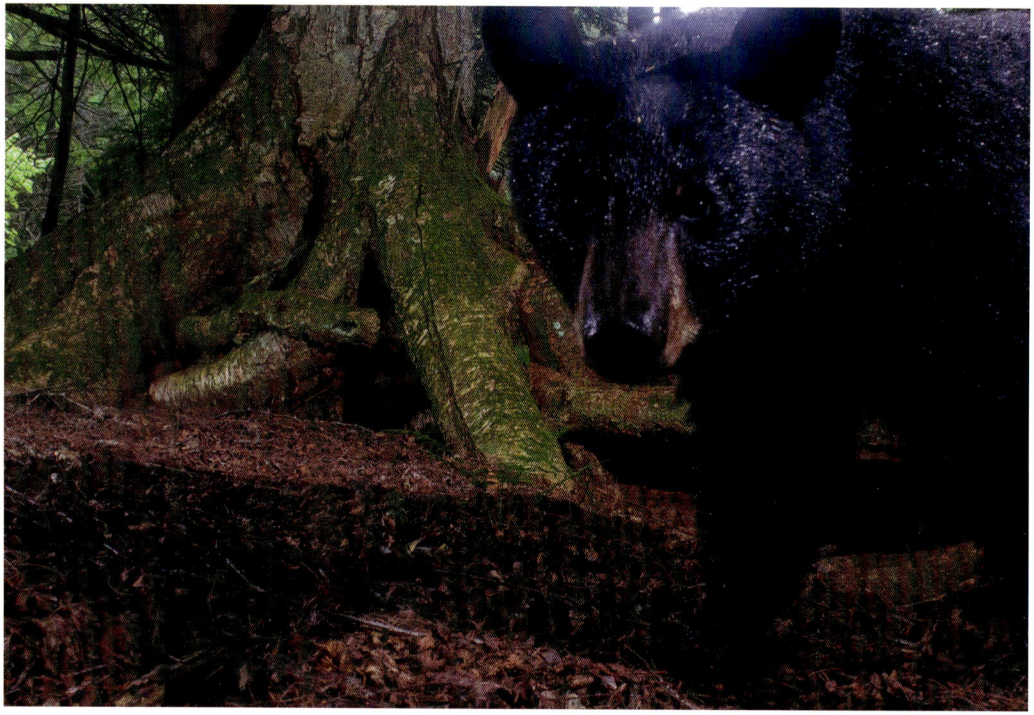

Scarlet blueberry shrubs in late September at the Dolly Sods Wilderness. Adjacent page

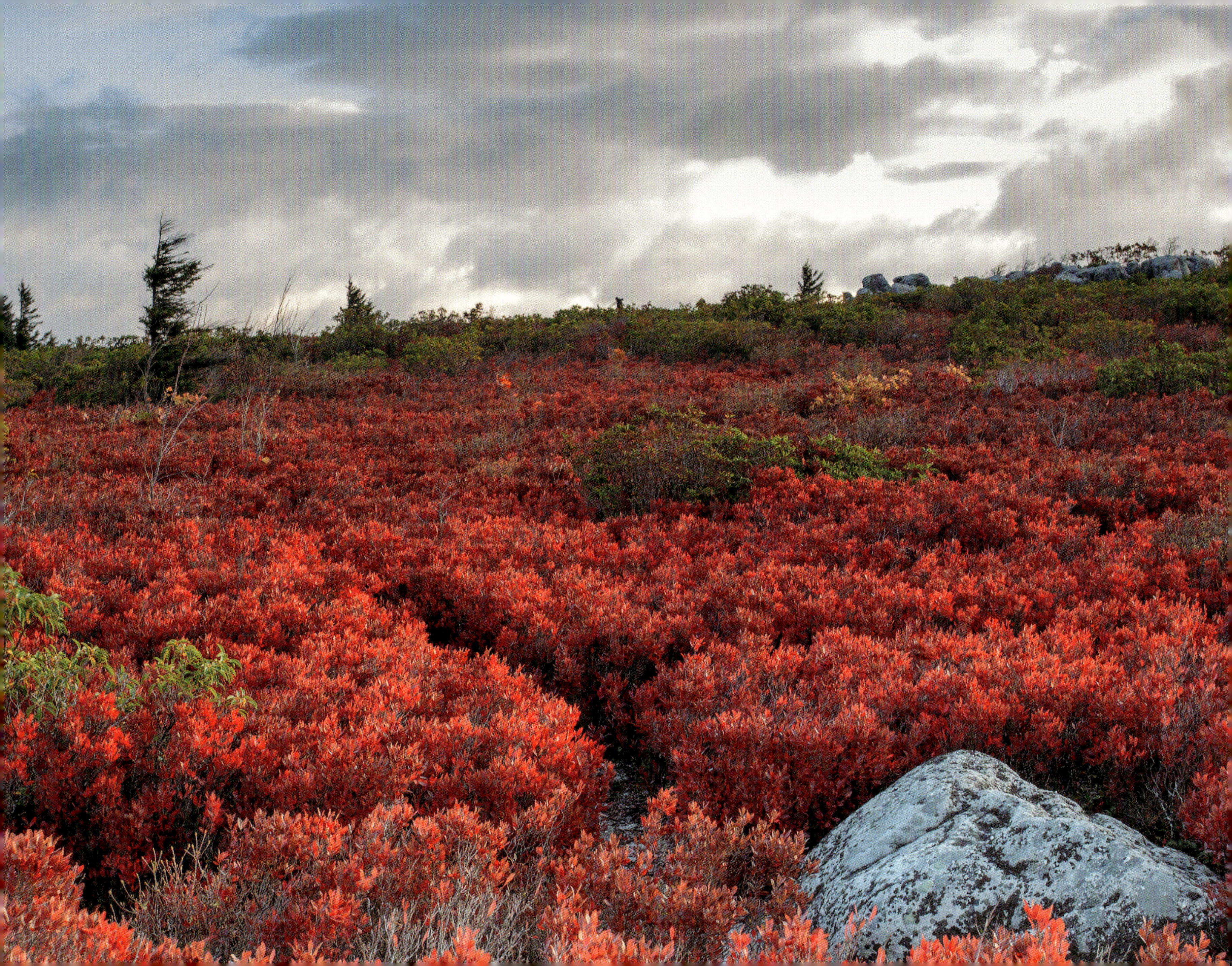

ACKNOWLEDGMENTS

A project of this scale could not have been possible without the support, friendship, and assistance of many wonderful people.

What began as a cold call to Dan Feller to discuss a story on bats I was researching led to a collaboration for which I am forever thankful. Besides allowing me to document so many species in which he specializes and sharing his ideas with me, Dan introduced me to many people who took this project down roads I could not have envisioned. Thank you, brother.

To my wife, Carolina, who was always so supportive, especially with all the time I spent away from home and for being my rock when I didn't think this project could be completed. When I was not home, spending long hours planning expeditions and creating this book, she was the sole provider for our wonderful children, Liliana and Adrian. Te amo.

To Deborah Landau and Bill Minarik, thank you for trusting me to tell your stories. They are important, beautiful, and inspirational. To inspire others to the natural world, I believe that the human element is a necessity. Both of your stories accomplish that.

To Rich Raesly, for the brotherhood, camaraderie, and beer, and for rescuing me when I was stuck in that mountain snowstorm. I still owe you!

To Robert Mulligan, for the friendship and assistance on many of these expeditions.

To Tim MacWelch, for the lovely foreword and for teaching me everything I know about survival.

And to Max Bolno, Ed Thompson, Chris Rieth, Tiffany Rieth, Denise Nooe, Kevin Hood, Matthew Olear, my parents (Mark and Angela Hendricks), Brad Hendricks, Chet Hendricks, Gloria Giraldo, the Medina family, Mike Powell, Alan Klotz, Jim McCann, Sunshine Brosi, Tom Dorsey, Jim McClelland, Steven David Johnson, Kodi Bowers, Chesapeake Bay Media for use of the bat story, and the Nature Conservancy chapters of Maryland/DC and West Virginia. Thank you.

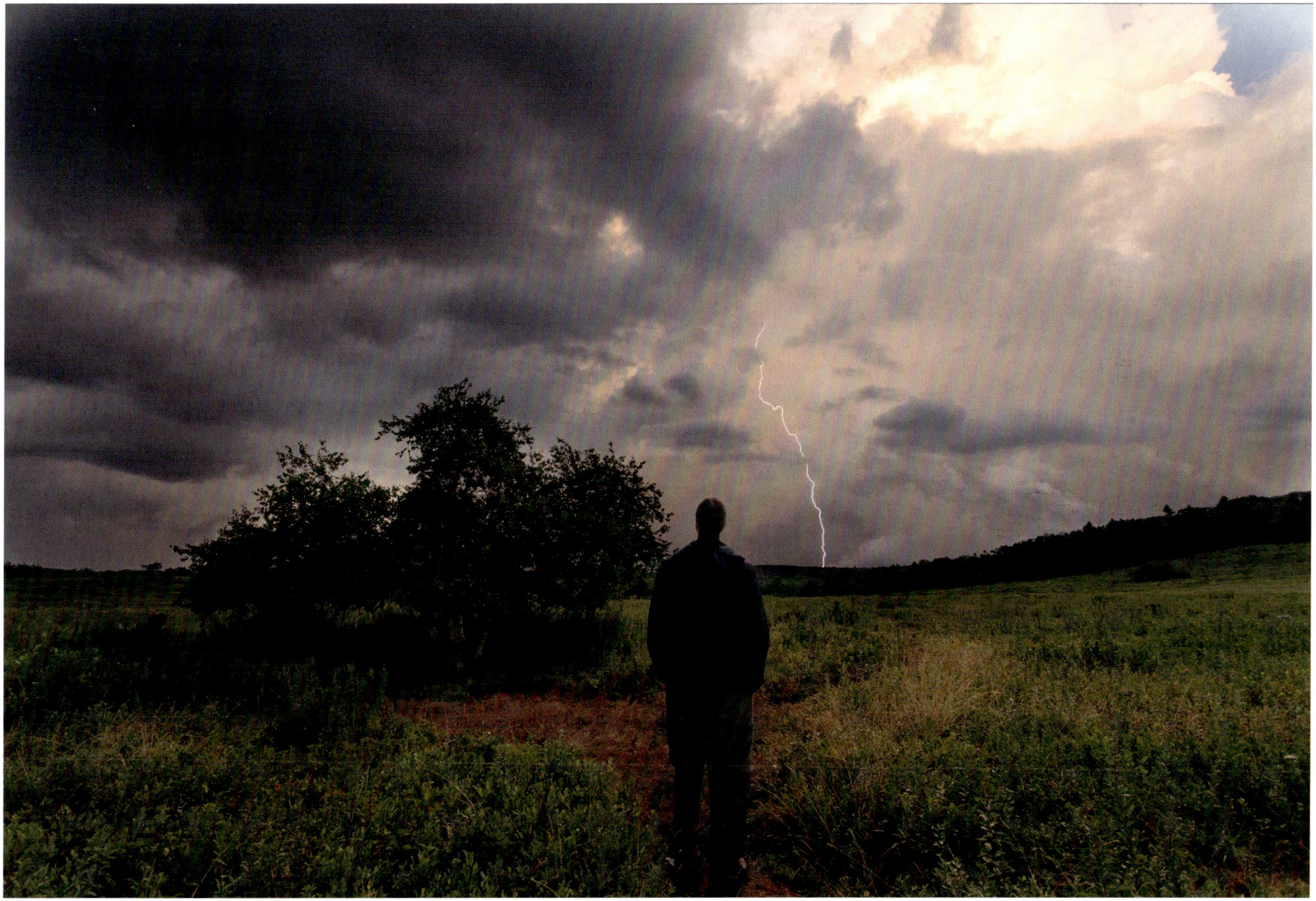

The author with the birch

REFERENCES

Crawford, Ben, and Meghan McCracken. *2,000 Miles Together: The Story of the Largest Family to Hike the Appalachian Trail*. United States: FFT Press, 2020.

Dallalio, Eric A., Adrianne B. Brand, and Evan H. Campbell Grant. "Climate-Mediated Competition in a High-Elevation Salamander Community." *Journal of Herpetology* 51, no. 2 (January 20, 2017): 190–96. https://doi.org/10.1670/15-157.

Harrison, Ralph L., and Lisa Cook. *The History of Pennsylvania: Elk Country*. Mechanicsburg, PA: Pennsylvania Forestry Association, 2008.

Schmidt, Martin F. *Maryland's Geology*. Atglen, PA: Schiffer, 2018.

Simpson, Ann, and Rob Simpson. *Falcon Pocket Guide: Nature Guide to Shenandoah National Park*. Guilford, CT: Falconguides, an imprint of Globe Pequot Press, 2013.

Stephenson, Steven L. *A Natural History of the Central Appalachians*. Morgantown: West Virginia University Press, 2013.

Weidensaul, Scott. *Mountains of the Heart: A Natural History of the Appalachians*. Golden, CO: Fulcrum, 2016.

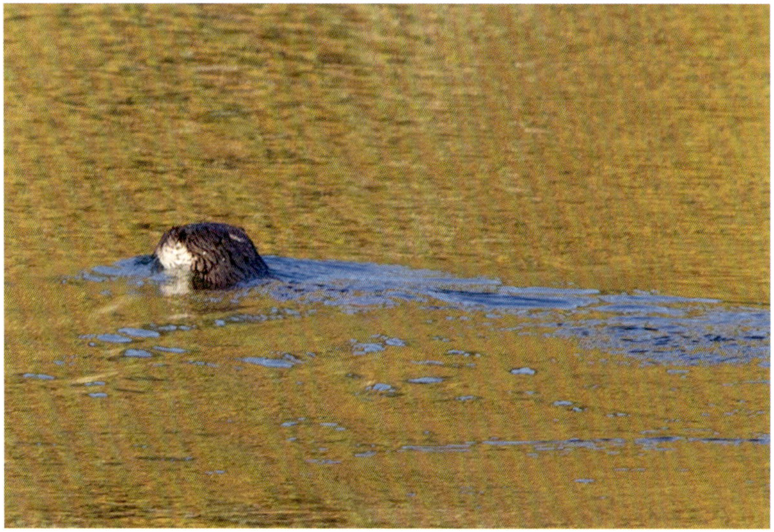

An otter swims through autumn reflections.

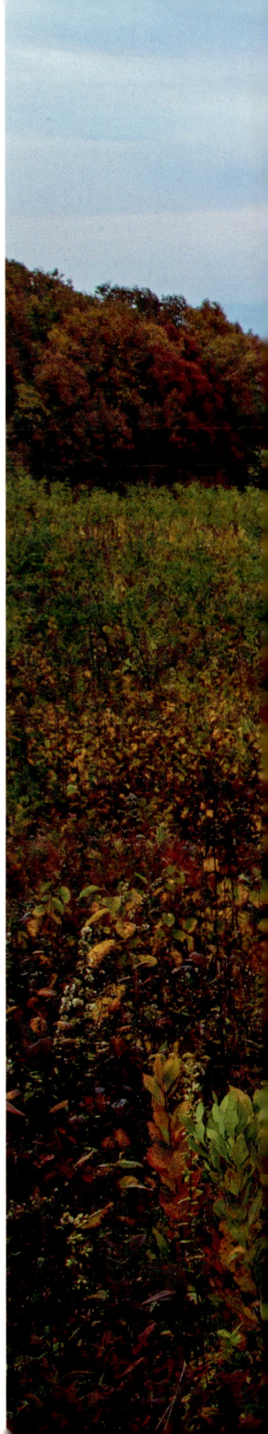

An autumn kaleidoscope of color with Old Rag Mountain in the background

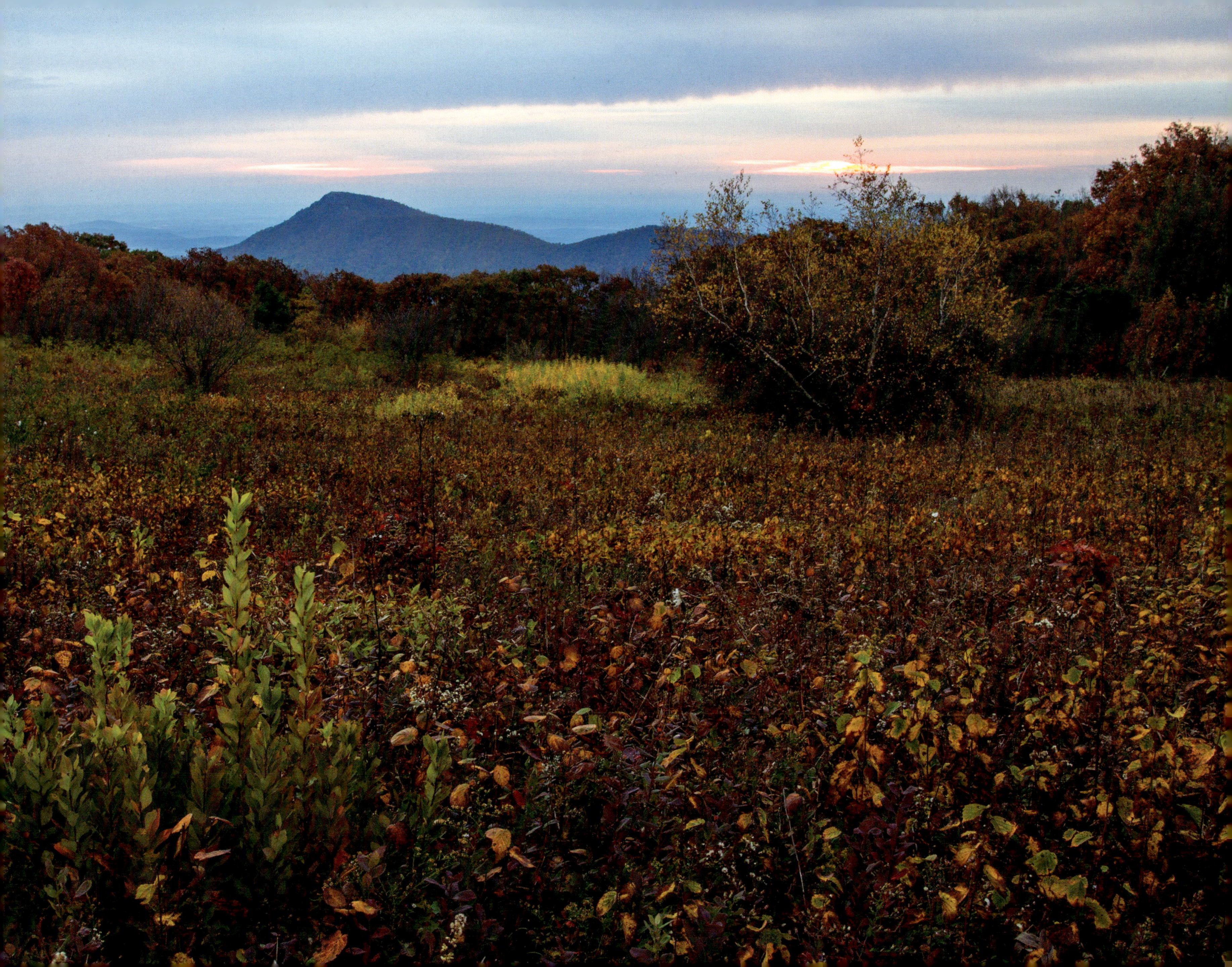

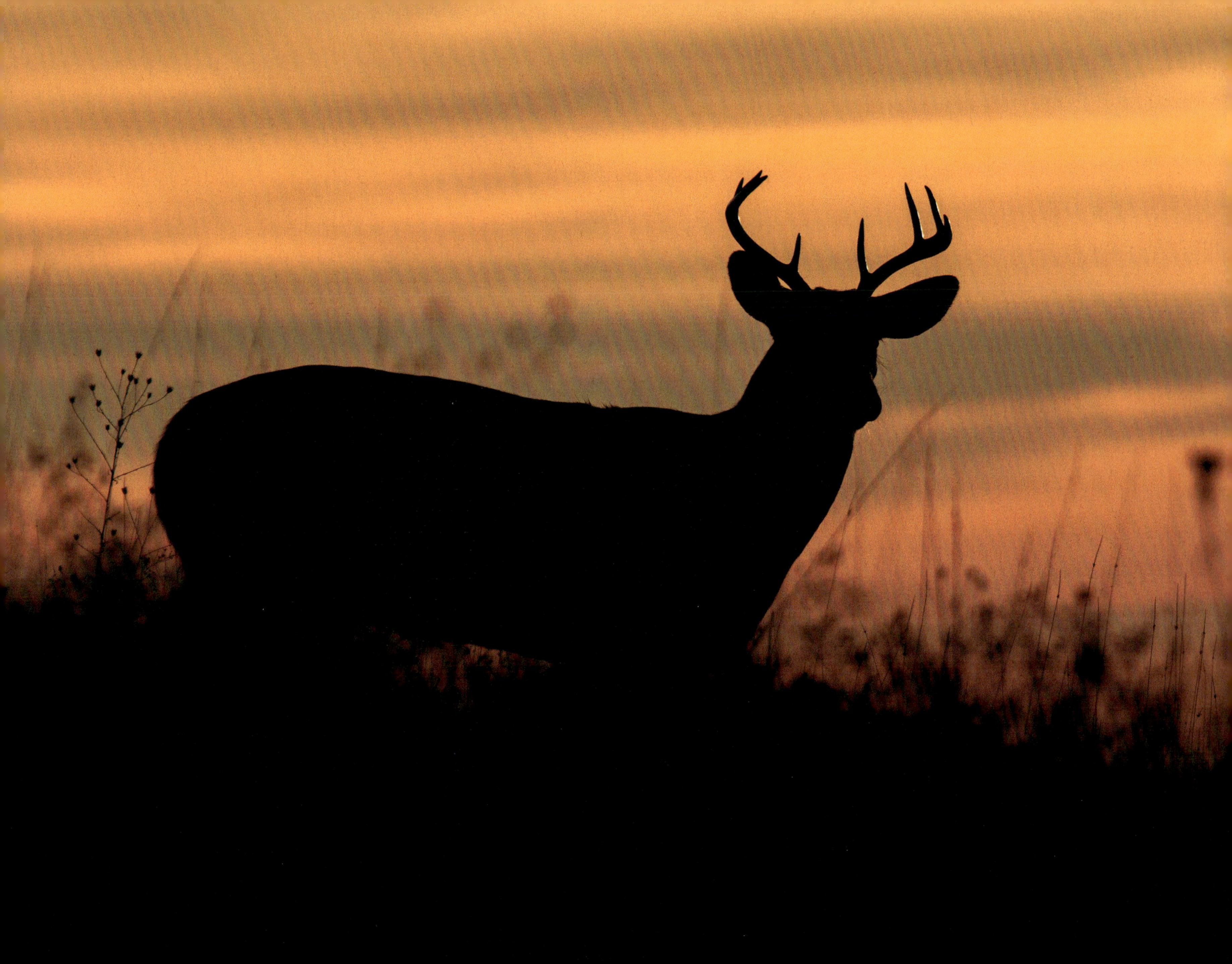

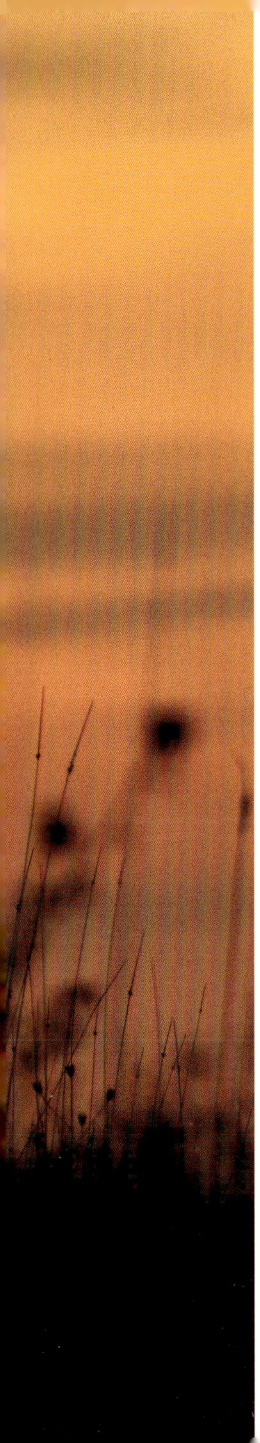

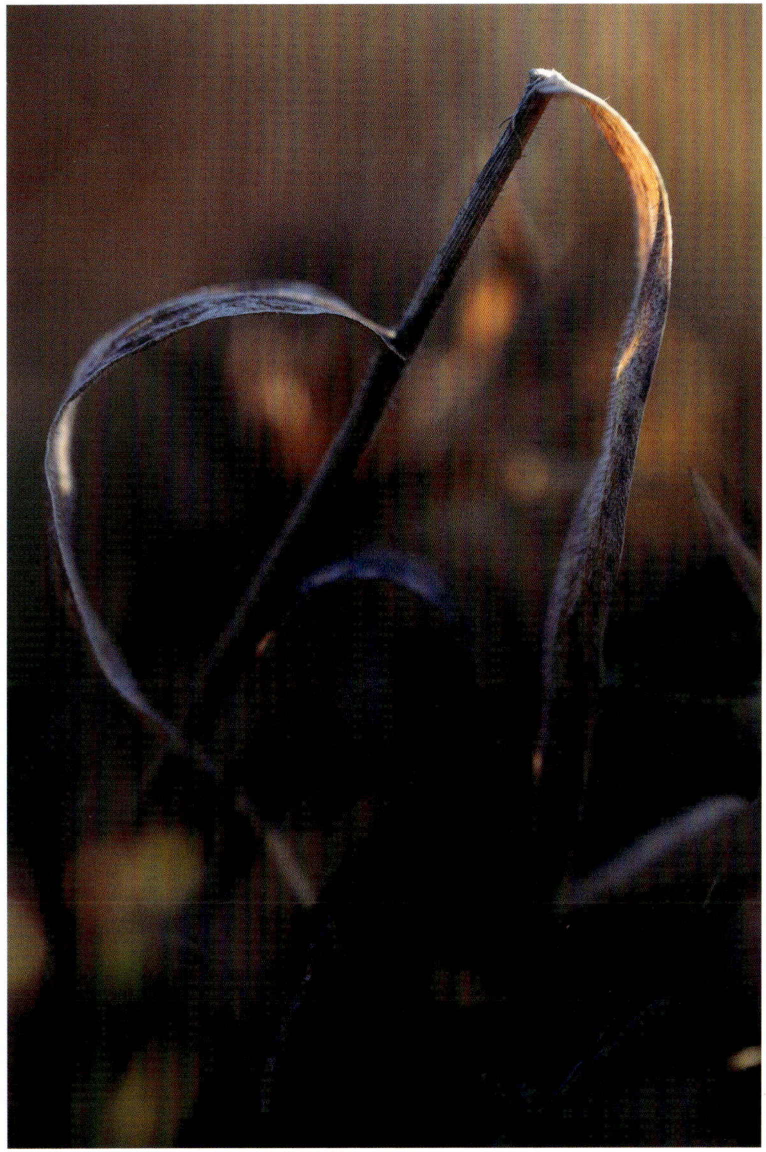

ABOUT THE AUTHOR

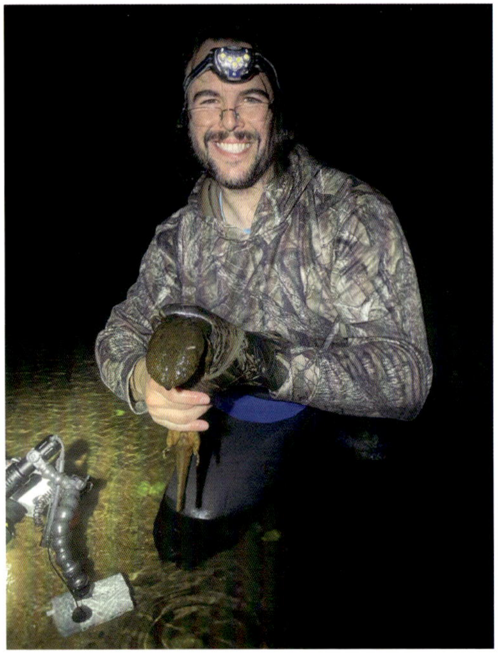

Mark Hendricks is an award-winning photographer, writer, and author. A former marine mammal biologist and animal rescuer, he turned to using his camera as a storytelling tool for conservation and envitonmental issues. His first book, *Natural Wonders of Assateague Island* (Schiffer Publishing), was a 2018 Foreword Reviews award winner. His images and articles have been featured in *National Geographic* (online), *The Guardian*, the *Washington Post*, *Living Bird*, and many others. He is a fellow in the International League of Conservation Writers, a member of the ethics committee for the North American Nature Photography Association (NANPA), and a faculty member at Towson University, where he serves as a lecturer in the Department of Psychology and adjunct in the Department of Biological Sciences.

www.markhendricksphoto.com